D1567111

Plant-Provided Food for Carnivorous Insects

A Protective Mutualism and Its Applications

Plants provide insects with a range of specific foods, such as nectar, pollen, and food bodies. In exchange, they may obtain various services from arthropods. The role of food rewards in the plant–pollinator mutualism has been broadly covered. This book addresses another category of food-mediated interactions, focussing on how plants employ foods to recruit arthropod "bodyguards" as a protection against herbivores.

Many arthropods with primarily carnivorous lifestyles require plant-provided food as an indispensable part of their diet. Only recently have we started to appreciate the implications of non-prey food for plant–herbivore–carnivore interactions. Insight into this aspect of multitrophic interactions is not only crucial to our understanding of the evolution and functioning of plant–insect interactions in natural ecosystems, it also has direct implications for the use of food plants and food supplements in biological control programs.

This edited volume provides essential reading for all researchers interested in plant–insect interactions.

Plant-Provided Food for Carnivorous Insects

A Protective Mutualism and Its Applications

Edited by

F. L. Wäckers
Netherlands Institute of Ecology

P. C. J. van Rijn
Netherlands Institute of Ecology

and

J. Bruin
University of Amsterdam

CAMBRIDGE UNIVERSITY PRESS
Cambridge, New York, Melbourne, Madrid, Cape Town, Singapore, São Paulo

CAMBRIDGE UNIVERSITY PRESS
The Edinburgh Building, Cambridge CB2 2RU, UK
Published in the United States of America by Cambridge University Press, New York

www.cambridge.org
Information on this title: www.cambridge.org/9780521819411

First published 2005

Printed in the United Kingdom at the University Press, Cambridge

A catalog record for this book is available from the British Library

Library of Congress Cataloging in Publication data

ISBN-13 978-0-521-81941-1 hardback
ISBN-10 0-521-81941-5 hardback

Contents

Contributors

J. Bruin
Section Population Biology, Institute for Biodiversity and Ecosystem Dynamics, University of Amsterdam, PO Box 94084, 1090 GB Amsterdam, The Netherlands

M. D. Eubanks
Department of Entomology and Plant Pathology, Auburn University, 301 Funchess Hall, Auburn, AL 36849, USA

G. M. Gurr
Faculty of Rural Management, University of Sydney, Orange, PO Box 883, Orange, NSW 2800, Australia

G. E. Heimpel
Department of Entomology, University of Minnesota, 1980 Folwell Avenue, St. Paul, MN 55108, USA

A. Janssen
Section Population Biology, Institute for Biodiversity and Ecosystem Dynamics, University of Amsterdam, PO Box 94084, 1090 GB Amsterdam, The Netherlands

M. A. Jervis
Cardiff School of Biosciences (BIOSI1), Cardiff University, PO Box 915, Cardiff, CF10 3TL, UK

J. Kean
Biocontrol and Biosecurity group, AgResearch Lincoln, PO Box 60, Lincoln, Canterbury, New Zealand

M. Keller
School of Agriculture and Wine, Adelaide University, Waite Campus, Private Bag 1, Glen Osmond, South Australia 5064, Australia

S. Koptur
Department of Biological Sciences, Florida International University, 11200 SW 8th Street, Miami, FL 33199, USA

D. A. Landis
204 Center for Integrated Plant Systems, Michigan State University, East Lansing, MI 48824, USA

W. J. Lewis
Crop Management and Research Laboratory, Agricultural Research Service, US Department of Agriculture, PO Box 748, Tifton, GA 31793, USA

D. M. Olson
Crop Management and Research Laboratory, Agricultural Research Service, US Department of Agriculture, PO Box 748, Tifton, GA 31793, USA

P. W. Price
Department of Biological Sciences, Northern Arizona University, North Campus 21, Flagstaff, AZ 86001, USA

J. Romeis
Agroscope FAL Reckenholz, Swiss Federal Research Station for Agroecology and Agriculture, Reckenholzstrasse 191, CH-8046 Zürich, Switzerland

M. W. Sabelis
Section Population Biology, Institute for Biodiversity and Ecosystem Dynamics, University of Amsterdam, PO Box 94084, 1090 GB Amsterdam, The Netherlands

E. Städler
Agroscope FAW Wädenswil, Swiss Federal Research Station for Fruit-Growing, Viticulture, and Horticulture, Schloss 334, CH-8820 Wädenswil, Switzerland

J. D. Styrsky
Department of Entomology and Plant Pathology, Auburn University, 301 Funchess Hall, Auburn, AL 36849, USA

K. Takasu
Faculty of Agriculture, Kyushu University, Fukuoka, 812–8581, Japan

J. Tylianakis
Soil, Plant and Ecological Sciences Division, Lincoln University, PO Box 84, Canterbury, New Zealand

P. C. J. van Rijn
Netherlands Institute of Ecology (NIOO-KNAW), Department for Multitrophic Interactions (MTI), PO Box 40, 6666 ZG Heteren, The Netherlands

F. L. Wäckers
Centre for Terrestrial Ecology, Netherlands Institute of Ecology (NIOO-KNAW), Boterhoeksestraat 48, 6666 GA Heteren, The Netherlands

T. K. Wilkinson
Department of Entomology, Michigan State University, East Lansing, MI 48824, USA

S. D. Wratten
Soil, Plant and Ecological Sciences Division, Lincoln University, PO Box 84, Canterbury, New Zealand

Foreword

Untangling Charles Darwin's "tangled bank" has motivated ecologists for well over a century. While Darwin could contemplate the many species and complex interactions of plants and animals along a roadside bank, he could hardly imagine the true dimensions of the intricacies in nature. Tangled interactions were soon perceived as more organized communities: the *Biozönose* of Karl Möbius in 1877 and the *microcosm* of Stephen Forbes in 1887. Food webs were gradually unraveled. An early example involved the cotton plant and the boll weevil complex, by Pierce, Cushman and Hood in 1912. Already, the staggering richness of trophic interrelationships was revealed, with over 300 species noted in the food web. These authors also documented the impact of cotton varieties on the attack on weevil larvae by parasitoids: an early example of three-trophic-level interactions.

An important next development was the dawn of chemical ecology which documented the ubiquity of multitrophic interactions. The aromas emanating from both plants and their herbivores were diagnosed as cues for higher trophic levels: phytochemicals could attract or repel herbivores and their natural enemies. But the field emphasized more the roles of plants and herbivores in influencing carnivores, and less the general ecology of the natural enemies themselves.

This book now remedies the former bias in the literature by placing natural enemies at center stage, and asking what the plant contributes to the intrinsic needs of the natural enemies: their full requirements of food, a place to live, and all their relationships with other species. *Plant-Provided Food for Carnivorous Insects* brings new vigor to the study of the carnivore's needs. The widely scattered literature is assembled and digested in an authoritative manner. Chapters are richly referenced: from Abe to Zoebelein, the reader is provided with an extensive bibliography. The common omnivory of perceived carnivores is revealed in all its complex and surreptitious ways. The plant angling with tasty lures for bodyguards, and plant food sources for adult herbivores are described. The benefits to plant fitness are evaluated, when provisioning for carnivores evolves as a strategy. Then, all this basic biology is used to understand the population dynamics of herbivores, its application in biological control, the role of habitat diversity, and application in the farm environment.

Such comprehensive treatment of the plant–carnivore mutualism, and other associations, is published at a time when the Earth needs more careful and gentle care. The encouragement of natural enemies of pests, with their acknowledged "friendly" environmental impact, requires sophisticated understanding of their ecology. This book provides a compelling argument for continuing detailed examination of carnivore biology, and a strong foundation of knowledge on which to build.

Peter W. Price

1

Food for protection: an introduction

FELIX L. WÄCKERS AND PAUL C. J. VAN RIJN

It has long been recognized that plants provide floral nectar and pollen to attract pollinators. In addition, plants also provide specific foods as part of a protection strategy. By producing extrafloral nectar or food bodies, plants attract predators that can act as bodyguards, clearing the plant of its antagonists. A wide range of arthropods with a primarily carnivorous lifestyle require plant-provided food as an indispensable part of their diet (Table 1.1). In some arthropod groups, the adult stages depend on nectar or pollen for survival and reproduction, whereas in other groups all stages feed on plant-provided food in addition to prey. Only recently have we started to appreciate the implications of non-prey food for plant–herbivore–carnivore interactions. Insight into these food-mediated interactions not only helps in understanding the functioning of multitrophic interactions in natural ecosystems, it also has direct implications for the use of food supplements in biological control programs. In this introductory chapter we first sketch a historical perspective on the topic of plant-provided foods. Subsequently, we present an outline of the book and briefly introduce the different chapters.

The scientific discovery of plant-provided foods

Humans have always shared the sweet tooth of many arthropods. However, for long we lacked the ability to obtain sugars directly from plants, and thus were entirely dependent on insects as intermediaries. Therefore, it is not surprising that nectar and honeydew in connection with insects attracted the attention of naturalists early on.

The Old Testament provides the first accounts of honeydew. The biblical "manna" (Exodus 16:13–36) is believed to be honeydew from the scale insect

Plant-Provided Food for Carnivorous Insects, ed. F. L. Wäckers, P. C. J. van Rijn, and J. Bruin.
Published by Cambridge University Press.

Coccus manniparus feeding on the shrub *Tamarix mannifera* (Bodenheimer 1947). In the Sinai, this honeydew is still collected as an alternative to honey under the local name of "menn" or "menu". Based on references by Al-Bīrūnī (973–1051) in his book on materia medica, Persian and Arab scholars of the medieval period already knew that honeydew originated from insects. Nevertheless, European naturalists argued for many centuries about the nature and origin of honeydew, before Leche (1765), basing himself on observations in Réamur's *Mémoires sur les insectes* (1734–42), described the production of honeydew by sap-feeding insects as well as the fact that ants tend them to obtain the sugar-rich solution.

Written records on floral nectar date back to antiquity as well. The Greek physician Dioscorides (50 BC – AD 10) wrote about floral nectar as the basis of honey production and the medicinal uses of the latter. The Roman naturalist Pliny the Elder (AD 23–79) in his *Naturalis Historiae* provides detailed accounts of nectar types secreted by flowers as well as their collection by bees as a basis for honey production:

> The honey that we see formed in the calix of flowers is of a rich and unctuous nature; that which is made from rosemary is thick, while that which is candied is little esteemed. Thyme honey does not coagulate, and on being touched will draw out into thin viscous threads, a thing which is the principal proof of its heaviness.

Both Pliny and Dioscorides already recognized that the phenomenon of toxic honey relates to its floral origin. They correctly attributed the toxicity of honey from the Black Sea region to the nectar from particular flowers (a.o. *Rhododendron* spp. and oleander (*Nerium oleander*)). The ecological role of nectar as a pollination reward was studied experimentally for the first time by Sprengel (1793), who did groundbreaking work on pollination ecology, recognizing and describing a range of pollination-related phenomena. Darwin (1855) built upon Sprengel's work by placing nectar production in the context of plant–pollinator co-evolution.

Hall (1762) is believed to have been the first to make the distinction between floral and extrafloral nectaries. In 1855, Glover reported that glands (extrafloral nectaries) of cotton (*Gossypium* spp.) secrete a sweet substance, which ants, bees, wasps, and plant bugs avail themselves of as food (Trelease 1879). Around that time, Darwin (1855) described that extrafloral nectaries on *Vicia* spp. are visited by bees, ants, and flies and suggested that they have a function other than pollination. Delpino (1873) recognized the specific functions of floral and extrafloral nectaries and proposed the terms "nuptial" and "extra-nuptial" nectaries as phrases that indicate their different ecological roles. He observed that "extranuptial nectar glands, by their secretion, attract to the plant that bears them, hordes of ants (rarely wasps)" and asserted that these "constitute a temporary and changing bodyguard". Around the same time, Thomas Belt (Belt 1874) argued that plants obtain a defensive benefit from insects visiting extrafloral nectaries and/or food bodies. The actual protective function of extrafloral nectaries was first demonstrated by Von Wettstein (1889). He excluded ants from bracteal nectaries on the flowering heads of two Compositae species, and was able to show that ant-tended plants suffered less damage to seeds by beetles and hemipteran bugs.

For almost a century following the seminal publications by Delpino and Belt, the protective function of plant-provided food was subject to intense debate (Bentley 1977). For many decades the concept of food as an indirect defense mechanism was discarded, before Janzen (1966) and others in the 1960s revived the idea. Through extensive experimental work, they were able to substantiate the fact that ants recruited to extrafloral nectaries and food bodies can benefit plant fitness.

Plant-provided food and biological control

The role of food supplements in plant–herbivore–carnivore interactions is not only an important topic in basic ecology, but is also directly linked to the applied discipline of biological pest control. Defensive food provision has evolved repeatedly and independently, suggesting that it constitutes a powerful mechanism through which plants can enhance the effectiveness of carnivores. We pursue the same objective in biological control programs. Here too, we aim at enhancing the efficacy with which carnivores control herbivorous pests.

The possibility of using predators for biological control of insect pests was recognized in China as far back as the fourth century AD. Hsi Han (AD 304) described in *Records of the Plants and Trees of the Southern Regions* how bags holding ant nests were traded in southern China. The bags were placed in citrus trees in order to protect the fruits from insect attacks. Farmers interconnected trees by means of bamboo bridges allowing ants to move between trees.

In biological control textbooks, this example is widely featured as the first known case of biological control. However, it is much less known that it also represents the first instance in which food supplements were used to enhance the efficacy of the biological control agent. Farmers provided food supplements, such as intestines and silkworm larvae, to help ants establish (Beattie 1985). The use of ants for biological control was not restricted to the Far East. In the New World, Indians independently developed methods to use ants as biological control agents. Rather than bringing the ants to the crop, they took the opposite approach by sowing cotton plants in the vicinity of ant nests (Cook 1905). Here again, food supplements played an integral role. The effectiveness of this practice can be explained by the fact that cotton features a range of extrafloral nectaries (Mound 1962) that are eagerly visited by ants (Rudgers 2002). By sowing cotton in the proximity of ant nests the Indians exploited this natural food-mediated association between cotton plants and predaceous ants.

In the twentieth century, awareness grew among biological control workers that the absence of non-prey food sources in agriculture or forestry could impose a serious constraint on the effectiveness of natural enemies (Illingworth 1921; Schneider 1940; Wolcott 1942; Hocking 1966). Hocking (1966) pointed out that lack of food availability can also prevent introduced parasitoids from establishing in classical biological control programs. Adding food sources to agroecosystems could be a simple and effective way to enhance the effectiveness of biological control programs.

Three types of approaches have been proposed to alleviate the shortage of food in modern monocultures. The first approach involves the diversification of agroecosystems, either through the use of non-crops in undergrowth or field margins (Van Emden 1965; Altieri and Whitcomb 1979) or through mixed cropping, e.g., with crops featuring flowers or extrafloral nectaries. A second approach involves the use of food sprays or other types of artificial food supplements to cater for the food needs of biological control agents (Hagen 1986). Finally, some crops produce suitable food supplements themselves. Examples of extrafloral nectar producing crops include *Prunus* spp. (cherry, plum, peach, almond), cassava, faba bean, zucchini, pumpkin, cashew, and cotton. These crop-produced foods may suffice as food sources for predators and parasitoids. In other cases, there may be room for plant breeding to improve the timing, quantity, and quality of food production, to better match the nutritional needs of biological control agents (Rogers 1985).

Cultivated cotton also provides a prominent example in which the potential for negative effects of food supplements became apparent. Cotton extrafloral nectaries are not only used by predators and parasitoids, but several major cotton pest species are known to feed on cotton extrafloral nectar as well. The generous use of broad-spectrum insecticides in the mid twentieth century not only temporarily eliminated herbivores from cotton fields, it also proved effective in clearing the field of predators and parasitoids. As a result, the indirect defensive function of extrafloral nectaries became obsolete. Under these conditions, nectar-bearing cotton varieties sometimes suffered higher levels of herbivore damage than nectariless varieties (Lukefahr *et al.* 1965; Adjei-Maafo and Wilson 1983). The replacement of broad-spectrum insecticides by more selective control methods rekindled the interest in cotton extrafloral nectar as a food source for beneficials (Rogers 1985; Schuster and Calderon 1986). Whereas the conditions of cotton production during the green revolution were obviously a far cry from the conditions of modern-day conservation biological control programs, this example nevertheless shows that we cannot ignore herbivore benefits when studying the impact of food supplements on biological control programs.

Outline of this book

This historical overview indicates how the provision of food by plants, and its impact on the effectiveness of predators and parasitoids, gradually gained interest. In the last two or three decades this interest seems to have accelerated. This has stimulated us to compose a book that reviews the current state of knowledge, and indicates directions of future research on this specific aspect of multitrophic (plant–herbivore–carnivore) interactions.

In the first section of this book, the spotlight is on the plants. What types of food supplements do they provide, why does food provision evolve and how does it affect plant–insect interactions? In the second section, the arthropods that feed on plant-provided food are at center stage. Why do they feed on this food, and how does it affect their behavior and life history? In the third section, we focus on the dynamics of the interactions between plants, carnivores, and herbivores. How do these interactions affect herbivore population levels, and what factors define the success of biological control?

Part I: Food provision by plants

Plants employ nutritional supplements to obtain a range of services. Best known are the mutualistic interactions in which sessile plants provide food in return for dispersal. This includes floral nectar to attract pollinators (Faegri and Van der Pijl 1979), and the fleshy fruit tissues and elaiosomes promoting seed dispersal. Other plant-provided foods, such as extrafloral nectar and food bodies, are likely to have evolved primarily to attract carnivores in order to obtain their protective services (Turlings and Wäckers 2004). As such they represent the most suitable models to study the evolution and functioning of food-for-protection strategies. Plant-provided foods are not only used by the intended consumers, they may also be exploited by arthropods from other guilds. Some of these unintended interactions are to the benefit of the plant, others to the detriment.

A separate category of food supplements comprises those that have evolved for functions other than arthropod nutrition. Pollen, for instance, primarily serves as a vehicle for gene transfer, but may serve a secondary function in attracting pollinators. Honeydew, the plant-derived waste product of phloem-feeding insects, can also be an important food source for (predaceous) arthropods. When ants tend sap-feeders to collect the honeydew, this sugar source can serve a (secondary) defensive function that may benefit both the sap-feeders and the plants.

Finally, some predators can feed directly on photosynthetic or reproductive plant tissue without special adaptations from the plant. It is obvious that these different food categories will differ in their implications for plant–carnivore interactions.

In this book, Wäckers (Chapter 2) presents an overview of the food sources provided by plants, and reviews their suitability in terms of their availability, detectability, accessibility, nutritional value, and mortality risks for the various arthropods that feed on them. The identified differences can be helpful in understanding the evolution and functioning of food supplements, and in selecting food supplements for use in biological control programs.

Koptur (Chapter 3) discusses the evolutionary origin of extrafloral nectar. A comparison is made between floral and extrafloral nectar, with respect to nectar composition, its consumers, and the ecological factors modifying its production. A strong emphasis is laid on ant–plant interactions.

Sabelis, van Rijn, and Janssen (Chapter 4) focus more closely on the evolutionary stability of extrafloral nectar production and of other nutritional rewards in the food-for-protection mutualism: how can they persist in the face of cheaters and other organisms ready to reap the benefits?

Part II: Arthropods feeding on plant-provided food

As most arthropod predators and parasitoids are able to feed on prey as well as on plant-provided food, they could actually be called omnivores. However, the need for plant food of these arthropods has long been overlooked. This has a number of reasons. Many species feed specifically on plant-provided food (nectar and/or pollen) and cause no visible damage to the plants. In some of these insects the stages that are carnivorous are not the stages that feed on plant substances. In other arthropods, carnivorous stages can also feed on plant tissue, but without obvious adaptations in their feeding apparatus.

In order to structure the great variety in plant-feeding among predators and parasitoids, we propose the following typology (see also Table 1.1).

- Life-history omnivory. Many holometabolic insects change their lifestyle during metamorphosis, and some of these insects shift from carnivory in the larval stage to herbivory (or nectarivory) in the adult stage. The larvae of some hymenopterans, such as parasitoids, most ants, and social wasps, only feed on animal prey (or hosts). Some of the nutrients obtained during larval stages are transferred to the adult stage, allowing the adults to survive and reproduce while feeding on nectar or honeydew only. An ontogenetic diet shift from herbivory to carnivory is much less common. Nymphs of stink bugs may start life as herbivores and later become predators or mixed feeders (McGavin 2000). The term "life-history omnivory" was first coined by Polis and Strong (1996).

- Temporal omnivory. Some predators and host-feeding parasitoids can supplement their carnivorous diet with plant food during part of their life cycle only. As an example, both juvenile and adult tiger beetles (Cicindelidae) forage on ground-dwelling prey, whereas adults also feed on plant seeds.
- Permanent omnivory. Many plant-inhabiting "predators" can feed on both prey and plant material in their juvenile as well as their adult phase. Typical are the heteropteran predators that can use their stinging mouthparts to feed on prey and plant tissue. Other predators only use pollen and/or nectar to supplement their diet (e.g., predatory mites and ladybirds).

The impact of plant-provided food on arthropods, and on their role in plant protection (herbivore suppression), depends on the type of omnivory. For life-history omnivores, such as parasitoids, the availability of the right material will typically extend survival of the adult insects, and will thereby expand their reproductive capacity. Olson, Takasu, and Lewis (Chapter 5) discuss the specific morphological and behavioral adaptations to this nectarivorous lifestyle and its ecological consequences. Heimpel and Jervis (Chapter 9) review the empirical evidence for nectar use by parasitoids under field conditions. They also address the impact of nectar feeding on parasitoid survival and reproduction, as well as on population establishment and pest control.

For real (temporal or permanent) omnivores, such as predatory bugs, the effect of plant-based feeding may be less clear, as it can be partly substituted by feeding on prey. The impact of plant food on the various life-history components (development, survival, reproduction) of these omnivores should therefore be studied at different, but fixed, prey densities. Another complicating factor is that plant feeding may go at the expense of the per capita prey consumption. Eubanks and Styrsky (Chapter 6) review the experimental studies on the various effects of omnivory, including its impact on herbivore suppression.

Not only predators and parasitoids shift their diets during development, also some herbivorous species change from tissue-feeding larvae to nectar or pollen-feeding adults. Romeis, Städler, and Wäckers (Chapter 7) review the foraging and feeding requirements of adult herbivorous butterflies, flies, and beetles. They discuss the implications of this adult feeding for herbivore reproductive fitness, herbivore–plant interactions, and pest management.

Table 1.1 *Types of omnivory among "carnivorous" arthropods*

Type	Plant-feeding stage	Arthropod examples can be found within:		Type of plant food utilized	Reference
Life-history omnivory	Adult	Neuroptera	Chrysopidae (green lacewings)	Nectar, pollen	Stelzl 1991
		Diptera	Syrphidae (hoverflies)	Nectar, pollen	Hickman *et al.* 1995
			Cecidomyiidae (gall midges)	Nectar	Opit *et al.* 1997
			Tachinidae (parasitoid flies)	Nectar	Gilbert and Jervis 1998
		Hymenoptera	a.o. Ichneumonidae, Braconidae (parasitoid wasps)	Nectar	Jervis 1998, Lewis *et al.* 1998, Wäckers 2001
			Vespidae (social wasps)	Nectar, fruit	Cuautle and Rico-Gray 2003
			Formicidae (ants)	Nectar	Beattie 1985
		Coleoptera	Meloidae (blister beetles)	Nectar, pollen	Adams and Selander 1979
	Juvenile	Heteroptera	Pentatomidae (stink bugs)	Plant juice	McGavin 2000
Temporal omnivory	Adult	Hymenoptera	a.o. Ichneumonidae, Braconidae (host-feeding parasitoids)	Nectar	Jervis 1998, Lewis *et al.* 1998
		Coleoptera	Cicindelidae (tiger beetles)	Seeds	Zerm and Adis 2001
			Cantharidae (soldier beetles)	Nectar, pollen	Traugott 2003
	Juvenile	Araneae	Araneidae (orb web spiders)	Pollen	Smith and Mommsen 1984

Permanent omnivory	Adult and juvenile	Acari: Mesostigmata	Phytoseiidae (predatory mites)	Nectar, pollen	Van Rijn and Tanigoshi 1999a, b
		Heteroptera	Pentatomidae (stink bugs)	Plant juice	Ruberson *et al.* 1986
			Miridae (mirid bugs)	Plant juice	Gillespie and McGregor 2000
			Geocoridae (big-eyed bugs)	Plant juice	Eubanks and Styrsky, Chapter 6
			Anthocoridae (flower bugs)	Pollen	Eubanks and Styrsky, Chapter 6
		Neuroptera	*Chrysopa*, Hemerobiidae (brown lacewings)	Nectar, pollen	Stelzl 1991, McEwen *et al.* 1993
		Thysanoptera	Aeolothripidae, Phlaeothripidae	Leaves, pollen	Kirk 1997
		Coleoptera	Coccinellidae (ladybirds)	Nectar	Pemberton and Vandenberg 1993
				Pollen	Cottrell and Yeargan 1998
			Carabidae (ground beetles)	Seeds	Goldschmidt and Toft 1997

Part III: Plant-provided food and biological control

Food-for-protection strategies have evolved independently in many plants, suggesting that food supplements can be a powerful tool to enhance the effectiveness of predators and parasitoids in the reduction of herbivores. In biological control programs, we rely on carnivorous arthropods to control herbivorous pest insects. Therefore, it seems an obvious step to emulate the food-for-protection strategies in our cropping systems.

The use of food plants and artificial food sprays has been advocated as a means to enhance biological and natural control. Some of the efforts seem successful, but in general the results remain variable and unpredictable (Bugg and Waddington 1994; Landis *et al.* 2000). The strategy is certainly promising, but we need to improve our understanding of the underlying mechanisms in order to increase the effectiveness of our efforts.

In theory, the enhanced performance of a carnivore supplied with plant food does not necessarily improve herbivore suppression. The positive effect on carnivore fitness can be canceled out by factors such as reduced attack rate or increased herbivore reproduction. Population feedback and modified distribution patterns may also reduce the effect on herbivore suppression. In a series of model exercises, Van Rijn and Sabelis (Chapter 8) investigate the conditions required for a positive relationship between food provision and plant protection. They consider food quality, life history, spatial structure, and food web structure.

The empirical verification of the food-for-protection hypothesis may be easier for parasitoids than for predators, for two reasons: the relative ease with which the strength of interaction between herbivore and parasitoid can be quantified (by means of percentage parasitism), and the fact that parasitoids often depend on a single food type (nectar). Heimpel and Jervis (Chapter 9) consider the assumptions underlying this nectar-limitation hypothesis, and discuss to what extent empirical evidence matches the predicted host parasitism and suppression.

When "companion" plants are grown, this may not only provide food for predators and parasitoids, but also other services such as shelter and alternative hosts. Wilkinson and Landis (Chapter 10) discuss the different spatial scales at which the plant resources can be present in the landscape, and how this eventually affects pest control.

Finally, Gurr and colleagues (Chapter 11) discuss how the implementation of food-for-protection strategies in agriculture can benefit from a directed approach that brings together general ecological theory, well-focussed empirical studies, and case-specific modeling.

References

Adams, C. L. and R. B. Selander. 1979. The biology of blister beetles of the Vittata Group of the genus *Epicauta* (Coleoptera, Meloidae). *Bulletin of the American Museum of Natural History* **162**: 139–266.

Adjei-Maafo, I. K. and L. T. Wilson. 1983. Factors affecting the relative abundance of arthropods on nectaried and nectariless cotton. *Environmental Entomology* **12**: 349–352.

Altieri, M. A. and W. H. Whitcomb. 1979. The potential use of weeds in manipulation of beneficial insects. *Horticulture Science* **14**: 12–18.

Beattie, A. J. 1985. *The Evolutionary Ecology of Ant–Plant Mutualisms*. Cambridge, UK: Cambridge University Press.

Belt, T. 1874. *The Naturalist in Nicaragua*. London: John Murray.

Bentley, B. L. 1977. Extrafloral nectaries and protection by pugnacious bodyguards. *Annual Review of Ecology and Systematics* **8**: 407–427.

Bodenheimer, F. S. 1947. The manna of Sinai. *Biblical Archeologist* **10**: 2–6.

Bugg, R. L. and C. Waddington. 1994. Using cover crops to manage arthropod pests of orchards: a review. *Agriculture Ecosystems and Environment* **50**: 11–28.

Cook, O. F. 1905. The social organization and breeding habits of the cotton-protecting kelep of Guatemala. *US Department of Agriculture Technical Series* **10**: 1–55.

Cottrell, T. E. and K. V. Yeargan. 1998. Effect of pollen on *Coleomegilla maculata* (Coleoptera: Coccinellidae) population density, predation, and cannibalism in sweet corn. *Environmental Entomology* **27**: 1402–1410.

Cuautle, M. and V. Rico-Gray. 2003. The effect of wasps and ants on the reproductive success of the extrafloral nectaried plant *Turnera ulmifolia* (Turneraceae). *Functional Ecology* **17**: 417–423.

Darwin, C. 1855. Nectar-secreting organs of plants. *Gardeners' Chronicle* **29**: 487.

Delpino, F. 1873. Ulteriori osservazioni e considerazioni sulla dichogamia nel regno vegetale. *Società Italiana di Scienze Naturali* **16**: 151–349.

Faegri, K. and L. Van der Pijl. 1979. *The Principles of Pollination Ecology*. Oxford, UK: Pergamon Press.

Gilbert, F. and M. Jervis. 1998. Functional, evolutionary and ecological aspects of feeding-related mouthpart specializations in parasitoid flies. *Biological Journal of the Linnean Society* **63**: 495–535.

Gillespie, D. R. and R. R. McGregor. 2000. The functions of plant feeding in the omnivorous predator *Dicyphus hesperus*: water places limits on predation. *Ecological Entomology* **25**: 380–386.

Goldschmidt, H. and S. Toft. 1997. Variable degrees of granivory and phytophagy in insectivorous carabid beetles. *Pedobiologia* **41**: 521–525.

Hagen, K. S. 1986. Ecosystem analysis: plant cultivars (HPR), entomophagous species and food supplements. In D. J. Boethel and R. D. Eikenbary (eds.) *Interactions of Plant Resistance and Parasitoids and Predators of Insects*. New York: John Wiley, pp. 153–197.

Hall, B. M. 1762. *Dissertatio Botanica sistens Nectaria florum*. Upsala: Dissertationes Academica (Linné).

Hickman, J. M., G. L. Lovei, and S. D. Wratten. 1995. Pollen feeding by adults of the hoverfly *Melanostoma fasciatum* (Diptera: Syrphidae). *New Zealand Journal of Zoology* **22**: 387–392.

Hocking, H. 1966. The influence of food on longevity and oviposition in *Rhyssa persuasoria* (L.) (Hymenoptera: Ichneumonidae). *Journal of the Australian Entomological Society* **6**: 83–88.

Illingworth, J. F. 1921. Natural enemies of sugar-cane beetles in Queensland. *Queensland Bureau Sugar Experimental Station Division of Entomology Bulletin* **13**: 1–47.

Janzen, D. H. 1966. Coevolution of mutualism between ants and acacias in Central America. *Evolution* **20**: 249–275.

Jervis, M. 1998. Functional and evolutionary aspects of mouthpart structure in parasitoid wasps. *Biological Journal of the Linnean Society* **63**: 461–493.

Kirk, W. D. J. 1997. Feeding. In T. Lewis (ed.) *Thrips as Crop Pests*. Wallingford, UK: CAB International, pp. 217–257.

Landis, D. A., S. D. Wratten, and G. M. Gurr. 2000. Habitat management to conserve natural enemies of arthropod pests in agriculture. *Annual Review of Entomology* **45**: 175–201.

Leche, J. 1765. Auszug aus dem täglichen Verzeichnisse der Witterungen das zu Abo, vom 1750sten bis zu Ende 1761 ist gehalten worden. III. Das Wasser, das aus der Luft herabfällt. *Der Königlischen schwedischen Akademie der Wissenschaften neue Abhandlungen aus der Naturlehre, Haushaltungskunst und Mechanik* **25**: 16–27.

Lewis, W., J. Stapel, A. Cortesero, and K. Takasu. 1998. Understanding how parasitoids balance food and host needs: importance to biological control. *Biological Control* **11**: 175–183.

Lukefahr, M. J., D. F. Martin, and J. R. Meyer. 1965. Plant resistance to five Lepidoptera attacking cotton. *Journal of Economic Entomology* **58**: 516–518.

McEwen, P. K., M. A. Jervis, and N. A. C. Kidd. 1993. Influence of artificial honeydew on larval development and survival in *Chrysoperla carnea* (Neu. Chrysopidae). *Entomophaga* **38**: 241–244.

McGavin, G. C. 2000. *Insects, Spiders and other Terrestrial Arthropods*. London: Dorling Kindersley.

Mound, L. A. 1962. Extra-floral nectaries of cotton and their secretion. *Empire Cotton Growing Review* **39**: 254–261.

Opit, G., B. Roitberg, and D. R. Gillespie. 1997. The functional response and prey preference of *Feltiella acarisuga* (Vallot) (Diptera: Cecidomyiidae) for two of its prey: male and female twospotted spider mites, *Tetranychus urticae* Koch (Acari: Tetranychiidae). *Canadian Entomologist* **129**: 221–227.

Pemberton, R. W. and N. J. Vandenberg. 1993. Extrafloral nectar feeding by ladybird beetles (Coleoptera, Coccinellidae). *Proceedings of the Entomological Society of Washington* **95**: 139–151.

Polis, G. A. and D. R. Strong. 1996. Food web complexity and community dynamics. *American Naturalist* **147**: 813–846.

Rogers, C. E. 1985. Extrafloral nectar: entomological implications. *Bulletin of the Entomological Society of America* **31**: 15–20.

Ruberson, J. R., M. J. Tauber, and C. A. Tauber. 1986. Plant feeding by *Podisus maculiventris* (Heteroptera: Pentatomidae): effect on survival, development, and preoviposition period. *Environmental Entomology* **15**: 894–897.

Rudgers, J. A. 2002. Evolutionary ecology of ant–wild cotton associations. Ph.D. thesis, University of California, Davis, California.

Schneider, F. 1940. Schadinsekten und ihre Bekämpfung in ostindischen Gambirkulturen. *Mitteillungen der schweizerischen entomologischen Gesellschaft* **18**: 77–207.

Schuster, M. F. and M. Calderon. 1986. Interactions of host plant resistant genotypes and beneficial insects on cotton ecosystems. In R. D. Boethel and R. D. Eikenbary (eds.) *Interactions of Plant Resistance and Parasitoids and Predators of Insects*. New York: John Wiley, pp. 84–97.

Smith, R. B. and T. P. Mommsen. 1984. Pollen feeding in an orb-weaving spider. *Science* **226**: 1330–1332.

Sprengel, C. K. 1793. *Das entdeckte Geheimnis der Natur im Bau und der Befruchtung der Blumen*. Berlin: Vieweg. (Reprinted 1972.)

Stelzl, M. 1991. Investigations on food of Neuroptera adults (Neuropteroidea, Insecta) in Central Europe, with a short discussion of their role as natural enemies of insect pests. *Journal of Applied Entomology / Zeitschrift für angewandte Entomologie* **111**: 469–477.

Traugott, M. 2003. The prey spectrum of larval and adult *Cantharis* species in arable land: an electrophoretic approach. *Pedobiologia* **47**: 161–169.

Trelease, W. 1879. Nectar, what it is and some of its uses. *US Department of Agriculture Report on Cotton Insects* **3**: 319–343.

Turlings, T. C. J. and F. L. Wäckers. 2004. Recruitment of predators and parasitoids by herbivore-injured plants. In R. T. Cardé and J. Millar (eds.) *Advances in Insect Chemical Ecology*. Cambridge, UK: Cambridge University Press, pp. 21–75.

Van Emden, H. F. 1965. The role of uncultivated land in the biology of crop pests and beneficial insects. *Scientific Horticulture* **17**: 121–136.

Van Rijn, P. C. J. and L. K. Tanigoshi. 1999a. The contribution of extrafloral nectar to survival and reproduction of the predatory mite *Iphiseius degenerans* on *Ricinus communis*. *Experimental and Applied Acarology* **23**: 281–296.

1999b. Pollen as food for the predatory mites *Iphiseius degenerans* and *Neoseiulus cucumeris* (Acari: Phytoseiidae): dietary range and life history. *Experimental and Applied Acarology* **23**: 785–802.

Von Wettstein, R. 1889. Über die compositen der Oesterreichisch-Ungarischen Flora mit zuckerabscheidenden Hüllschuppen. *Oesterreiche Akademie der Wissenschaften, Sitzungsberichte Teil I, Minerologie, Krystallographie, Botanik* **97**: 570–589.

Wäckers, F. L. 2001. A comparison of nectar and honeydew sugars with respect to their utilization by the hymenopteran parasitoid *Cotesia glomerata*. *Journal of Insect Physiology* **47**: 1077–1084.

Wolcott, G. N. 1942. The requirements of parasites for more than hosts. *Science* **96**: 317–318.

Zerm, M. and J. Adis. 2001. Spatio-temporal distribution of larval and adult tiger beetles (Coleoptera: Cicindelidae) from open areas in Central Amazonian floodplains (Brazil). *Studies on Neotropical Fauna and Environment* **36**: 185–198.

Part I FOOD PROVISION BY PLANTS

2

Suitability of (extra-)floral nectar, pollen, and honeydew as insect food sources

FELIX L. WÄCKERS

Introduction

Although arthropod predators and parasitoids are usually associated with their carnivorous qualities, they often require plant-provided foods as well, at least during part of their life cycle. The level in which predators or parasitoids depend on these plant-provided foods varies (see Wäckers and van Rijn, Chapter 1). Temporal omnivores and permanent omnivores are *facultative* consumers of plant-derived food, using it as complement to their prey. This category includes mites (Bakker and Klein 1992b), spiders (Ruhren and Handel 1999), hemipterans (Bugg *et al.* 1991), beetles (Larochelle 1990; Pemberton and Vandenberg 1993; Pfannenstiel and Yeargan 2002), lacewings (Limburg and Rosenheim 2001), wasps (Beggs 2001), and ants (Porter 1989). Life-history omnivores, on the other hand, are *obligatory* consumers of plant-provided foods during certain stages (usually the adult stage). This means that they are entirely dependent on non-prey food for their survival and metabolic upkeep. Examples are syrphid flies (Lunau and Wacht 1994), some lacewings (Canard 2001), and many parasitoids (Jervis *et al.* 1996) and ants (Porter 1989; Tobin 1994).

Ants play a key role in the evolution of a range of food-mediated mutualisms, including extrafloral nectar, food bodies, elaiosomes, Lycaenid dorsal gland secretions and certain honeydews (see also Koptur, Chapter 3). The degree in which ants depend on these foods varies widely. The dietary requirements of ants range from species that are primarily predaceous, to species that rely almost entirely on honeydew and extrafloral nectar. Although it has long been held that the majority of ant species are

Plant-Provided Food for Carnivorous Insects, ed. F. L. Wäckers, P. C. J. van Rijn, and J. Bruin.
Published by Cambridge University Press.

predominantly carnivorous (Sudd and Franks 1987; Hölldobler and Wilson 1990), Tobin (1994) argues that the dominant species are largely primary consumers, for which the bulk of their diet consists of plant-derived carbohydrates. An important dichotomy might occur between the nutrition of immature and mature stages. Ants tend to pass on protein-rich food to their larvae, whereas the adults survive mostly on a diet of plant-derived carbohydrates (Haskins and Haskins 1950; Vinson 1968; Boevé and Wäckers 2003). Foraging castes retain the majority of sugar-rich foods, while passing the bulk of protein-rich food to castes remaining in the nest (Markin 1970; Schneider 1972). The importance of carbohydrates to ant nutrition was also demonstrated by Porter (1989). He showed that fire ant colonies kept on insect prey only had a retarded growth and reproduction rate in comparison to colonies fed both prey and sugar-water.

Plant-provided food can have a strong effect on life-history parameters of predators and parasitoids. Facultative consumers can use foods like (extra-) floral nectar, pollen, or honeydew to bridge periods of low prey availability (Limburg and Rosenheim 2001). When combined with predation, these food supplements can increase predator fitness over prey feeding alone (Porter 1989; Van Baalen *et al.* 2001). To the category of obligatory consumers, plant-provided foods are indispensable. Their longevity and fecundity are often seriously compromised in the absence of these food sources (Leatemia *et al.* 1995; Jervis *et al.* 1996; Wäckers 2003b). Lack of suitable food can also have an impact on the insect's behavior, either by affecting the overall activity level, or by causing the insect to cease search for prey or hosts in favor of food foraging (Wäckers 1994; Takasu and Lewis 1995). Through its impact on the fitness and behavior of predators or parasitoids, plant-provided food can be an essential element in the evolution and functioning of multitrophic interactions. There is strong theoretical and empirical evidence that the availability of suitable food supplements can play a key role in the population dynamics of predator–prey (Van Rijn *et al.* 2002) as well as parasitoid–host systems (Krivan and Sirot 1997; Wäckers 2003a).

Despite the obvious importance of non-prey foods to predator and parasitoid ecology, we lack comprehensive comparisons between food categories with regard to their suitability and their role in insect nutrition. This chapter offers such a comparison between floral nectar, extrafloral nectar, pollen, and honeydew. Although honeydew is not directly provided by the plant, it is nevertheless included because this sugar-rich excretion of sap-feeding insects is an important alternative to nectar for many insect groups. Additional plant-derived foods are presented briefly in Box 2.1.

Box 2.1 Additional types of plant-derived food

Fruits

Especially when damaged or rotting, fruits provide a source of readily accessible sugars. They are used by a broad range of insects, including Diptera (Eijs *et al.* 1998), Lepidoptera (DeVries *et al.* 1999), and Hymenoptera (Jander 1998). Ripe fruits may contain considerable levels of carbohydrates. Whereas cultivated fruits are often sucrose-rich, wild fruits are frequently dominated by fructose and glucose (Milton 1999). Fruits are likely of particular importance to the diet of insects that use (fermenting) fruits as oviposition substrate (Eijs *et al.* 1998), as well as their natural enemies.

Plant sap

Damage to plant tissue usually results in the exuding of plant sap. Insects may actively inflict the damage for the purpose of plant-sap feeding (Wellenstein 1952); Eubanks and Styrsky, Chapter 6, or they may use sap at tissues that have been damaged otherwise. A plant's phloem sap usually contains 5–30% sugars (Zimmerman and Ziegler 1975). Although sucrose is the principal phloem sugar in most plants, some plants (e.g., Oleaceae) primarily transport galactose-containing oligosaccharides such as raffinose and stachyose, and the phloem of others can contain high concentrations of sugar alcohols, such as mannitol, sorbitol, or dulcitol. Phloem sap can be inferior as a food source, as it may contain substantial amounts of secondary metabolites (Van Dam *et al.* 2003), whereas the sugar content is often low relative to that of nectar or honeydew. In some plants, however, sites of plant damage may act as an analog to extrafloral nectaries. When *Eucalyptus* plants are damaged by herbivores, they bleed copious amounts of sugar-rich sap that is eagerly collected by ants (Basden 1965; Steinbauer 1996). This way of recruiting ants is particularly effective as an indirect defensive mechanism, since the site of herbivory coincides with the site of ant recruitment.

Leaked assimilates

Tissue damage is not always necessary to make phloem sugars accessible. Undamaged plants passively leak small amounts of carbohydrates and amino acids to the leaf surface. The amounts depend on the plant and its leaf characteristics, as well as the presence of a water film (rain, dew) on the leaf surface (Tuckey 1971). The amount of carbohydrates (sucrose, glucose, and fructose) present on an uninhabited bean leaf surface is in the range 0.2–2 μg (Mercier and Lindow 2000). The role of these sugars in insect nutrition has not been evaluated adequately (Tuckey 1971). However, in addition to the low quantities exuded, insects likely have limited access to these sugars, as they are rapidly broken down by bacterial and fungal epiphytes (Leveau and Lindow 2001).

Fungal fluids

Several rust fungi (Basidiomycetes; Uredinales) employ sugars to attract insects for the purpose of spore dispersal. They produce a sugar-rich fluid from the ostioles of the pycnidium amidst maturing sexual spores. They may also produce specific odors to lure insects to this pycnial fluid (Buller 1950; Ingold 1971). Insects visiting this sugar source facilitate the transfer of spores to other rust colonies where fertilization can occur. Some rust fungi, like *Puccinia monoica*, suppress the flower formation in its host plants (*Arabis* spp.) and instead induce the production of pseudo-inflorescences. These fake flowers often produce a spore-laden "nectar" in amounts exceeding the nectar from uninfested plants (Roy 1993).

Food bodies

Food bodies include Beltian bodies, Müllerian bodies, Beccarian bodies, and pearl bodies, all of which consist of solid epidermal tissues (Rickson 1980). Like extrafloral nectar, food bodies are believed to have evolved as a food reward in plant–ant mutualism. Whereas their function might be similar to that of extrafloral nectar, food bodies are usually rich in proteins and/or lipids, and relatively poor in carbohydrates. This applies especially to food bodies of plants that have strong mutualistic interactions with ants (myrmecophytes) (Heil *et al.* 1998). This nutritional composition enables the plant to retain ant colonies even when the availability of insect protein is low. Some ant species rely entirely on food bodies from their particular host plant for their protein supplements (Carroll and Janzen 1973). Even though food bodies are collected by some non-mutualists (Letourneau 1990), the range of potential consumers is not as broad as in the case of the easily accessible and digestible extrafloral nectar. This makes food bodies less vulnerable to consumption by unintended consumers.

Elaiosomes

In addition to their defensive role, ants also play an important role in the dissemination of seeds. Their tendency to harvest seeds and to transport them to their (underground) nests makes ants effective seed dispersers (Horvitz and Schemske 1986; Jolivet 1998). Some plant species stimulate this interaction by producing seed appendages, the so-called elaiosomes (Milewski and Bond 1982). Elaiosomes are usually lipid rich, and contain lower levels of carbohydrates and proteins (Bresinsky 1963). Ants collect these seeds preferentially, and transport them to their nest where the nutritious appendages are consumed. Subsequently the hard seeds may be discarded in underground waste dumps. The scarring of the seeds, the moist and nutrient-rich conditions in the ant nest, and the clustering of seeds all can benefit germination and seedling growth (Beattie 1985).

Gall secretions

Several galls produce sugar-rich secretions that are commonly frequented by ants (Bequaert 1924). Whereas the sugar secretion is believed to be produced by the epidermal and underlying plant tissue, it is usually the galling insect that appears to be the beneficiary of the ant recruitment, as ants protect the galling insect by aggressively chasing away their parasitoids (Washburn 1984; Fernandes *et al.* 1999) or inquilines (Seibert 1993).

Lycaenid dorsal gland secretions

The recruitment of ants as bodyguards is not restricted to plants (extrafloral nectar) or aphids (honeydew). Caterpillars of various species of blues (Lepidoptera: Lycaenidae) also excrete a glucose-rich fluid from a dorsal gland, which is eagerly collected by ants (Wada *et al.* 2001). Ants return the favor by defending the caterpillars against their enemies. With some blues, this mutualism goes so far that the caterpillars are "adopted" into the ant nest where they receive bed and board (Fiedler *et al.* 1996). There are obvious ecological parallels between the dorsal gland secretions and the honeydew in ant-tended aphids. Still, the latter represents a waste product that only in some plant/sap-feeder systems has evolved a secondary defensive function, while dorsal gland secretions are believed to have a primary function in ant-mutualism.

Parameters affecting food suitability

The rest of this chapter focusses on pollen, floral nectar, extrafloral nectar, and honeydew, as they often represent the main categories of plant-derived food supplements for arthropods. To assess the suitability of these food supplements and to elucidate their potential role in predator and parasitoid food ecology, I discuss each category with regard to the parameter's availability, apparency, accessibility, and nutritional composition. In addition, the foraging risks associated with visiting the food sites are addressed.

Availability

The availability of food sources can vary considerably in time and space. Plant growth and the induction of phenological stages, e.g., flowering, are often synchronized by the same abiotic triggers, such as daylength, seasonal precipitation, or temperature. This can result in distinct periods of high or low availability of plant-provided foods. Whereas the availability of floral nectar and pollen are obviously linked to the (brief) period of flowering, honeydew and extrafloral nectar are often also available during vegetative stages of plant growth, thereby extending food availability. Such differences in abundance and distribution between food sources are addressed.

Gaps in availability of suitable food sources are most likely to occur in low-diversity ecosystems (including agroecosystems). These systems may show distinct peaks in food availability during flowering of the dominant plant species, or during particular sap-feeder outbreaks, but food availability is often low during the remainder of the season. Food scarcity may be aggravated when these ecosystems are dominated by nectariless plants. The fact that more diverse ecosystems typically provide a broader range of food supplements usually helps in equalizing the extremes in food availability.

Apparency

The rate at which insects encounter food sources is not only determined by the availability of these sources, but also by their detectability (Vet *et al.* 1991; Wäckers and Swaans 1993). Food sources differ considerably with respect to their apparency, ranging from flowers employing marked olfactory and visual advertisements to inconspicuous extrafloral nectaries and honeydew. The detectability depends on the stimuli emitted by the food source, the transmitting medium, and the receptor sensitivity of the receiving organism. In combination these parameters describe the field over which an insect may perceive a food source. Most insects primarily use chemical (olfaction, gustation) and visual (color, shape, movement) cues during food foraging (Kevan and Baker 1998). The three-dimensional field within which an insect can visually or chemically detect an object increases by the cube of the range over which receptors can perceive an object. For instance, a duplication in object size increases the visual field of detectability by a factor of eight.

Although the apparency of food sources is of obvious importance in food finding, sensory detection does not warrant that the food source is located succesfully. Food-associated cues may be perceived without triggering a behavioral response, or alternatively the insect may be repelled by the perceived stimuli (Wäckers 2004). The nature of an insect's behavioral response towards food stimuli is determined by its innate preferences and its physiological state (Wäckers 1994). Rewarding as well as unrewarding experiences may modify these innate responses (Takasu and Lewis 1993; Wäckers *et al.* 2002).

Even when perception of a food source evokes a positive response, this does not warrant that the insect will be successful in tracking down the stimulus source. For instance, olfactory information is usually transmitted in meandering plumes of odor pockets (Elkinton and Cardé 1984). Successful location of the odor source requires upwind orientation within the odor plume. Many insects are limited in their ability to fly upwind. These physical restrictions to food detectability apply to small insects, as well as to poor or non-flying species.

Accessibility

Once a food source has been located successfully, consumption may be constrained by its accessibility. Food accessibility depends both on the mouthpart morphology of the insect and the characteristics of the food source. Only a few insects possess mouthparts adapted to pollen feeding (Kirk 1984; Flechtmann and McMurtry 1992; Romeis *et al.*, Chapter 7), whereas morphological adaptations to nectar-feeding are more common, especially among Lepidoptera and Diptera (Faegri and Van der Pijl 1979). Most predators and parasitoids, however, have short mouthparts that are relatively unsuited to reach concealed food sources. The unspecialized labrum of most hymenopteran parasitoids (see Jervis (1998) for exceptions), limits their feeding to easily accessible liquids, such as honeydew, extrafloral nectar, or floral nectar from unspecialized flowers (Hocking 1966; Bugg *et al.* 1987). The mouthpart morphology will also determine the nectar concentration that can be consumed. Insects that take up sugar through a narrow tube-like proboscis, such as Lepidoptera, will often feed on less concentrated (less viscous) sugar sources (Watt *et al.* 1974; May 1985). The mouthparts of most Hymenoptera, Diptera, and Coleoptera allow them to imbibe more concentrated sugar solutions (Baker and Baker 1983a). Some Lepidoptera, (parasitic) Diptera, and Hymenoptera can utilize crystalline sugar by dissolving it in excreted saliva (Stoffolano 1995; Wäckers 2000).

Food source accessibility under laboratory conditions does not guarantee accessibility in the field. Low humidity and exposure to sun and wind can result in higher viscosity of nectar sources. In addition, other consumers may deplete food sources or monopolize and aggressively defend them (Beggs 2001). Besides inaccessibility of food sources, lack of consumption may also be due to rejection of the feeding site (Henning *et al.* 1992; Wäckers 2004), or the food source itself (Feinsinger and Swarm 1978). Therefore studies on food accessibility will have to be supplemented with data on food acceptance.

Composition/nutritional suitability

Once an accessible food source has been located, food suitability still hinges on food ingestion (gustatory acceptance) and the successful metabolization of consumed compounds. Nectar, honeydew, and pollen differ considerably with regard to their composition, e.g., the level and type of carbohydrates, amino acids, proteins, lipids, and secondary plant metabolites. The particular chemical composition affects both food acceptance and nutritional suitability. The decision to ingest or reject food sources is often governed by the presence of feeding stimulants and feeding deterrents (Chapman 1995). The stimulatory effect of the common sugar sucrose is almost universal, but insects vary with regard to their gustatory sensitivity and response towards other sugars (Wäckers 1999). Several insects prefer a sugar solution with amino acids over nectar solutions containing sugar only (Ricks and Vinson 1970; Inouye and Waller 1984; Lanza and Krauss 1984; Potter and Bertin 1988; Alm *et al.* 1990; Rusterholz and Erhardt 1998). Wada *et al.* (2001) showed that the amino acid glycine enhances the electrophysiological response of sugar receptors in the ant *Camponotus japonicus*, and increases its response to glucose.

Rejection of food sources may be based on the presence of feeding deterrents (Stoffolano 1995), on the absence of appropriate (concentrations of) feeding stimulants, or on their masking by other food constituents. Various compounds can act as feeding deterrents, including certain sugars (Dethier 1976; Allsopp *et al.* 1998) and amino acids (Shiraishi and Kuwabra 1970; Potter and Bertin 1985). Even stimulatory compounds may become deterrent at high concentrations (Dethier 1976; Hern and Dorn 1999). Several instances of feeding deterrency have been attributed to secondary metabolites (Feinsinger and Swarm 1978; Stephenson 1982a; Detzel and Wink 1993). However, it has to be considered that deterrency is species dependent. Specialized insects may actually be stimulated by compounds that are deterrent to generalists (Schoonhoven 1972). Also the response of the individual to feeding deterrents can vary relative to its physiological state (Simpson and Raubenheimer 2001).

Following ingestion, compounds have to pass through several physiological steps before they are assimilated successfully. First, the molecule has to be absorbed through the gut wall, which often requires enzymatic breakdown. Once absorbed, molecules often need to be metabolized before they can be assimilated or stored. Each of these steps may represent physiological hurdles that restrict the nutritional value of a food item.

Insects may be able to recognize and avoid unsuitable food sources. Positive correlations between nutritional suitability and gustatory response have been reported for several nectar-feeders (Von Frisch 1934; Wäckers 2001). However, gustatory responses are not always adapted to the recognition of unsuitable food items, as illustrated by numerous reports of insect poisoning following ingestion of toxic nectar, pollen, or honeydew (Barker 1990; Detzel and Wink 1993; Adler 2001).

Toxicity of plant-provided foods may be due to primary metabolites, such as toxic sugars or non-protein amino acids, or to plant secondary metabolites. The occurrence of toxins in foods intended for the recruitment of pollinators has long puzzled ecologists and has given rise to a range of intriguing hypotheses proposing various ecological functions (Adler 2001). Rather than seeking an adaptive explanation, Ehrlén and Erickson (1993) and Adler (2001) propose that the occurrence of secondary plant metabolites in plant-provided food supplements may be due to inadvertent "leakage" of these compounds from the phloem into fruits or nectar. This would thus represent a pleiotropic constraint of having phloem-transported defensive chemicals.

Foraging risks

Not only are foraging insects at risk of encountering toxic or insufficient food (starvation), the act of foraging in itself also poses risks. These risk factors include adverse abiotic conditions (wind, rain, drought) (DeBach *et al.* 1955; Dyer and Landis 1996; Schwörer and Völkl 2001) as well as biotic factors (encountering pathogens, or predators) (Stamp and Bowers 1988; Weisser *et al.* 1994; Völkl and Kroupa 1997; Rosenheim 1998). It is difficult to quantify the relative contribution of various risk factors to insect mortality under field conditions. Out of the range of potential mortality factors, predation risk seems to have received the most attention.

Rosenheim (1998) reviewed a number of field studies that quantify parasitoid predation risks. Although this gives a good general overview, the available studies do not specifically differentiate between food foraging and other activities. With respect to food foraging, two broad risk categories can be distinguished. The first involves risks associated with traveling between patches. Movement increases both exposure to passively hunting predators such as web spiders (Völkl and Kraus 1996) and visibility towards visually hunting predators (e.g., birds). The second category includes specific risks associated with particular feeding sites. In this chapter I focus on the latter risk category, even though traveling risks are likely to be also important in shaping insect foraging decisions. Where feeding-site inherent risks can be identified, they are addressed per food source.

In the remainder of this chapter, I apply the above criteria to assess the suitability of nectar, extrafloral nectar, pollen, and honeydew as insect food sources. These food categories are very broad, and treating them as singular classes may invite sweeping generalizations that ignore the often substantial variation within each group. Overemphasis on variation and exceptions, on the other hand, may obscure the forest for the trees. I have tried to strike a balance by identifying general attributes that characterize the different categories, while also addressing exceptions and sources of variation within each category.

Food suitability is determined by consumer features, as much as it depends on attributes of the food source. In order for an insect to be able to exploit a particular food source, its behavior, morphology, and physiology need to match the appearance, build, and chemical composition of the food source. I have tried to present examples from different insect taxa. Nevertheless, in keeping with the scope of this book and affected by my own research background, this chapter has a bias towards natural enemies, in particular parasitoids.

Floral nectar

Function and consumers

Floral nectar is one of the primary rewards by which plants recruit pollinators. Early on in the evolution of entomophilous flowers, the nutritious pollen served as the sole food reward (Labandeira 1997). During the co-evolution between plants and pollinators, floral nectar evolved as an additional food lure that in many cases has become the primary reward. In comparison to pollen, the nitrogen-poor nectar is usually cheaper to produce and it is both easier and faster to consume and digest by a wider group of pollinators. Moreover, the switch to the new food reward is believed to have been facilitated by the fact that many insects were pre-adapted to the exploitation of nectar, as they were already using "older" sugar sources such as honeydew (Downes and Dahlem 1987).

The phrase "floral nectar" is primarily based on the location of the nectary (within the flower) and the timing of secretion (during anthesis) rather than being defined by its function in pollination. This demarcation excludes nectar produced on the outside of inflorescences, e.g., bracts, even when these nectaries are involved in pollination (Ford and Forde 1976), as well as floral nectar produced after anthesis has been completed (the so-called post-floral nectar) (Daumann 1932). On the other hand, it includes floral nectaries that have lost their pollination function, such as the nectaries consisting of entire but reduced flowers found among Pedaliaceae, tropical Fabaceae, and Tiliaceae (Van der Pijl 1951). These reduced flowers likely represent an indirect defensive function analogous to extrafloral nectaries as they are intensively visited by ants. A predominantly defensive function of floral nectar has also been described in other plant species (Domínguez *et al.* 1989; Yano 1994).

Floral nectar is utilized by a wide range of insects from various taxa, especially by Hymenoptera, Lepidoptera, Coleoptera, and Diptera (Kevan and Baker 1998; Romeis *et al.*, Chapter 7). The mutualism between flowers and their particular pollinators has resulted in the evolution of clearly distinguishable pollination syndromes, that often feature adaptations both from the plants and the pollinators (Kugler 1970). Insects that contribute little or nothing to pollination may freeload on the floral rewards by seeking out flowers and collecting pollen and nectar. This category includes many predators (Bugg *et al.* 1991; Bakker and Klein 1992b; Ruhren and Handel 1999) and parasitoids (Kevan 1973; Jervis *et al.* 1993).

Whereas ants are the principal consumers of extrafloral nectar, they visit floral nectaries less frequently (Janzen 1977; Rico-Gray *et al.* 1998). Ants that do exploit floral nectar often act as nectar robbers or otherwise contribute little to pollination (Inouye 1983). Only a few examples of ant-pollinated flowers have been described (Beattie 1985).

Availability

The availability of floral nectar can vary considerably in time and space, as both the composition of the local vegetation and its phenology are determined by geographic, seasonal, and environmental factors. Flowering activity may be triggered by temperature, daylight, and/or precipitation. This can result in synchronized periods of high or low flower availability (Abe and Kamo 2003). For example, in central Europe flowering plants are scarce in early spring and late autumn. Moreover, the plant species flowering during these periods are characterized by pollen-rich flowers, adding to the relative shortage in nectar supply. In contrast, early summer is characterized by copious nectar supply and lower pollen availability (Gleim 1985).

Given the limited lifespan and activity radius of most insects, the individual foraging insect is primarily confronted with local and short-term variation. This variation can be substantial, as plants have evolved various mechanisms to minimize the costly production of floral nectar. Blooming of individual flowers is often limited to a few days. During this short blooming period, most plants do not secrete floral nectar continuously, but rather according to fixed daily rhythms. Many plants show a peak in nectar production during morning hours, whereas some produce mainly in the afternoon or at night (Kleber 1935). The diurnal peak in nectar secretion usually coincides with the time of day (or night) when the plant's main pollinator is active. Some plants actually resorb residual nectar during the remainder of the day (Faegri and Van der Pijl 1979; Burquez and Corbet 1991).

The amount of nectar excreted per flower can vary broadly between plant species. Expressed in sugar dry weight this may range from less than 10 μg to as much as 163 mg sugar per flower (Percival 1965). The daily amount of nectar secreted is lowest in generalist open bowl flowers and highest in specialized zygomorphic and stereomorphic flowers (Kevan and Baker 1998). The volume and duration of floral nectar secretion and the composition of the excreted nectar are primarily plant characteristics, but several factors may affect the quantity and quality of the secreted nectar. These include the age of the nectary, irradiance, temperature, relative humidity, soil conditions, nutrients, and water balance (Beutler 1935; Burquez and Corbet 1991), the state of pollination (Gori 1983), and nectar removal (Cruden and Hermann 1983). The amount of nectar actually available to the flower visitor will be limited by exploitative competition among nectar feeders (Comba *et al.* 1999; Hansen *et al.* 2002; Lee and Heimpel 2003).

Apparency

Among plant-provided food supplements, floral nectar is likely easiest to detect for food-foraging insects due to the fact that zoophilous flowers advertise their floral rewards with conspicuous visual and olfactory signals. Flower coloration and shape, as well as the floral odor blend are often adjusted to the sensory preferences of the primary pollinator (Dobson 1994).

To the foraging insect, the visual apparency of flowers will depend on floral size, shape, and color, and on the contrast to the background. Even insects with color vision perceive achromatic contrast from a larger distance than color contrasts (Giurfa *et al.* 1996). The distance from which a flower can be detected is therefore determined by both the surface area that produces achromatic contrast with the background (Ne'eman and Kevan 2001) and the visual acuity of the insect's compound eye (Wehner and Srinivasan 1989). The resolution of a typical insect eye is almost two orders of magnitude poorer than that of human eyes. Given sufficient green contrast, a flower with an 8-cm diameter is visible from about 90 cm, while in absence of this contrast this is reduced to 15 cm (Giurfa *et al.* 1996). Movement, as experienced when approaching a flower, also enhances visual acuity (Wehner and Srinivasan 1989). Many insects are (innately) attracted to specific colors, of which yellow appears to have a particularly broad appeal (Kevan and Baker 1998). When deprived of sugar feeding, the parasitoid *Cotesia rubecula* seeks out yellow surfaces and shows feeding-specific behaviors after landing on this color. Following sugar satiation, parasitoids no longer exhibit these behaviors (Wäckers 1994). Given that the vast majority of palearctic flowers are either yellow or white (Weevers 1952; Kevan 1972), and that pollen is also predominantly yellow (Osche 1983), a limited set of innate color preferences would suffice to cover the bulk of floral nectar sources.

The detectability of floral odors will depend both on the specific blend of emitted volatiles and the rate of emission. Once an insect perceives a floral odor blend, its behavioral response will depend on the insect's innate preferences and its previous experience (Dobson 1994; Takasu and Lewis 1996; Wäckers *et al.* 2002). The range of floral odors is vast (Knudsen *et al.* 1993). Floral fragrances are typically bouquets of several odors from different chemical classes, including isoprenoids, benzenoids, aminoids, and derivates of fatty acids (Dobson 1994). Flower visitors may exhibit innate responses to a few common floral volatiles, rather than to the endless variety of possible odor blends. Linalool is an example of such a common floral scent (Knudsen *et al.* 1993), attractive to honeybees (Henning *et al.* 1992) as well as to the parasitoid *Aphidius ervi* (Du *et al.* 1998). In a study on innate responses of three hymenopteran parasitoids to flower odors, food-deprived individuals exhibited a positive response to only four out of the eleven flowers tested, while they were repelled by two flowers (Wäckers 2004). These results indicate that inexperienced parasitoids might forage by means of innate odor repellency as much as by innate attraction.

Accessibility

Flower morphology has been recognized as one of the most important factors limiting floral nectar use (Faegri and Van der Pijl 1979; Jervis *et al.* 1993). In contrast to the generally exposed extrafloral nectar and honeydew, floral nectaries are often hidden within complex floral structures. This strategy not only protects floral rewards from drought and rain, it also enables the plant to restrict nectar access to a limited subset of consumers. Some organisms circumvent this obstacle course and use their biting mouthparts to gnaw their way to the hidden nectar. These so-called nectar robbers are found among short-tongued bees, wasps, ants, beetles, and birds (Inouye 1983). For those visitors that access flowers along the intended route, nectar accessibility is determined by the match between floral architecture and the morphology and behavior of the flower visitor (Patt *et al.* 1997; Jervis 1998; Kevan and Baker 1998). Once the nectar feeder has successfully managed to collect the nectar, it can improve its flower handling through (operant) learning (Lewis 1986).

Most hymenopteran parasitoids (Gilbert and Jervis 1998) and many predators have short mouthparts that restrict their feeding to exposed nectaries. Wäckers *et al.* (1996) and Wäckers (2004) investigated the ability of three nectar-feeding parasitoid species to access floral nectar from a range of flowering plants. Nectar uptake, as indicated by an increase in weight, was recorded in *Aegopodium podagraria* and *Daucus carota* (both Apiaceae), as well as in *Vicia sepium* (Fabaceae), all of which offer easily accessible nectaries. Parasitoids confined to the other six plant species were not successful in obtaining nectar. Jervis *et al.* (1993) assessed correlations between parasitoid size (head width) and flower visitation. Larger ichneumonids were found to be mostly excluded from exploiting the narrow flowers of Compositae and Fabaceae. This disadvantage is compensated by size-related advantages during food search. Bigger insects are usually better suited for resource-directed orientation, such as upwind flight in flower odor plumes or directional location after visual fixation upon the flower image.

Being restricted to exposed nectaries implies that the available nectar is often relatively concentrated. The resulting high viscosity can further restrict accessibility. However, parasitoids can deal with a wide range of nectar concentrations (Siekmann *et al.* 2001) and to some extent even manage to feed when presented with crystalline sucrose (Wäckers 2000).

Under field conditions numerous insects compete for limited nectar resources. As a result, individual flowers may hold little or no nectar (Comba *et al.* 1999). This exploitative competition is likely to be most fierce in the case of accessible nectaries. When aggressive competitors monopolize nectar sources and attack subsequent flower visitors, this direct competition impedes access to even the most exposed nectary. Kikuchi (1963) established dominance rankings between flower visitors, showing that bumblebees chase away syrphids, which in turn outcompete butterflies. The fact that the exploitation of floral nectar by ants is subordinate to their use of extrafloral nectar (Beattie 1985) implies that the impact of ants on flower visitors is often relatively minor.

Plants have evolved various mechanisms to prevent ants from impeding plant–pollinator interactions. Myrmecophilous plants may employ floral ant repellents that (temporarily) segregrate pollinators from defensive mutualists (Willmer and Stone 1997; Ghazoul 2001). In addition, various morphological barriers have been described that prevent crawling insects from entering the inflorescence, including sticky belts in *Viscaria vulgaris* (Faegri and Van der Pijl 1979), glandular hairs and waxy corrolas (Feinsinger and Swarm 1978; Beattie 1985). It is believed that these morphological traits primarily serve to keep ants from accessing flowers (Beattie 1985). The fact that ants are often unwelcome to floral nectar can be explained by their low effectiveness as pollinators (Beattie 1985), composited with their high effectiveness in robbing floral rewards (Faegri and Van der Pijl 1979) and/or castrating flowers (Izzo and Vasconcelos 2002). Feinsinger and Swarm (1978) and Guerrant and Fiedler (1981) demonstrated that it is usually not nectar quality that keeps ants from visiting flowers. They showed that nectar collected from plants not visited by ants is readily accepted when offered outside the flower. However, the nectar of two of the species tested by Feinsinger and Swarm (1978) did prove to be ant repellent, showing that nectar consumption can be restricted by its chemical composition as well.

Composition/nutritional suitability

Nectar is a sugar-rich excretion derived from phloem sap alone, or from a combination of phloem and xylem sap (Fahn 1988). In addition to carbohydrates, nectar may contain various amino acids, and lesser amounts of proteins, lipids, vitamins, and secondary plant metabolites as well as other organic compounds and minerals (Baker and Baker 1982). Besides positive nutritional effects, certain nectar constituents may actually reduce insect survival (Sols *et al.* 1960; Detzel and Wink 1993; Adler 2001; Wäckers 2001; Romeis and Wäckers 2002).

Carbohydrates

Sugar concentrations in nectar can vary considerably with regard to the types of saccharides and their relative proportions. In the majority of plants sugar concentrations range between 20% and 40%, with extremes as low as 8% in *Fritillaria imperialis* and as high as 76% in *Origanum vulgare* (Kugler 1970). Environmental conditions may affect nectar concentration both indirectly through their effects on the nectar-producing plant, and directly through evaporation, hygroscopy, or rain dilution (Dafni 1992).

Floral nectar is generally dominated by the monosaccharides fructose and glucose and the disaccharide sucrose (Baker and Baker 1982). Fungi and bacteria introduced by nectar visitors can alter the nectar composition (Baker and Baker 1983a). In general, the proportions of the three main sugars are rather constant within a species, but broad differences exist between flowering species. Percival (1961) and Baker and Baker (1982) showed that the sucrose/hexose ratios of flowering plants can vary from less than 0.1 to more than 0.999. The sucrose/hexose ratio can be correlated with particular pollinator groups, indicating different sugar preferences among insects. Preferences for sucrose over glucose and fructose have been described for Hymenoptera (Fonta *et al.* 1985; Cornelius *et al.* 1996; but see Koptur and Truong 1998) and Lepidoptera (Rusterholz and Erhardt 1997; Romeis and Wäckers 2000), whereas flies are mainly associated with hexose-dominated nectars (see also Romeis *et al.*, Chapter 7). Insects are generally able to assimilate the three common nectar sugars and use them in energy metabolism. In addition to the three main nectar sugars, nectar may contain low concentrations of other carbohydrates such as raffinose, galactose, mannose, or xylose (Baker and Baker 1983a; Jackson and Nicolson 2002; Wäckers 2003b). Some of these minor nectar sugars can be deterrent, nutritionally unsuitable, or even toxic to insects (Finch and Coaker 1969; De la Fuente *et al.* 1986; Wäckers 2001; Romeis and Wäckers 2002). In mixtures, unsuitable sugars sometimes suppress the nutritional benefit of suitable nectar sugars (Wäckers 2001).

Amino acids

Floral nectar often contains a broad range of amino acids at relatively low levels. The overall concentration in nectar typically lies within the range of 30 to 400 μg/ml (Baker and Baker 1975; Gottsberger *et al.* 1984; Rusterholz and Erhardt 1998). In some nectars the amino acid fraction is dominated by essential amino acids (Dafni *et al.* 1988), whereas in other nectars more than 90% are non-essential amino acids (Rusterholz and Erhardt 1998). Several insects prefer nectar mimics with amino acids over an alternative with sugars only (Inouye and Waller 1984; Lanza and Krauss 1984; Potter and Bertin 1985; Alm *et al.* 1990; Rusterholz and Erhardt 1998; but see Romeis and Wäckers 2000 for exceptions). Baker and Baker (1975) showed that overall amino acid levels correlate with pollinator type. By far the highest amino acid levels are found in nectar from carrion and dung-fly pollinated flowers. Nectar from flowers pollinated by non-pollen-feeders present about twice the amino acid content as compared to flowers pollinated by pollen-feeding insects (Kevan and Baker 1983). The amino acid composition of floral nectar can vary with a range of factors, including soil nitrogen levels (Gardener and Gillman 2001), CO_2 levels (Rusterholz and Erhardt 1998), time of day (Willmer 1980), age of the flower, and floral damage inflicted by nectar robbers (Willmer 1980; Gottsberger *et al.* 1990). Insect visitors may increase amino acid levels through their saliva, or by introducing pollen grains (Willmer 1980).

Secondary plant metabolites

Extensive studies, in which hundreds of nectars were sampled and analyzed, have shown that secondary metabolites are rather common in nectar (Baker *et al.* 1978). However, only a few studies have shown that secondary metabolites in nectar affect pollinators (Waller 1972; Carey and Wink 1994; Adler 2001). Nectar may contain (considerably) lower levels of secondary metabolites in comparison to other plant tissues (Detzel and Wink 1993). However, the picture can also be reversed. Peumans *et al.* (1997) showed that lectin levels in nectar from *Allium porum* are six times those found in foliage. Although toxic effects of nectar, or various of its components, have been well documented, it would be erroneous to conclude that floral nectar in general is an inferior food. Floral nectar composition has been particularly scrutinized because of the fact that it is the primary source for honey production. Moreover, many of the reports on unsuitable nectars are based on laboratory studies and do not necessarily translate to the field. Many examples of bee-toxic nectars involve flowers that are not visited by bees and whose actual pollinators are not affected (Adler 2001).

Foraging risks

Flower visits, be it for the purpose of pollen- or nectar-feeding, can represent particular mortality risks. For one, the plant itself might kill its pollinators, by trapping them with glue, webs, watery pits, or closing flowers (Schubert 1972). In addition, flowers may also harbor predators that hide out in the inflorescence to ambush and kill visitors (Boyko 1939; Morse 1986; Jervis 1990). Crab spiders are especially adapted to this predation strategy. Some species actively blend in with the coloration of the floral background rendering them invisible to approaching flower visitors (Chittka 2001), while the coloration of others creates a color contrast against the floral background, increasing attractiveness to bees with innate preference for color patterning (Heiling *et al.* 2003). Crab spiders primarily hunt during the day, but nocturnal flower visitors are not necessarily out of harm's reach. Hunting spiders of the families Heteropodidae and Ctenidae often reside on night-blooming species. They recognize, locate, and attack flower-approaching insects on the basis of their wing vibrations (Wasserthal 1993). Wasserthal (1993) proposes that the extremely long proboscis of some hawkmoth species combined with an oscillating hovering flight has evolved as a strategy to avoid this floral predation. While crab spiders may seek out flowers, the risk of ant attack during flower visits is relatively low as compared to more intensively ant-tended food sources such as honeydew and extrafloral nectar.

Flowers can also harbor pathogens or parasites. In the case of social insects, where parasites and disease can spread within the colony, infections may be more costly than predation. A spiroplasma species lethal to honeybees was recorded in nectar from the tulip poplar (*Liriodendron tulipifera*) and southern magnolia (*Magnolia grandiflora*) (Clark 1978). Durrer and Schmid-Hempel (1994) reported that a trypanosome parasite can be horizontally transmitted between bumblebees through the shared use of flowers. Phoretic mites may use flowers to transfer between pollinators (Schwarz and Huck 1997), or use pollinators to transfer between flowers (Seeman 1996). It remains unclear whether these phoretic mites constitute a fitness cost to the pollinator.

Overall, flower-dwelling antagonists may represent a considerable cost to the use of floral food. Some flower visitors are able to avoid perceivable risks (Dukas 2001). Over evolutionary time flower-associated mortality risks are likely to have shaped the pattern of flower use and may have favored the use of less risky food alternatives.

Extrafloral nectar

Function and consumers

Extrafloral nectar glands include a wide range of nectar-excreting structures that can be found on various plant parts, such as leaves, stems, bracts, and fruits. Unlike their floral counterparts, they are usually not involved in pollination (but see Ford and Forde 1976 for an exception). Instead, they are believed to function as an indirect plant defense: production of extrafloral nectar allows plants to recruit predators or parasitoids that may return the favor by safeguarding plants against herbivory (Bentley 1977a; Koptur 1992; Turlings and Wäckers 2004; see also Koptur, Chapter 3). In some instances, extrafloral nectar can also have a direct negative impact on herbivores, trapping them in the sticky exudate (Putman 1958; Wootton and Sun 1990).

In several plant systems it has been established that ants attracted by extrafloral nectar both reduce plant damage (O'Dowd and Catchpole 1983; Heil *et al.* 2001) and increase plant reproductive fitness (Rico-Gray and Thien 1989; Oliveira 1997; Wagner 1997). Extrafloral nectar can make a significant contribution to the diet of ants. Fisher *et al.* (1990) reported that six ant species investigated derive between 11% and 48% of their diet from extrafloral nectar. Retana *et al.* (1987) found (extrafloral) nectar to be the main food source for the ant *Camponotus foreli.*

Although ants are often the primary visitors of extrafloral nectar, extrafloral nectaries are also frequented by a range of other carnivorous arthropods (Bugg *et al.* 1989; Koptur 1994; Whitman 1996). The provision of these food supplements may enhance herbivore suppression by spiders (Ruhren and Handel 1999), predatory mites (Bakker and Klein 1992a), coccinelids (Stephenson 1982b), predatory wasps (Cuautle and Rico-Gray 2003), or parasitoids (Lingren and Lukefahr 1977; Stapel *et al.* 1997). The nutritional benefits of extrafloral nectar consumption have been demonstrated widely. Extrafloral nectar not only represents an excellent source of energy, it may also sustain development of immatures (Van Rijn and Tanigoshi 1999a; Limburg and Rosenheim 2001) and increase fecundity in mature insects (Lingren and Lukefahr 1977). The free amino acids and proteins presented by extrafloral nectar are usually insufficient to cover the protein requirements of carnivorous insects, but they can allow predators to bridge periods with low prey densities (Van Rijn and Tanigoshi 1999a; Limburg and Rosenheim 2001).

Availability

Extrafloral nectaries have been described in over 1000 species from 93 plant families (Bentley 1977a; Koptur 1992). They occur in numerous dicotyledonous species, as well as in such diverse monocotyledonous taxa as ferns, lilies, orchids, sedges, and grasses. They are found in virtually all plant types including herbs, vines, shrubs, and trees; annuals as well as perennials; successional as well as climax species (Koptur 1992; Whitman 1996). While extrafloral nectar plants are most abundant in the tropics (Zimmerman 1932), they also occur in the temperate zones (Koptur 1992). Extrafloral nectaries are not restricted to one part of the plant, but can be found on various vegetative parts, as well as on flowers and fruiting structures (Bentley 1977a; Koptur 1992). The often copious nectar volume secreted by extrafloral nectaries may exceed floral nectar production markedly. Individual leaves of *Ricinus communis* excrete 3.6 mg of sugar per day, representing 1‰ of the leaf's daily assimilate production (Wäckers *et al.* 2001). Bracteal nectaries of *Gossypium hirsutum* excrete up to 12 mg of highly concentrated nectar per fruit per day (Wäckers and Bonifay 2004). One hectare of cotton represents a daily production of 3.8 liters extrafloral nectar (Butler *et al.* 1972). Nectar production is even higher in bull's horn acacias (*Acacia sphaerocephala*) that have been reported to produce 1 ml per nectary per day (Janzen 1966). As in floral nectar, the production of extrafloral nectar can be affected by a range of abiotic factors, including light conditions, soil nutrient levels, and humidity.

In addition to the often high rates of excretion, extrafloral nectaries also usually produce nectar over extended periods, functioning much longer (3–6 weeks) than floral nectaries (1–3 days on average) (Rogers 1985; Koptur 1992). In *Gossypium* spp., where floral nectar production is limited to a single day of flowering, nectar production by bracteal nectaries exceeds 5 weeks (Wäckers and Bonifay 2004). The fact that extrafloral nectaries are often associated with vegetative growth further prolongs the period over which extrafloral nectar is present.

Some extrafloral nectaries produce nectar in a distinct diurnal pattern (Bentley 1977b; Gaume and McKey 1999; Heil *et al.* 2000), whereas others have an almost continuous production (Bentley 1977a). Most plants excrete some level of extrafloral nectar irrespective of the presence of herbivores. Such strict constitutive nectar production may be synchronized, for example, with the most susceptible stages of plant growth (Bentley 1977a), with the periods of activity of damaging herbivores (Tilman 1978), or with the daily activity pattern of ants (Pascal and Belin-Depoux 1991). Further synchronization is achieved in those plants that actively increase nectar secretion in response to herbivory (Koptur 1989; Wäckers and Wunderlin 1999; Heil *et al.* 2001; Wäckers *et al.* 2001) or ant attendance (i.e., nectar removal) (Koptur 1992; Heil *et al.* 2000). This sophisticated two-way induction allows plants to match their defensive investment with the presence of both the herbivore and the carnivore. However, not all types of extrafloral nectar induction may be adaptive to the plant (Wäckers and Bezemer 2003).

Apparency

For insects that forage by vision or olfaction, the exploitation of extrafloral nectar is likely restricted by the fact that these food rewards are usually far less conspicuously advertised than floral nectar. As most extrafloral nectars themselves do not have obvious odors, Koptur (1992) concluded that olfaction is probably not involved in their detection. In studies with *Cotesia rubecula* (F. Wäckers, unpublished data) no long-range volatile attraction to *Vicia faba* extrafloral nectaries could be demonstrated, but short-range attraction of *Microplitis croceipes* to cotton extrafloral nectaries has been reported (Stapel *et al.* 1997). The conspicuous coloration of some extrafloral nectaries is thought to represent a visual signal advertising the extrafloral nectar reward to the intended insect visitors. It is remarkable, though, that the color markings of extrafloral nectaries are typically (dark) red or black. Unless these visual marks reflect ultraviolet (UV) light, their hue would lie outside the color spectrum of most insects (Briscoe and Chittka 2001). Even though the studies addressing color vision in ants are few, the studies available did not find red-sensitive receptors (Briscoe and Chittka 2001; but see Kretz 1979). The often dark coloration of extrafloral nectaries also represents a brightness contrast against the lighter surrounding plant tissue. Again, dark coloration does not seem to be the most suitable attractant, since many insects prefer lighter over dark substrates. However, ants may be attracted to the small dark spots if they mistake them for aphids. Alternatively the dark coloration may play into the tendency of ants to seek out (dark) crevices (Huxley 1980).

Accessibility

Given that ants, as well as most other predators and parasitoids, lack elongated mouthparts, extrafloral nectar has to be easily accessible in order to effectively cater to this group. The far majority of extrafloral nectaries are indeed fully exposed, as they typically occur on the surface of leaves, stipules, stems, or fruits (Koptur 1992). In a few instances, extrafloral nectaries might be embedded in tissues of other organs (Elias 1983). Zimmerman (1932) distinguished six basic groups of extrafloral nectaries: formless, flattened, pit, hollow, scale-like, and elevated nectaries. In all of these categories, the nectary is basically exposed. Access might be restricted for some elevated extrafloral nectaries, like the stalked nectaries found on for instance *Sambucus nigra* and *Ricinus communis*.

In the field, the access to extrafloral nectar is likely primarily limited by the presence of dominant ants. Their tendency to aggressively monopolize extrafloral nectar limits the use of this food source by other species (Blüthgen *et al.* 2000).

However, not all ants are equally aggressive (Bentley 1977a). Furthermore, the intensity of ant tending may vary with factors such as proximity of the ant nest (Inouye and Taylor 1979), climate conditions (Wirth and Leal 2001), period of the day, and general ant activity (Orivel and Dejean 2002).

Composition/nutritional suitability

Like floral nectar, extrafloral nectar is typically dominated by sucrose and its hexose components glucose and fructose (Koptur 1994). The fact that these common sugars are acceptable to the majority of insects makes extrafloral nectar a suitable food source for a broad range of consumers. Additional sugars, such as galactose, raffinose, xylose, and arabinose, have been found in some types of extrafloral nectar (Bowden 1970; Bory and Clair Maczulajtys 1986; Caldwell and Gerhardt 1986; Olson and Nechols 1995). As these sugars are less suitable for potential consumers (Dethier 1976; Wäckers 2001; Romeis and Wäckers 2002; Boevé and Wäckers 2003) it will be interesting to investigate the ecological implications of their presence in extrafloral nectar.

In general, overall sugar concentrations of extrafloral nectaries vary much more than floral nectaries from the same plant, even when protected from rain (Koptur 1994). Although herbivory can induce nectar secretion, it does not necessarily affect the sugar concentration or sugar composition of the secreted nectar (Wäckers *et al.* 2001). Compared to floral nectar, extrafloral nectar often has a higher overall sugar concentration (Gracie 1991; Koptur 1994; Wäckers *et al.* 2001), as well as increased fructose and glucose levels (Tanowitz and Koehler 1986; Koptur 1994). These characteristics can be explained by the exposed nature of most extrafloral nectaries, which both expedites microbial breakdown of sucrose and increases evaporation. The high sugar concentration may also serve an ecological function (Wäckers *et al.* 2001). High sugar concentrations reduce intake by visiting ants, prolong ant visits (Josens *et al.* 1998), and help prevent nectar use by a range of non-intended visitors (Wäckers 2001). The latter applies especially to Lepidoptera, whose mouthpart morphologies restrict them to feeding on nectar with relatively low sugar concentrations. The shift from sucrose to hexose sugars may also reduce the attractiveness of extrafloral nectar to Lepidoptera (Rogers 1985; Romeis and Wäckers 2000).

In addition to carbohydrates, extrafloral nectar may contain variable amounts of amino acids, as well as low concentrations of lipids and vitamins (Baker *et al.* 1978; Smith *et al.* 1990). The amino acid composition usually differs between extrafloral and floral nectar of the same plant (Baker *et al.* 1978). Certain amino acids (e.g., cysteine) are more frequently found in extrafloral nectar, while glutamic acid appears to be more prevalent in floral nectar (Baker *et al.* 1978). In species that feature both extrafloral and floral nectar, the overall level of amino acids is often lower in the extrafloral variant (Pate *et al.* 1985; Wunnachit *et al.* 1992; Koptur 1994; but see Pickett and Clark 1979). The reduction in amino acid levels relative to phloem sap is usually far more pronounced (Baker *et al.* 1978; Pate *et al.* 1985). Plants may increase the amino acid complement of extrafloral nectar in response to herbivory (Smith *et al.* 1990), making the food reward more appealing to ants and thus enhancing their recruitment (Lanza 1988; Wagner and Kay 2002). Nevertheless, extrafloral nectar by itself usually falls short from providing a well-balanced diet. Low amino acid levels or the absence of certain essential amino acids may be adaptive to the plant, as they force predators to seek out supplementary protein sources, thereby stimulating predation (Hagen 1986).

Relative to the substantial number of accounts of toxic floral nectars, few cases of insect poisoning have been attributed to the consumption of extrafloral nectar (Barker 1990). Part of this bias may stem from the fact that insect toxicity has received most attention in honeybee research, a field that has traditionally concentrated on floral nectar. However, the fact that ant repellency has been described in floral but not in extrafloral nectar (Feinsinger and Swarm 1978; Stephenson 1982a) may indicate that the latter may overall be more suitable, at least for ants.

Not only do we know little about negative consequences of extrafloral nectar consumption, the studies that have investigated the presence of plant secondary metabolites in extrafloral nectar are equally scant. Calatayud *et al.* (1994) reported the presence of flavonoid and cyanogenic glycosides as well as cyanide in extrafloral nectar of cassava. Our own studies have shown that extrafloral nectaries of cotton species can contain low levels of gossypol and other terpenoids (F. Wäckers, unpublished data). Overall, extrafloral nectar appears to be a broadly suitable food source. Additional comparative studies will be needed to compare its suitability relative to floral nectar.

Foraging risks

Crab spiders are usually associated with flowers (Morse 1986), but they have also been reported to occasionally lie in wait near extrafloral nectaries (Knoll 1930). The picture is inverse for ants, which are more frequently found in association with extrafloral nectaries (Koptur 1992; Rico-Gray *et al.* 1998). Ants vary with regard to the aggressiveness with which they defend attended sugar sources (Beattie 1985). The most aggressive species often monopolize the most remunerative nectar sources (Blüthgen *et al.* 2000). Although this tendency enhances the defensive function of extrafloral nectar, it creates considerable mortality risks for other nectar visitors. These risks apply in particular to small and/or poorly flying insects since large and mobile insects are more likely to evade ant attacks. Insects may be able to visually recognize and avoid ant-tended plants, as was demonstrated for the nymphalid butterfly *Eunica bechina* (Freitas and Oliveira 1996). Such early recognition would allow insects to reduce foraging risks. Ant predation is less of an issue in agricultural systems, as frequent soil tillage suppresses soil-nesting ant species.

Pollen

Function and consumers

During the process of pollination, pollen serves to transfer a plant's male genetic information from the stamen to the female pistil. In self-pollinating plants this transfer takes place within the flower. Cross-pollinating plants have to ensure that the pollen finds its way to flowers of another plant. Based on the mode of pollen transfer two categories of cross-pollinated plants can be distinguished: wind-pollinated (anemophilous) plants, such as beech, hazel, and various grasses, and biotically pollinated plants. The former typically produce large quantities of small pollen grains (Faegri and Van der Pijl 1979; Harder 1998), whereas the latter often produce smaller quantities of larger pollen grains (Percival 1955; Faegri and Van der Pijl 1979). Besides their reduced size, anemophilous pollens may show other adaptations that promote wind dispersal, such as the air bladders found primarily among conifers (Runions *et al.* 1999). Animal-carried pollens also frequently feature adaptations that facilitate transport, such as rough surface structures and the presence of a lipid-rich outer layer (pollenkitt) (Knoll 1930), both ensuring adhesion of pollen to flower visitors.

For the pollinating insect, pollen may be the primary nutritional motive to visit flowers. Pollen is believed to pre-date nectar as a floral reward (Faegri and Van der Pijl 1979), and several "pollen flowers" rely solely on pollen as a reward for their visitors (Percival 1955). Various specific adaptations for the facilitation of animal transport have been described, but there is still little or no unambiguous evidence for the existence of nutritional adaptations that would enhance the function of pollen as food (Roulston and Cane 2000).

Among brood-rearing insects such as bees and bumblebees, pollen collection is often more important than the collection of nectar (Faegri and Van der Pijl 1979). Pollen is also consumed by certain Coleoptera, Diptera, Lepidoptera (Romeis *et al.*, Chapter 7), Thysanoptera (Kirk 1985; Van Rijn and Sabelis, Chapter 8), and Collembola (Scott and Stojanovich 1963). Several predaceous arthropods supplement their diet with pollen-feeding. This group includes adult syrphid flies (Haslett 1989; Wratten *et al.* 1995), heteropteran bugs (Alomar and Wiedemann 1996), ladybird beetles (Triltsch 1997; Cottrell and Yeargan 1998), green lacewings (Sheldon and MacLeod 1971), and predatory mites (Van Rijn and Tanigoshi 1999b). In contrast to the prevalence of pollen-feeding among predators, active pollen-feeding seems to be rather uncommon among hymenopteran parasitoids. Observations of direct pollen-feeding by this group are scarce (Györfi 1945; Patt *et al.* 1997) and usually it remains unclear whether the observed parasitoids derive actual nutritional benefits from their pollen meal. Patt *et al.* (1997) described that *Edovum puttleri* and *Pediobius foveolatus* consume pollen, either directly from the anthers or from other floral parts onto which pollen had dropped. Jervis (1998) reported that Mutillidae and Scoliidae show mouthpart specializations for pollen-feeding.

Availability

Anemophilous pollen is broadly available, also during periods in which other food sources are scarce (Gleim 1985). The pollen-to-ovule ratio is often high ($10^6:1$) in dioecious anemophilous plants. Since wind typically carries these pollen in large quantities, the availability of anemophilous pollen is not restricted to the inflorescence. Following wind dispersal, the shelf-life of pollen depends on environmental conditions. Under suitable conditions pollen grains may remain viable for several days (Dafni and Firmage 2000). Exceptional cases have been described where pollen from some Rosaceae and Liliaceae remained viable for over 100 days (Leduc *et al.* 1990). Exposure to moisture and UV usually decreases pollen viability (Feder and Shrier 1990; Shivanna *et al.* 1991). The factors that affect pollen viability are likely to influence their nutritional expiration date as well. Under dry and cool conditions pollen may remain suitable as a food source for years (Van Rijn and Tanigoshi 1999b). However, under natural conditions, pollen usually takes up moisture from the environment resulting in a rapid decline in food quality. The perishabililty of pollen differs between plant species (Hagedorn and Moeller 1968). Van Houten and van Rijn (unpublished data) distributed pollen on cucumber plants in a greenhouse (20°C and 75–80% relative humidity), recollected it at 3-day intervals, and measured the performance of different phytoseiid mites when provided with this pollen as food source. Birch pollen (*Betula pubescens*) lost its suitability as a food source within 6 days on the plants. Two-week-old cattail pollen (*Typha latifolia*), on the other hand, still increased oviposition rates of adult mites and allowed half of the juveniles to mature.

The availability of entomophilous pollen is more strictly linked to the period and pattern of flowering. The daily pattern of flower accessibility determines pollen-feeding opportunities. Flowers vary with respect to the time of pollen presentation (e.g., morning, midday, afternoon, night). Also the duration of pollen presentation varies, ranging from 16 hours (*Helleborus abchasicus*) to

4–6 hours (e.g., *Sonchus oleraceus*), while the individual inflorescence may only provide pollen for a mere 10 minutes per day (Percival 1955). Among the flowers tested in the latter study, the daily amount of pollen produced per flower ranged between 0.005 mg (*Sonchus oleraceus*) and 54.7 mg (*Cucurbita pepo*). The duration of anther dehiscence per flower varied from a few hours (*Sinapis arvensis*) to 26 days (*Helleborus abchasicus*) (Percival 1955).

Apparency

Insects foraging for pollen from insect-pollinated (entomophilous) flowers have the same visual and olfactory flower signals at their disposal as described for floral nectar. In addition, they may use signals derived from the pollen itself, such as pollenkitt volatiles or pollen coloration (Knoll 1930; Dobson and Bergström 2000; Lunau 2000). Volatiles emitted by pollen are species specific and chemically distinct from odors emitted from other floral parts (Dobson *et al.* 1990). Given that pollen odors are usually quantitatively weak as compared to other floral odors, they likely act at short distances (Dobson and Bergström 2000). Strong pollen odors are mainly found in plants pollinated by pollen-feeding insects (Faegri and Van der Pijl 1979). In addition to odors acting as attractants, insect-repellent pollen odors have been described as well. The repellent compounds are thought to have an antimicrobial function or to serve in the defense of the anthers against destructive pollen-feeders (Dobson *et al.* 1990).

Even though anemophilous plants do not need to advertise the presence of pollen, their pollen nevertheless often release some volatiles (Dobson and Bergström 2000). They are also usually yellow as a result of the flavonoid pigments in the pollen wall, which protect the pollen against UV and overheating (Harborne and Grayer 1993). This enhances the visibility of the often large quantities of exposed anemophilous pollen accumulated on the anthers (Lunau 2000). Once dispersed, anemophilous pollen becomes less detectable, as the grains are small and possess little pollenkitt. This can be compensated by the high number of grains, increasing the chance of random encounter.

Accessibility

As with floral nectar, accessibility of pollen from entomophilous flowers is largely determined by floral architecture. Given that nectaries are usually positioned beyond the anthers, pollen is generally more accessible than floral nectar from the same flower. For smaller insects, pollen uptake can be restricted by the size of grains, typically ranging between 15 and 60 μm, with extremes at 3 μm (forget-me-not) and 250 μm (pumpkin) (Erdtman 1952). A few specialized pollen-feeders like flower thrips are not affected by this problem as they pierce individual grains (Kirk 1984). Insects equipped with chewing mouthparts may feed on the pollenkitt. This feeding mode has been described for micropterygid moths of the genus *Sabatinca* (Thien *et al.* 1985). Insects equipped for liquid feeding (Lepidoptera) may use exuded saliva to soak pollen. The soluble compounds that are thus extracted are consumed when the saliva is reingested (Boggs 1987). Water-soluble nutrients present in the pollenkitt also leak out when pollen grains come in contact with moisture (e.g., rain, dew, or nectar). Finch (1974) demonstrated the presence of fructose, glucose, and the sugar alcohol inositol in water extracts of various pollen. This provides insects that lack specific adaptation to handle pollen grains with access to part of the pollen nutrients. Even those insects that do not feed directly on pollen may be exposed to pollen constituents when they consume pollen-contaminated nectar (Erhardt and Baker 1990) or honeydew. These sugar solutions may also elicit (pseudo-) germination of the pollen (Stanley and Linskens 1974), thereby facilitating the release of pollen nutrients.

Composition/nutritional suitability

Unlike sugar-dominated nectar and honeydew, pollen is primarily a source of nitrogenous compounds (proteins and free amino acids). Protein levels range from 2.5% to as high as 61% (Roulston and Cane 2000). In addition, pollen usually contains some sterols, lipids, and carbohydrates (mainly starch) (Dobson 1988; Nepi and Franchi 2000). Most pollen contains the full complement of the 10 essential amino acids in the free state (Solberg and Remedios 1980). The relative level of free pollen amino acids corresponds with the amino acid composition of pollen protein (Stanley and Linskens 1974).

There has been a long and ongoing discussion as to whether the pollen of entomophilous plants has been adapted to the nutritional needs of their pollinators (Roulston and Cane 2000). Petanidou and Vokou (1990) reported that entomophilous pollens have a higher energetic value than anemophilous pollen and interpreted this as an adaptation to their function as a pollinator reward. However, Collin and Jones (1980) did not find such a caloric dichotomy. Solberg and Remedios (1980) also failed to find obvious differences between amino acid profiles of pollen from wind- versus insect-pollinated plants. Baker and Baker (1983b) reported that anemophilous pollen contain more starch, whereas entomophilous pollen contain more lipid. Roulston and Buchmann (2000) challenged their conclusions, using phylogeny-based statistics, and pointed out that comparisons between different ecological groups are constrained by the paucity of available data.

Insects often exhibit distinct preferences for pollen from certain plant species (Campana and Moeller 1977; Crailsheim *et al.* 1992). The relative amounts of pollen collected may also be affected by the morphological characteristics of the grains. When bees were given a choice between pollen from different plants, cotton pollen was collected least. This poor representation was not due to a low preference for cotton pollen, but could be explained by the fact that spines on cotton pollen physically interfere with pollen aggregation by honeybees (Vaissière and Vinson 1994).

The efficacy of pollen digestion can vary with the pollen feeder and the pollen type (Roulston and Cane 2000). Herbivores may be adapted to use the pollen of their particular host plant. Kirk (1985) showed that specialized thrips achieved a higher fecundity when fed their host pollen as compared to pollen from non-hosts. Generalist species, on the other hand, accepted and utilized a broader spectrum. The honeybee *Apis mellifera* extracts 50–98% of the ingested pollen grains (Crailsheim *et al.* 1992). Similar figures have been reported for other organisms, including those that do not regularly consume pollen (Roulston and Cane 2000).

Bioassays with pollen from various plant sources have shown that there can also be large differences in the effect of pollen on insect life-table parameters. Genissel *et al.* (2002) showed that the floral origin of pollen had a clear effect on the reproductive success of the bumblebee *Bombus terrestris*. This variation could be correlated to the pollen's protein content. Similarly, life-history studies with predatory mites (Yue and Tsai 1996; Van Rijn and Tanigoshi 1999b; Broufas and Koveos 2000) and thrips (Hulshof and Vänninen 2002) showed distinct differences between pollen diets with respect to their effect on the consumer's development and reproductive success. Brodbeck *et al.* (2002) analyzed the pollen used in the latter study and proposed that the differences in amino acid composition may be responsible for the variation in thrips performance.

Some pollens are toxic to honeybees (Stanley and Linskens 1974; Roulston and Cane 2000). Pollen toxicity has been ascribed to the presence of mannose sugars (Crane 1978) known for their bee toxicity (De la Fuente *et al.* 1986). In a range of other cases toxicity has been explained by the presence of secondary plant metabolites in pollen. Secondary plant compounds, such as alkaloids, phenolics, and terpenes, have been detected in pollen of both anemophilous (Meurer *et al.* 1988; Carisey and Bauce 1997) and entomophilous flowers (Detzel and Wink 1993; Kretschmar and Baumann 1999). The relatively high levels of secondary metabolites in some pollen could explain reports of toxicity following pollen-feeding (Detzel and Wink 1993; Yue *et al.* 1994). Specialist herbivores have usually developed mechanisms to overcome the specific secondary chemistry of their host plant and may even use these compounds as feeding stimulants (Schoonhoven 1972). The presence of secondary metabolites in pollen may therefore explain the finding that specialized herbivores spend more time feeding on pollen of their particular host plant (Kirk 1985).

Foraging risks

When insects collect pollen from entomophilous flowers, the same risks apply as described for floral nectar. Since flower-associated predators mainly seek out insect-pollinated flowers, pollen-feeding on anemophilous flowers is probably less risky. Flower-associated risks are avoided when insects feed on pollen after it has dropped from the flowers onto the vegetation below or when wind dispersed pollen accumulates in certain plant structures, such as leaf axils. Pollen-feeding predators may aggregate strongly in these patches, thus increasing the risk for other pollen feeders to be attacked. Van Rijn *et al.* (2002) demonstrated that western flower thrips visiting pollen patches suffer increased mortality due to attacks by the predatory mite *Iphiseius degenerans*. These risks may be low when the pollen is spread out over larger areas of the plant's surface. Pollen might also be contaminated with insect pathogens (Cury 1951), but it remains unclear how prevalent such contaminations are.

Honeydew

Function and consumers

Honeydew is a generic term for sugar-rich excretions produced by sap-feeding Sternorrhynchae. With the exception of the parenchym-feeding Adelgidae, all honeydew-producers are phloem-feeders. It is generally accepted that phloem sap is nutritionally imbalanced for sap-feeders (Sandström and Moran 1999). The carbohydrate/amino acid ratio of the ingested phloem sap usually exceeds the nutritional requirements of the phloem feeders. To obtain sufficient (essential) amino acids, phloem-feeding insects have to excrete excess carbohydrates through the production of honeydew. As such, honeydew is a waste product rather than having a primary function in mutualistic interactions. Nevertheless, honeydew can serve a (secondary) defensive function when honeydew is collected by ants (Stadler and Dixon 1999; Völkl *et al.* 1999). Ants return the favor by protecting sap-feeders against predators and parasitoids and prevent honeydew accumulation that would impede the development of the sap-feeder colony. However, not all sap-feeders benefit when tended by ants (Stadler and Dixon 1999) and the majority of honeydew-producers is actually not ant-tended (Bristow 1991). The ants' decision whether or not to tend a sap-feeder colony is often based on honeydew quantity and composition, as well as the presence of alternative sugar sources (Völkl *et al.* 1999; Engel *et al.* 2001; Fischer *et al.* 2001).

Honeydew is not only appreciated by ants. Hosts of other organisms exploit this often plentiful sugar source (Zoebelein 1956; Beggs 2001). Zoebelein (1956) recorded 246 insect species consuming honeydew in a natural ecosystem. These included important pollinators such as bees and a broad range of predators, as well as 59 species of parasitic Hymenoptera. In agricultural systems honeydew may be consumed by predators and parasitoids (Hagen 1986; Evans and England 1996; Lee and Heimpel 2003; Wäckers and Steppuhn 2003), as well as pest organisms (Swirski *et al.* 1980; Beerwinkle *et al.* 1993).

Availability

Honeydew requires the presence of sap-feeding Sternorrhynchae. Many honeydew-producers are typical *r*-selected species, which combine sessile generations with a high rate of increase. As a result the population structure, and consequently the availability of honeydew, often shows strong geographic and temporal variability (Kunkel and Kloft 1985). Honeydew, like extrafloral nectar, can be especially important as a source of sugars in early spring and late autumn when floral nectar is scarce (Leius 1960). Because most crops lack nectar sources honeydew is often the predominant food source in agricultural systems. Using high-performance liquid chromatography (HPLC) analyses, it was shown that 80% of the parasitoid *Cotesia glomerata* and 55% of *Microplitis mediator* collected in a cabbage field contained honeydew-specific sugars (Wäckers and Steppuhn 2003). The amount of honeydew varies among sap-feeding species and is also dependent on plant parameters and environmental factors (Kunkel and Kloft 1985). At the high end of the spectrum, a full-grown lime tree can produce 31 l of honeydew in a year, with a dry weight of 8.5 kg (Dixon 1971). Some sap-feeders can actively increase the quantity of excreted honeydew when tended by ants (Takeda *et al.* 1982).

Only a minority of sap-feeders is actually ant-tended. When ants are present they often outcompete other potential honeydew consumers. This decreases the availability of honeydew from ant-tended relative to honeydew from non-tended sap-feeders. Due to the low prevalence of ants in most agricultural systems, honeydew from otherwise ant-tended species might become available to other consumers.

Apparency

Honeydew is likely the most prevalent source of exogenous sugars available to sugar-feeding insects, but its exploitation can be restricted by the fact that honeydew can be difficult to detect (Wäckers and Swaans 1993). Johnson *et al.* (1987) argue that this is one of the reasons that honeydew is not a major carbohydrate source for Lepidoptera, as most butterflies use vision to find flowers and aphids do not provide adequate visual stimuli. Aphid species whose coloration contrasts with the green background could provide a visual cue to honeydew-foraging insects (Dixon 1959). Some aphid parasitoids are able to perceive honeydew volatiles and use this olfactory information as a cue to locate aphid colonies (Bouchard and Cloutier 1985).

The generally low apparency can be understood based on the fact that honeydew is primarily a waste product, unlike the actively advertised floral nectar. Consequently, there is usually little benefit to the honeydew-producer in drawing attention to this sugar secretion. To the contrary, since honeydew can serve predators and parasitoids as a kairomone leading to its producer (Budenberg 1990; Romeis and Zebitz 1997; Wäckers 2000), sap-feeders should be subject to a strong selection pressure to minimize honeydew detectability. The low apparency might not be a severe constraint to ant species, as the numerous ant scouts are likely to find aphid colonies or honeydew patches even if they were to explore randomly.

In addition to the fact that honeydew is often inconspicuous in itself, aphid infestation may also result in relatively low levels of volatiles released by the infected plant (Turlings *et al.* 1998). Food-deprived *Cotesia glomerata* parasitoids showed no response to aphid-infested plants, either in flight chamber or in olfactometer experiments (Wäckers and Swaans 1993).

Accessibility

Since honeydew is typically released as little droplets that are often sprayed onto plant substrate, it is usually easily accessible as a food source. Hence honeydew is thought to be especially valuable to the many predators and parasitoids whose access to floral nectar is restricted by their unspecialized mouthparts (Wäckers *et al.* 1996; Jervis 1998). However, the small size of the droplets promotes evaporation and some common honeydew sugars, such as raffinose and melezitose, crystallize rapidly. As a result, honeydew on the plant surface is often highly viscous or crystallized, which can severely interfere with its consumption (Wäckers 2000).

Ant-tended aphids often do not actively shoot away honeydew droplets. Instead they wait for ants to be present before releasing the honeydew. They may even resorb honeydew if not collected (Way 1963). Thus in heavily tended colonies honeydew is often not accessible to insects other than ants. In the absence of (sufficient) ants, though, these aphids eventually release honeydew without instant removal.

Some aphids (e.g., *Brevicoryne brassicae*) cover excreted honeydew with a wax layer. This likely serves a hygienic purpose, but it may also limit honeydew accessibility to potential consumers.

Composition/nutritional suitability

Although honeydew might be the predominant sugar source, especially in agricultural systems, it is often nutritionally inferior to nectar or sucrose (Leius 1961; Avidov *et al.* 1970; Wäckers 2000). Low suitability may be based on plant-derived compounds (Wink and Römer 1986; Vrieling *et al.* 1991; Romeis *et al.* 2003) or on aphid-synthesized compounds (Wäckers 2000).

Honeydew composition may vary depending on a broad range of parameters, including the species of sap-feeder (Hendrix *et al.* 1992), the presence of bacterial symbionts in the digestive tract (Sasaki *et al.* 1990; Wilkinson *et al.* 1997), the developmental stage of the aphid (Sasaki and Ishikawa 1991; Arakaki and Hattori 1998), the presence of aphid-tending ants (Fischer and Shingleton 2001; Yao and Akimoto 2001), and whether or not the aphid is parasitized (Cloutier 1986). Variables from the plant's side include plant species (Hendrix *et al.* 1992; Sandström and Moran 2001) and its growth conditions (Crafts-Bradner 2002).

Carbohydrates

Even though the sugar composition of honeydew reflects the original composition of the phloem sap, the individual sugar compounds and their relative quantities can be altered by the phloem-feeder. Ingested phloem sugars such as sucrose and maltose are broken down by digestive enzymes in the gut of the phloem-feeder. The sap-feeders may also synthesize more complex sugars, such as the trisaccharides melezitose and erlose (fructomaltose) and the disaccharides trehalose and trehalulose (Byrne and Miller 1990; Mittler and Meikle 1991; Hendrix *et al.* 1992; Hendrix and Salvucci 2001). The resulting sugar spectrum may range from honeydews that are almost entirely composed of the phloem sugar sucrose and its hexose components fructose and glucose, to honeydews that are low in hexoses and are dominated by insect-synthesized oligosaccharides (Kunkel and Kloft 1985; Hendrix *et al.* 1992; Völkl *et al.* 1999; Wäckers and Steppuhn 2003). Honeydew-specific sugars can be a reliable tool in identifying consumption of honeydew (Heimpel *et al.* 2004). The often distinct differences between honeydew profiles can even indicate which honeydew has been consumed (Wäckers and Steppuhn 2003).

It has been suggested that by synthesizing honeydew sugars, sap-feeders cater to the particular gustatory preferences of ants (Kiss 1981). Ants may indeed prefer melezitose and raffinose over sucrose (Völkl *et al.* 1999). However, the majority of ants does not show such preferences (Cornelius *et al.* 1996; Blüthgen and Fiedler 2004) and other insects may show poor gustatory responses to honeydew-specific sugars (Wäckers 1999; Romeis and Wäckers 2000; Beach *et al.* 2003). Furthermore, longevity studies indicate that honeydew oligosaccharides may have a negative effect on the nutritional suitability of honeydew (Wäckers 2001; Romeis and Wäckers 2002; Boevé and Wäckers 2003).

Amino acids

In comparison to nectar, honeydew often contains relatively high levels of amino acids (Hendrix *et al.* 1992; Sandström and Moran 2001; Crafts-Bradner 2002; Fischer *et al.* 2002). Two mechanisms may affect the amino acid composition of the excreted honeydew. First, some sap-feeders induce the plant to increase phloem amino acid levels (Sandström and Moran 2001). Second, some sap-feeders feature symbiotic bacteria in their gut that can synthesize amino acids (Sasaki and Ishikawa 1991).

Although the overall amino acid quantity in honeydew may be relatively high, this does not guarantee nutritional suitability. The bulk of honeydew amino acids are typically non-essential, because sap-feeders selectively assimilate essential amino acids from the ingested phloem sap (Sandström and Moran 2001). The fraction of non-essential amino acids in honeydew may be strongly dominated by only a few compounds. Especially prevalent are glutamine, glutamic acid, asparagine, aspartic acid, and serine (Byrne and Miller 1990; Hendrix *et al.* 1992; Sandström and Moran 2001; Crafts-Bradner 2002; Fischer *et al.* 2002). Sap-feeders can adjust the amino acid composition in response to ant-tending. Yao and Akimoto (2002) demonstrated that honeydew of tended aphids contains a higher diversity and a higher total concentration of amino acids than honeydew of ant-excluded aphids.

Secondary plant metabolites

Phloem-feeding allows sap-feeders to minimize exposure to secondary plant chemistry, as many toxic compounds are xylem transported (Hartley and Jones 1997). Nevertheless, a range of secondary metabolites also occurs in phloem sap and these are consequently ingested by phloem-feeders. Many compounds are unaffected by the passage through the insects' digestive system, meaning that phloem sap composition has a direct impact on honeydew composition. This applies in particular to plant lectins, protease inhibitors, and other proteins, as phloem-sucking insects typically possess low proteolytic activity in the gut (Rahbé *et al.* 1995; but see Salvucci *et al.* 1998). Also a range of other plant compounds, including cardenolides (Malcolm 1990), alkaloids (Wink and Römer 1986; Vrieling *et al.* 1991), phenolics (Peng and Miles 1991), flavonoids (Calatayud *et al.* 1994), and glucosinolates (Weber *et al.* 1986) have been reported to occur in honeydew.

Comparative studies have shown that honeydew may differ from phloem sap with regard to the composition and concentration of secondary plant metabolites (Wink and Römer 1986). The overall levels of secondary metabolites in honeydew may exceed levels in plant tissue, indicating effective excretion

(Malcolm 1990; Vrieling *et al.* 1991). Other sap-feeders may sequester secondary metabolites to obtain protection against their antagonists (Wink and Witte 1991; Mendel *et al.* 1992). Some even increase their toxicity through chemical degradation of compounds into more toxic breakdown products (Francis *et al.* 2001).

Ant-tending, or the lack thereof, may be a rough indication as to whether honeydew is nutritionally suitable to other consumers. However, this method may fall short in two respects. For one, nutritional and gustatory requirements of ants may differ from those of other honeydew consumers (Wäckers 1999). Moreover, gustatory responses may not always be correlated to nutritional suitability (Dethier *et al.* 1956; Nettles and Burks 1971).

Foraging risks

As with extrafloral nectar, ant predation likely represents the main risk associated with honeydew-feeding. This foraging risk is aggravated by the fact that the most aggressive ants monopolize the highest-quality honeydew sources (Blüthgen *et al.* 2000). Ant-inflicted mortality has been quantified for aphid parasitoids. In ant-tended colonies females of the aphid parasitoid *Trioxys angelicae* are attacked as soon as they are encountered by the ants. Few survive for more than 5 minutes (Völkl 1992). Certain aphid parasitoids can forage and parasitize undisturbed in the lion's den, escaping ant predation through cryptic behavior and/or ant mimicry (Völkl and Mackauer 1993). Not only do these parasitoids avoid predation, their offspring actually benefit from the ants' protection (Völkl 1992).

Although not all ant species are equally aggressive, some aphid parasitoids have been shown to take the cautious approach, exhibiting an innate retreating behavior when encountering ants (Völkl 2001). Following experience with non-aggressive ants, the parasitoid *Pauesia picta* learns to become less vigilant, thereby increasing its reproductive success (Völkl 2001).

Whereas an encounter with aggressively guarding ants may impose a serious fitness penalty to honeydew consumers, there is also the option to obtain honeydew sweets without sweat from those producers that are not tended by ants (Bristow 1991). However, the lack of interest from the site of the ants may be indicative of a lower honeydew suitability. This implies that the honeydew consumer may be faced with a choice between long-term and short-term fitness costs. In (annual) agricultural systems ants are scarce, and here honeydew from otherwise ant-tended species is safely available to other consumers.

Long-exposed honeydew can represent a different risk, when toxin-producing fungi use honeydew as a substrate. An example of the latter was described by Vitzthum (1930), who reported that oak honeydew infested with *Aspergillus calyptratus* represents a particular danger to honeydew-collecting bees.

Discussion and implications

This chapter illustrates the substantial variation between and among food categories with regard to their availability, detectability, accessibility, and composition, as well as the mortality risks associated with food exploitation. Each of these traits affects the suitability of food sources, and as such can be a factor in shaping the evolution of food rewards in plant–insect and insect–insect interactions. Identifying and quantifying these parameters not only helps to understand the complex functioning of food supplements in plant–insect and insect–insect interactions, it also has direct implications for the use of food supplements in biological control programs.

This chapter has mainly focussed on food source characteristics, but it has to be emphasized that suitability is a function of both food source and its potential consumer. A flower that is attractive and accessible to a hummingbird is unlikely to be optimally suited for visiting beetles. Food sources may vary broadly with regard to their appearance, build, and chemical composition, yet consumers may show an even broader variation in foraging behavior, mouthpart morphology, and physiology. An effective exploitation of food sources requires that consumer attributes fit the food source characteristics. Therefore, the broad variation in consumer characteristics needs to be taken into account when assessing food suitability.

Mutualistic interactions, including food-mediated interactions, have often been interpreted within the perspective of pairwise interacting species (Stanton 2003). Whereas the direct pairwise interaction between food producer and intended consumer is obviously key to understanding the evolution of food source characteristics, food-mediated interactions do not occur in isolation. Under natural conditions organisms function within a complex web of (trophic) interactions. Even though complexity is likely common and important in the evolution and functioning of food sources, it has received relatively little attention to date. Sources of complexity include multiple functions of food rewards and the exploitation of food by unintended visitors.

Multiple functions

The fact that food rewards can be multifunctional is evident in the case of pollen, whose role as a food reward is derived from its primary function in gene transfer during pollination. But also floral and extrafloral nectar might serve combined functions in pollination and in defense (Wäckers and Bonifay 2004). Such a dual function can be illustrated by so-called "post-floral nectaries" (Daumann 1932). These nectaries are located within the flower where they act as a pollination reward. Yet they continue nectar secretion for several days or even weeks after pollination has been effected. During the latter phase, they attract ants that can provide protection against seed predators (Keeler 1981). Although the defensive and reproductive functions of these post-floral nectaries are separated in time, they may overlap in other instances, for example when predators or parasitoids act as pollinators (Beattie 1985; Kunze 1999), or when nectar simultaneously attracts pollinators and predators (Ford and Forde 1976; Knox *et al.* 1985; Yano 1994).

When multiple functions are distinctly separated in time, food rewards may show a switch in features to suit both functions (Gracie 1991). When multiple functions overlap, this may result in the evolution of hybrid features that are not necessarily optimal for either function, but instead yield the highest combined fitness benefit to the food producer. This underlines that the evolution of food source features can only be fully understood if all functions are considered.

Unintended visitors

A second factor adding complexity to the evolution and functioning of food rewards lies in exploitation of food sources by unintended consumers (see also Sabelis *et al.*, Chapter 4 and Romeis *et al.*, Chapter 7). These uninvited guests can entail considerable fitness costs to the food provider (Bronstein and Ziv 1997). When the impact of non-mutualists is restricted to food consumption, fitness costs do not need to exceed the costs of food production. Additional costs arise when unintended consumers interfere with the food-mediated mutualism, either through food depletion (exploitative competition) (Comba *et al.* 1999; Hansen *et al.* 2002), or by physically attacking/repelling the mutualists (Kikuchi 1963; Beattie 1985; Blüthgen *et al.* 2000). A particular category of fitness costs applies when food provision promotes the activity of organisms that either damage the producer directly (herbivores in the case of plant-provided food; see Romeis *et al.*, Chapter 7), transmit diseases to the producer (Roy 1993; Shykoff and Bucheli 1995; Biere and Honders 1996), or parasitize/predate the producer's mutualists (Stephens *et al.* 1998). Overall, unintended consumers can form a significant – yet largely unexplored – additional force in shaping the evolution of food rewards (Bronstein and Ziv 1997; Wagner and Kay 2002).

As pointed out, potential consumers often vary broadly with regard to their foraging behavior, mouthpart morphology, gustatory preferences, and food digestion. This variation opens up the possibility for the evolution of selective food-supply strategies or selective foods. Such selectivity would allow the food provider to cater to its mutualists, while excluding unintended consumers. The induction of food body production in response to ant presence (Risch and Rickson 1981) may be an example of a selective supply strategy. High viscosity of extrafloral nectar (Wäckers *et al.* 2001) and the presence of feeding deterrents in some floral nectar (Feinsinger and Swarm 1978; Adler 2001) may represent examples of selective foods. Overall, these examples show that we need to consider intended as well as unintended consumers if we want to understand the evolution and functioning of food traits.

Implications for agricultural systems

It is increasingly recognized that the use of food supplements can be a powerful tool to enhance the effectiveness of predators and parasitoids in agriculture or forestry. Sustaining biological control through the provision of food supplements is often one of the objectives in functional biodiversity programs and conservation biological control. Seed mixtures are commercially available containing plants that are thought to be particularly attractive and suitable as food supplements for beneficials. Moreover, several countries offer sizable subsidies for the use of such seed mixtures in field borders, non-cropping areas, or undergrowth (Anonymous 1999, 2001). Despite the substantial investments in these programs, there is often surprisingly little information available to substantiate the suitability of the included plants as insect food sources. The criteria discussed in this chapter could serve as a guideline for the selection of food sources, both natural and artificial (food sprays). By selecting those foods whose availability, appearance, build, and chemical composition match the behavior, morphology, and physiology of target organisms, we can optimize the enhancement of biological control. In agriculture, as in natural ecosystems, there is the pitfall that the provision of food sources may also benefit herbivorous pest insects. By identifying and exploiting differences between herbivores and their enemies with respect to their foraging behavior, food handling, and food utilization this drawback may be avoided.

The parameters identified in this chapter provide various tools to tailor food supply to achieve such selectivity. Food availability, for instance, can be attuned to those periods and locations in which predators and parasitoids are present. Timing of food availability can be especially effective in those instances in which the nectar-feeding stages of the herbivore and its enemy do not coincide, for instance diurnal parasitoids attacking larvae of a nocturnal moth.

Selectivity may also be based on attractiveness. Herbivores and their enemies often differ with regard to their sensory capacities or foraging behavior. Butterflies, for instance, usually respond to bright colors with scents often playing a subordinate role. Moreover, butterflies often require flowers to be relatively large and to feature a landing platform. Their parasitoids, on the other hand, respond to odors as well as to colors (Wäckers 1994, 2004). Due to their smaller size, parasitoids do not require flowers to be large nor to feature a landing platform.

The often distinct difference in mouthpart morphology and size between herbivores and their enemies mean that the accessibility of food sources may be another source of selectivity in food exploitation between both consumer categories (Baggen *et al.* 1999).

Herbivores and their enemies may also differ with regard to their gustatory response, nutritional requirements, and metabolic efficiency (Wäckers 1999, 2001; Romeis and Wäckers 2000, 2002). We can take advantage of these differences when selecting foods of a particular chemical composition. Finally, food-associated mortality risks may also be consumer-specific and thus may contribute to specificity.

Exploring these differences is not only important in order to identify strategies that selectively sustain natural enemies. It may also help to avoid the potential pitfall of reverse selectivity, i.e. food sources that are exploited by herbivores without benefitting predators or parasitoids (Winkler *et al.* 2003).

Acknowledgments

The valuable comments of Diane Wagner were much appreciated.

References

Abe, T. and K. Kamo. 2003. Seasonal changes of floral frequency and composition of flower in two cool temperate secondary forests in Japan. *Forest Ecology and Management* **175**: 153–162.

Adler, L. S. 2001. The ecological significance of toxic nectar. *Oikos* **91**: 409–420.

Allsopp, M. H., S. W. Nicolson, and S. Jackson. 1998. Xylose as a nectar sugar: the response of Cape honeybees, *Apis mellifera capensis* Eschscholtz (Hymenoptera: Apidae). *African Entomology* **6**: 317–323.

Alm, J., T. E. Ohnmeiss, J. Lanza, and L. Vriesenga. 1990. Preference of cabbage white butterflies and honey bees for nectar that contains amino acids. *Oecologia* **84**: 53–57.

Alomar, O. and R. N. Wiedemann. 1996. *Zoophytophagous Heteroptera: Implications for Life History and Integrated Pest Management*. Lanham, MD: Entomological Society of America.

Anonymous. 1999. *Subsidieregeling Agrarisch Natuurbeheer*. The Hague, the Netherlands: Ministerie van Landbouw, Natuurbeheer en Visserij.

Anonymous. 2001. *Verordnung über die Direktzahlungen an die Landwirtschaft*. Bern, Switzerland: Bundesamt für Landwirtschaft.

Arakaki, N. and M. Hattori. 1998. Differences in the quality and quantity of honeydew from first instar soldier and ordinary morph nymphs of the bamboo aphid, *Pseudoregma koshunensis* (Takahashi) (Homoptera: Aphididae). *Applied Entomology and Zoology* **33**: 357–361.

Avidov, Z., M. Balshin, and U. Gerson. 1970. Studies on *Aphytis coheni*, a parasite of the california red scale, *Aonidiella aurantii*, in Israel. *Entomophaga* **15**: 191–207.

Baggen, L. R., G. M. Gurr, and A. Meats. 1999. Flowers in tri-trophic systems: mechanisms allowing selective exploitation by insect natural enemies for conservation biological control. *Entomologia Experimentalis et Applicata* **91**: 155–161.

Baker, H. G. and I. Baker. 1975. Studies of nectar-constitution and pollinator–plant coevolution. In L. E. Gilbert and P. H. Raven (eds.) *Co-Evolution of Animals and Plants*. Austin, TX: University of Texas Press, pp. 100–140.

1982. Chemical constituents of nectar in relation to pollination mechanisms and phylogeny. In M. H. Nitecki (ed.) *Biochemical Aspects of Evolutionary Biology*. Chicago, IL: University of Chicago Press, pp. 131–171.

1983a. Floral nectar sugar constituents in relation to pollinator type. In C. E. Jones and R. J. Little (eds.) *Handbook of Experimental Pollination Biology*. New York: Van Nostrand Reinhold, pp. 117–141.

1983b. Some evolutionary and taxonomic implications of variation in the chemical reserves of pollen. In D. L. Mulcahy and E. Ottaviano (eds.) *Pollen: Biology and Implications for Plant Breeding*. New York: Elsevier, pp. 43–52.

Baker, H. G., P. A. Opler, and I. Baker. 1978. A comparison of the amino acid complements of floral and extrafloral nectar. *Botanical Gazette* **139**: 322–332.

Bakker, F. M. and M. E. Klein. 1992a. How cassava plants enhance the efficacy of their phytoseiid bodyguards. Proc. 8th Int. Symp. *Insect–Plant Relationship*, Dordrecht, pp. 353–354.

1992b. Transtrophic interactions in cassava. *Experimental and Applied Acarology* **14**: 293–311.

Barker, R. J. 1990. Poisoning by plants. In R. A. Morse and R. Nowogrodzki (eds.) *Honey Bee Pests, Predators and Diseases*. Ithaca, NY: Cornell University Press, pp. 275–296.

Basden, R. 1965. The occurrence and composition of Manna in *Eucalyptus* and *Angophora*. *Proceedings of the Linnean Society of New South Wales* **90**(2): 152–156.

Beach, J. P., L. Williams, D. L. Hendrix, and L. Price. 2003. Different food sources affect the gustatory response of *Anaphes iole*, an egg parasitoid of *Lygus* spp. *Journal of Chemical Ecology* **29**: 1203–1222.

Beattie, A. J. 1985. *The Evolutionary Ecology of Ant–Plant Mutualisms*. Cambridge, UK: Cambridge University Press.

Beerwinkle, K. R., T. N. Shaver, and J. D. Lopez Jr. 1993. Field observations of adult emergence and feeding behavior of *Helicoverpa zea* (Lepidoptera: Noctuidae) on dallisgrass ergot honeydew. *Environmental Entomology* **22**: 554–558.

Beggs, J. 2001. The ecological consequences of social wasps (*Vespula* spp.) invading an ecosystem that has an abundant carbohydrate resource. *Biological Conservation* **99**: 17–28.

Bentley, B. L. 1977a. Extrafloral nectaries and protection by pugnacious bodyguards. *Annual Review of Ecology and Systematics* **8**: 407–427.

1977b. The protective function of ants visiting the extrafloral nectaries of *Bixa orellana* (Bixaceae). *Journal of Ecology* **65**: 27–38.

Bequaert, J. 1924. Galls that secrete honeydew: a contribution to the problem as to whether galls are altruistic adaptations. *Bulletin of the Brooklin Entomological Society* **19**(4): 101–124.

Beutler, R. 1935. Nectar. *Bee World* **24**: 106–162.

Biere, A. and S. Honders. 1996. Host adaptation in the anther smut fungus *Ustilago violacea* (*Microbotryum violaceum*): infection success, spore production and alteration of floral traits on two host species and their F_1- hybrid. *Oecologia* **107**: 307–320.

Blüthgen, N. and K. Fiedler. 2004. Preferences for sugars and amino acids and their conditionality in a diverse nectar-feeding ant community. *Journal of Animal Ecology* **73**: 155–166.

Blüthgen, N., M. Verhaagh, W. Goitia, *et al.* 2000. How plants shape the ant community in the Amazonian rainforest canopy: the key role of extrafloral nectaries and homopteran honeydew. *Oecologia* **125**: 229–240.

Boevé, J. L. and F. L. Wäckers. 2003. Gustatory perception and metabolic utilization of sugars by *Myrmica rubra* ant workers. *Oecologia* **136**: 508–514.

Boggs, C. L. 1987. Ecology of nectar and pollen feeding in Lepidoptera. In F. Slansky and J. G. Rodriguez (eds.) *Nutritional Ecology of Insects, Mites, Spiders, and Related Invertebrates*. New York: John Wiley, pp. 369–391.

Bory, G. and D. Clair Maczulajtys. 1986. Nectar composition and role of the extrafloral nectar in *Ailanthus glandulosa*. *Canadian Journal of Botany* **64**: 247–253.

Bouchard, Y. and C. Cloutier. 1985. Role of olfaction in host finding by the aphid parasitoid *Aphidius nigriceps* (Hymenoptera: Aphidiidae). *Journal of Chemical Ecology* **11**: 801–808.

Bowden, B. 1970. The sugars in the extrafloral nectar of *Andropogon gayanus* var *bisquamulatus*. *Phytochemistry* **9**: 2315–2318.

Boyko, A. K. 1939. Larvae of *Senotainia triguspis* Meig. causing heavy losses of bees. *Doklady Akademiia Nauk USSR* **24**: 306–309.

Bresinsky, A. 1963. Bau, Entwicklungsgeschichte und Inhaltstoffe der Elaiosomen. *Bibliotheca Botanica* **126**: 1–54.

Briscoe, A. D. and L. Chittka. 2001. The evolution of color vision in insects. *Annual Review of Entomology* **46**: 471–510.

Bristow, C. M. 1991. Why are so few aphids ant-tended? In C. R. Huxley and D. F. Cutler (eds.) *Ant–Plant Interactions*. Oxford, UK: Oxford University Press, pp. 104–119.

Brodbeck, B. V., J. Funderburk, J. Stavisky, P. C. Andersen, and J. Hulshof. 2002. Recent advances in the nutritional ecology of Thysanoptera, or the lack thereof. Proc. 7th Int. Symp. *Thysanoptera*, Canberra, pp. 145–153.

Bronstein, J. L. and Y. Ziv. 1997. Costs of two non-mutualistic species in a yucca/yucca moth mutualism. *Oecologia* **112**: 379–385.

Broufas, G. D. and D. S. Koveos. 2000. Effect of different pollens on development, survivorship and reproduction of *Euseius finlandicus* (Acari: Phytoseiidae). *Environmental Entomology* **29**: 743–749.

Budenberg, W. J. 1990. Honeydew as a contact kairomone for aphid parasitoids. *Entomologia Experimentalis et Applicata* **55**: 139–148.

Bugg, R. L., L. E. Ehler, and L. T. Wilson. 1987. Effect of common knotweed (*Polygonum aviculare*) on abundance and efficiency of insect predators of crop pests. *Hilgardia* **55**: 1–52.

Bugg, R. L., R. T. Ellis, and R. W. Carlson. 1989. Ichneumonidae (Hymenoptera) using extrafloral nectar of faba bean (*Vicia faba* L., Fabaceae) in Massachusetts. *Biological Agriculture and Horticulture* **6**: 107–114.

Bugg, R. L., F. L. Wäckers, K. E. Brunson, J. D. Dutcher, and S. C. Phatak. 1991. Cool-season cover crops relay intercropped with cantaloupe: influence on a generalist predator *Geocoris punctipes* (Hemiptera: Lygaeidae). *Journal of Economical Entomology* **84**: 408–416.

Buller, A. H. R. 1950. *Researches on Fungi*, vol. 7, *The Sexual Process in the Uredinales*. Toronto, Ontario: University of Toronto Press.

Burquez, A. and S. A. Corbet. 1991. Do flowers reabsorb nectar? *Functional Ecology* **5**: 369–379.

Butler, G. D., G. M. Loper, S. E. McGregor, J. L. Webster, and H. Margolis. 1972. Amounts and kinds of sugars in the nectars of cotton (*Gossypium* spp.) and the time of their secretion. *Agronomy Journal* **64**: 364–368.

Byrne, D. N. and W. B. Miller. 1990. Carbohydrate and amino acid composition of phloem sap and honeydew produced by *Bemisia tabaci*. *Journal of Insect Physiology* **36**: 433–439.

Calatayud, P. A., Y. Rahbé, B. Delobel, *et al.* 1994. Influence of secondary compounds in the phloem sap of cassava on expression of antibiosis towards the mealybug *Phenacoccus manihoti*. *Entomologia Experimentalis et Applicata* **72**: 47–57.

Caldwell, D. L. and K. O. Gerhardt. 1986. Chemical analysis of peach extrafloral exudate. *Phytochemistry* **25**: 411–413.

Campana, B. J. and F. E. Moeller. 1977. Honey bees: preference for and nutritive value of pollen from five plant sources. *Journal of Economic Entomology* **70**: 39–41.

Canard, M. 2001. Natural food and feeding habits of lacewings. In P. McEwen, T. R. New, and A. E. Whittington (eds.) *Lacewings in the Crop Environment*. Cambridge, UK: Cambridge University Press, pp. 116–129.

Carey, D. B. and M. Wink. 1994. Elevational variation of quinolizidine alkaloid contents in a lupine (*Lupinus argenteus*) of the Rocky Mountains. *Journal of Chemical Ecology* **20**: 849–857.

Carisey, N. and E. Bauce. 1997. Impact of balsam fir flowering on pollen and foliage biochemistry in relation to spruce budworm growth, development and food utilization. *Entomologia Experimentalis et Applicata* **85**: 17–31.

Carroll, C. R. and D. H. Janzen. 1973. Ecology of foraging by ants. *Annual Review of Ecology and Systematics* **4**: 231–257.

Chapman, R. F. 1995. Chemosensory regulation of feeding. In R. F. Chapman and G. de Boer (eds.) *Regulatory Mechanisms in Insect Feeding*. New York: Chapman and Hall, pp. 101–136.

Chittka, L. 2001. Camouflage of predatory crab spiders on flowers and the colour perception of bees (Aranida: Thomisidae / Hymenoptera: Apidae). *Entomologia Generalis* **25**: 181–187.

Clark, T. B. 1978. Honey bee spiroplasmosis, a new problem for beekeepers. *American Bee Journal* **118**: 18–23.

Cloutier,C. 1986. Amino acid utilization in the aphid *Acyrthosiphon pisum* infected by the parasitoid *Aphidius smithi*. *Journal of Insect Physiology* **32**: 263–267.

Collin, L. J. and C. E. Jones. 1980. Pollen energetics and pollination modes. *American Journal of Botany* **67**: 210–215.

Comba, L., S. A. Corbet, L. Hunt, and B. Warren. 1999. Flowers, nectar and insect visits: evaluating British plant species for pollinator-friendly gardens. *Annals of Botany* **83**: 369–383.

Cornelius, M. L., J. K. Grace, and J. R. Yates III. 1996. Acceptability of different sugars and oils to three tropical ant species (Hymen., Formicidae). *Anzeiger für Schädlingskunde, Pflanzenschutz und Umweltschutz* **69**: 41–43.

Cottrell, T. E. and K. V. Yeargan. 1998. Effect of pollen on *Coleomegilla maculata* (Coleoptera: Coccinellidae) population density, predation, and cannibalism in sweet corn. *Environmental Entomology* **27**: 1402–1410.

Crafts-Bradner, S. J. 2002. Plant nitrogen status rapidly alters amino acid metabolism and excretion in *Bemisia tabaci*. *Journal of Insect Physiology* **48**: 33–41.

Crailsheim, K., L. H. W. Schneider, N. Hrassnigg, *et al.* 1992. Pollen consumption and utilization in worker honeybees (*Apis mellifera carnica*): dependence on individual age and function. *Journal of Insect Physiology* **38**: 409–419.

Crane, E. 1978. Sugars poisonous to bees. *Bee World* **59**: 37–38.

Cruden, R. W. and S. M. Hermann. 1983. Studying nectar? Some observations on the art. In B. Bentley and T. Elias (eds.) *The Biology of Nectaries*. New York: Columbia University Press, pp. 223–241.

Cuautle, M. and V. Rico-Gray. 2003. The effect of wasps and ants on the reproductive success of the extrafloral nectaried plant *Turnera ulmifolia* (Turneraceae). *Functional Ecology* **17**: 417–423.

Cury, R. 1951. Micoses das abelhas e fungos das colmeias. *Biologico (Brazil)* **17**: 214–220.

Dafni, A. 1992. *Pollination Ecology*. Oxford, UK: Oxford University Press.

Dafni, A. and D. Firmage. 2000. Pollen viability and longevity: practical, ecological and evolutionary implications. In A. Dafni, M. Hesse, and E. Pacini (eds.) *Pollen and Pollination*. Vienna: Springer-Verlag, pp. 113–133.

Dafni, H., Y. Lensky, and A. Fahn. 1988. Flower and nectar characteristics of nine species of Labiatae and their influence on honeybee visits. *Journal of Apicultural Research* **27**: 103–114.

Daumann, E. 1932. Über postflorale Nektarabscheidung. *Beihefte zum botanischen Zentralblatt* **49**: 720–734.

De la Fuente, M., P. F. Penas, and A. Sols. 1986. Mechanism of mannose toxicity. *Biochemical and Biophysical Research Communications* **140**: 51–55.

DeBach, P., T. W. Fischer, and J. Landi. 1955. Some effects of meteorological factors on all stages of *Aphytis lingnanensis*, a parasite of the California red scale. *Ecology* **36**: 743–753.

Dethier, V. G. 1976. *The Hungry Fly*. Cambridge, MA: Harvard University Press.

Dethier, V. G., D. R. Evans, and M. V. Rhoades. 1956. Some factors controlling the ingestion of carbohydrates by the blowfly. *Biological Bulletin* **111**: 204–222.

Detzel, A. and M. Wink. 1993. Attraction, deterrence or intoxication of bees (*Apis mellifera*) by plant allelochemicals. *Chemoecology* **4**: 8–18.

DeVries, P. J., T. R. Walla, and H. F. Greeney. 1999. Species diversity in spatial and temporal dimensions of fruit-feeding butterflies from two Ecuadorian rainforests. *Biological Journal of the Linnean Society* **68**: 333–353.

Dixon, A. F. G. 1971. The role of aphids in wood formation. II. The effect of the lime aphid *Eucallipterus tiliae* L. (Aphididae) on the growth of lime, *Tilia* × *vulgaris* Hayne. *Journal of Applied Ecology* **8**: 393–399.

Dixon, J. J. 1959. Studies on the oviposition behaviour of Syrphidae. *Transactions of the Royal Entomological Society of London* **11**: 57–80.

Dobson, C. H. 1994. Floral volatiles in insect biology. In E. Bernays (ed.) *Insect–Plant Interactions*. Boca Raton, FL: CRC Press, pp. 47–81.

Dobson, H. E. M. 1988. Survey of pollen and pollenkitt lipids: chemical cues to flower visitors? *American Journal of Botany* **75**: 170–182.

Dobson, H. E. M. and G. Bergström. 2000. Ecology and evolution of pollen odors. In A. Dafni, M. Hesse and E. Pacini (eds.) *Pollen and Pollination*. Vienna: Springer-Verlag, pp. 63–89.

Dobson, H. E. M., G. Bergström, and I. Groth. 1990. Differences in fragrance chemistry between flower parts of *Rosa rugosa*. *Israel Journal of Botany* **39**: 143–156.

Domínguez, C. A., R. Dirzo, and S. H. Bullock. 1989. On the function of floral nectar in *Croton suberosus* (Euphorbiaceae). *Oikos* **56**: 109–114.

Downes, W. L. and G. A. Dahlem. 1987. Keys to the evolution of Diptera: role of Homoptera. *Environmental Entomology* **16**: 847–854.

Du, Y. J., G. M. Poppy, W. Powell, *et al.* 1998. Identification of semiochemicals released during aphid feeding that attract the parasitoid *Aphidius ervi*. *Journal of Chemical Ecology* **24**: 1355–1368.

Dukas, R. 2001. Effects of perceived danger on flower choice by bees. *Ecology Letters* **4**: 327–333.

Durrer, S. and P. Schmid-Hempel. 1994. Shared use of flowers leads to horizontal pathogen transmission. *Proceedings of the Royal Society of London Series B* **258**: 299–302.

Dyer, L. E. and D. A. Landis. 1996. Effects of habitat, temperature, and sugar availability on longevity of *Eriborus terebrans* (Hymenoptera: Ichneumonidae). *Environmental Entomology* **25**: 1192–1201.

Ehrlén, J. and O. Erickson. 1993. Toxicity in fleshy fruits: a non-adaptive trait? *Oikos* **66**: 107–113.

Eijs, I. E. M., J. Ellers, and G.-J. van Duinen. 1998. Feeding strategies in drosophilid parasitoids: the impact of natural food resources on energy reserves in females. *Ecological Entomology* **23**: 133–138.

Elias, T. S. 1983. Extrafloral nectaries: their structure and distribution. In B. L. Bentley and T. S. Elias (eds.) *The Biology of Nectaries*. New York: Columbia University Press, pp. 174–203.

Elkinton, J. S. and R. T. Cardé. 1984. Odor dispersion. In W. J. Bell and R. T. Cardé (eds.) *Chemical Ecology of Insects*. London: Chapman and Hall, pp. 73–91.

Engel, V., M. K. Fischer, F. L. Wäckers, and W. Völkl. 2001. Interactions between extrafloral nectaries, aphids and ants: are there competition effects between plants and homopteran sugar sources? *Oecologia* **129**: 577–584.

Erdtman, G. 1952. *Pollen Morphology and Plant Taxonomy*. Stockholm: Almquist and Wicksell.

Erhardt, A. and I. Baker. 1990. Pollen amino acids: an additional diet for a nectar feeding butterfly? *Plant Systematics and Evolution* **169**: 111–121.

Evans, E. W. and S. England. 1996. Indirect interactions in biological control of insects: pests and natural enemies in alfalfa. *Ecological Applications* **6**(3): 920–930.

Faegri, K. and L. Van der Pijl. 1979. *The Principles of Pollination Ecology*. Oxford, UK: Pergamon Press.

Fahn, A. 1988. Secretory tissues in vascular plants. *New Phytologist* **108**: 229–258.

Feder, W. A. and R. Shrier. 1990. Combination of UV-B and ozone reduces pollen-tube growth more than either stress alone. *Environmental and Experimental Botany* **30**: 451–454.

Feinsinger, P. and L. A. Swarm. 1978. How common are ant-repellent nectars? *Biotropica* **10**: 238–239.

Fernandes, G. W., M. Fagundes, R. L. Woodman, and P. W. Price. 1999. Ant effects on three-trophic level interactions: plant, galls, and parasitoids. *Ecological Entomology* **24**: 411–415.

Fiedler, K., B. Hölldobler, and P. Seufert. 1996. Butterflies and ants: the communicative domain. *Experientia* **52**: 14–24.

Finch, S. 1974. Sugars available from flowers visited by the adult cabbage root fly, *Erioischia brassicae* (Bch.) (Diptera, Anthomyiidae). *Bulletin of Entomological Research* **64**: 257–263.

Finch, S. and T. H. Coaker. 1969. Comparison of the nutritive values of carbohydrates and related compounds to *Erioischia brassicae*. *Entomologia Experimentalis et Applicata* **12**: 441–453.

Fischer, M. K. and A. W. Shingleton. 2001. Host plant and ants influence the honeydew sugar composition of aphids. *Functional Ecology* **15**: 544–550.

Fischer, M. K., K. H. Hoffmann, and W. Volkl. 2001. Competition for mutualists in an ant–homopteran interaction mediated by hierarchies of ant attendance. *Oikos* **92**: 531–541.

Fischer, M. K., W. Völkl, R. Schopf, and K. H. Hoffmann. 2002. Age-specific patterns in honeydew production and honeydew composition in the aphid *Metopeurum fuscoviride*: implications for ant-attendance. *Journal of Insect Physiology* **48**: 319–326.

Fisher, B. L., S. da Silva Lobo Sternberg, and D. Price. 1990. Variation in the use of orchid extrafloral nectar by ants. *Oecologia* **83**: 263–266.

Flechtmann, C. H. W. and J. A. McMurtry. 1992. Studies of cheliceral and deutosternal morphology of some Phytoseiidae (Acari: Mesostigmata) by scanning electron microscopy. *International Journal of Acarology* **18**: 163–169.

Fonta, C., M. Pham-Delègue, R. Marilleau, and C. Masson. 1985. Rôle des nectars de tournesol dans le comportement des insectes pollinisateurs et analyse qualitative et quantitative des éléments glucidiques de ces sécrétions. *Acta Oecologia* **6**: 175–186.

Ford, H. A. and N. Forde. 1976. Birds as possible pollinators of *Acacia pycnantha*. *Australian Journal of Botany* **24**: 793–795.

Francis, F., J. Lognay, J. P. Wathelet, and E. Haubruge. 2001. Effects of allelochemicals from first (Brassicaceae) and second (*Myzus persicae* and *Brevicoryne brassicae*) trophic levels on *Adalia bipunctata*. *Journal of Chemical Ecology* **27**: 243–256.

Freitas, A. V. L. and P. S. Oliveira. 1996. Ants as selective agents on herbivore biology: effects on the behaviour of a non-myrmecophilous butterfly. *Journal of Animal Ecology* **65**: 205–210.

Gardener, M. C. and M. P. Gillman. 2001. Analyzing variability in nectar amino acids: composition is less variable than concentration. *Journal of Chemical Ecology* **27**: 2545–2558.

Gaume, L. and D. McKey. 1999. An ant–plant mutualism and its host-specific parasite: activity rhythms, young leaf patrolling, and effects on herbivores of two specialist plant-ants inhabiting the same myrmecophyte. *Oikos* **84**: 130–144.

Genissel, A., P. Aupinel, C. Bressac, J. N. Tasei, and C. Chevrier. 2002. Influence of pollen origin on performance of *Bombus terrestris* micro-colonies. *Entomologia Experimentalis et Applicata* **104**: 329–336.

Ghazoul, J. 2001. Can floral repellents pre-empt potential ant–plant conflicts? *Ecology Letters* **4**: 295–299.

Gilbert, F. and M. A. Jervis. 1998. Functional, evolutionary and ecological aspects of feeding-related mouthpart specializations in parasitoid flies. *Biological Journal of the Linnean Society* **63**: 495–535.

Giurfa, M., M. Vorobyev, P. Kevan, and R. Menzel. 1996. Detection of coloured stimuli by honeybees: minimum visual angles and receptor specific contrasts. *Journal of Comparative Physiology A* **178**: 699–709.

Gleim, K. H. 1985. *Die Blütentracht, Nahrungsquellen des Bienenvolkes*. Germany: Sankt Augustin, Delta-Verlag.

Gori, D. F. 1983. Post-pollination phenomena and adaptive floral changes. In C. E. Jones and R. J. Little (eds.) *Handbook of Experimental Pollination Biology*. New York: Van Nostrand Reinhold, pp. 31–49.

Gottsberger, G., T. Arnold, and H. F. Linskens. 1990. Variation in floral nectar amino acids with aging of flowers, pollen contamination, and flower damage. *Israel Journal of Botany* **39**: 167–176.

Gottsberger, G., J. Schrauwen, and H. F. Linskens. 1984. Amino acids and sugars in nectar, and their putative evolutionary siginificance. *Plant Systematics and Evolution* **145**: 55–77.

Gracie, C. 1991. Observation of dual function of nectaries in *Ruellia radicans* (Nees) Lindau (Acanthaceae). *Bulletin of the Torrey Botanical Club* **118**: 1101–1106.

Guerrant, E. O. and P. L. Fiedler. 1981. Flower defenses against nectar pilferage by ants. *Biotropica* **13**: 25–33.

Györfi, J. 1945. Beobachtungen über die Ernährung der Schlupfwespenimagos. *Erdészeti Kisérletek* **45**: 100–112.

Hagedorn, H. H. and F. E. Moeller. 1968. Effect of the age of pollen used in pollen supplements on their nutritive value for the honey bee. I. Effect on thoracic weight, development of hypopharyngeal glands, and brood rearing. *Journal of Apicultural Research* **7**: 89–95.

Hagen, K. S. 1986. Ecosystem analysis: plant cultivars (HPR), entomophagous species and food supplements. In D. J. Boethel and R. D. Eikenbary (eds.) *Interactions of Plant Resistance and Parasitoids and Predators of Insects*. New York: John Wiley, pp. 153–197.

Hansen, D. M., J. M. Olesen, and C. G. Jones. 2002. Trees, birds and bees in Mauritius: exploitative competition between introduced honey bees and endemic nectarivorous birds? *Journal of Biogeography* **29**: 721–734.

Harborne, J. B. and R. J. Grayer. 1993. Flavonoids and insects. In J. B. Harborne (ed.) *The Flavonoids: Advances in Research since 1986*. London: Chapman and Hall, pp. 589–618.

Harder, L. D. 1998. Pollen-size comparisons among animal pollinated angiosperms with different pollination characteristics. *Biological Journal of the Linnean Society* **64**: 513–525.

Hartley, S. E. and C. G. Jones. 1997. Plant chemistry and herbivory, or why the world is green. In M. J. Crawley (ed.) *Plant Ecology*. Oxford, UK: Blackwell Science, pp. 284–324.

Haskins, C. P. and E. F. Haskins. 1950. Notes on the biology and social behavior of the archaic ponerine ants of the genera *Myrmeca* and *Promyrmeca*. *Annals of the Entomological Society of America* **43**: 461–491.

Haslett, J. R. 1989. Interpreting patterns of resource utilization: randomness and selectivity in pollen feeding by adult hoverflies. *Oecologia* **78**: 433–442.

Heil, M., B. Fiala, B. Baumann, and K. E. Linsenmair. 2000. Temporal, spatial and biotic variations in extrafloral nectar secretion by *Macaranga tanarius*. *Functional Ecology* **14**: 749–757.

Heil, M., B. Fiala, W. Kaiser, and K. E. Linsenmair. 1998. Chemical contents of *Macaranga* food bodies: adaptations to their role in ant attraction and nutrition. *Functional Ecology* **12**: 117–122.

Heil, M., T. Koch, A. Hilpert, *et al.* 2001. Extrafloral nectar production of the ant-associated plant, *Macaranga tanarius*, is an induced, indirect, defensive response elicited by jasmonic acid. *Proceedings of the National Academy of Sciences of the USA* **98**: 1083–1088.

Heiling, A. M., M. E. Herberstein, and L. Chittka. 2003. Pollinator attraction: crab-spiders manipulate flower signals. *Nature* **421**: 334–334.

Heimpel, G. E., J. C. Lee, Z. Wu, *et al.* 2004. Gut sugar analysis in field-caught parasitoids: adapting methods originally developed for biting flies. *International Journal of Pest Management* **50**: 193–198.

Hendrix, D. L. and M. E. Salvucci. 2001. Isobemisiose: an unusual trisaccharide abundant in the silverleaf whitefly, *Bemisia argentifolii*. *Journal of Insect Physiology* **47**: 423–432.

Hendrix, D. L., Y. Wei, and J. E. Leggett. 1992. Homopteran honeydew sugar composition is determined by both the insect and plant species. *Comparative Biochemistry and Physiology B* **101**: 23–27.

Henning, J. A., Y. S. Peng, M. A. Montague, and L. R. Teubler. 1992. Honey-bee (Hymenoptera, Apidae) behavioral response to primary alfalfa (Rosales, Fabaceae) floral volatiles. *Journal of Economic Entomology* **85**: 233–239.

Hern, A. and S. Dorn. 1999. Sexual dimorphism in the olfactory orientation of adult *Cydia pomonella* in response to alpha-farnesene. *Entomologia Experimentalis et Applicata* **92**: 63–72.

Hocking, H. 1966. The influence of food on longevity and oviposition in *Rhyssa persuasoria* (L.) (Hymenoptera: Ichneumonidae). *Journal of the Australian Entomological Society* **6**: 83–88.

Hölldobler, B. and E. O. Wilson. 1990. *The Ants*. Cambridge, MA: Harvard University Press.

Horvitz, C. C. and D. W. Schemske. 1986. Seed dispersal of a neotropical myrmecochore: variation in removal rates and dispersal distance. *Biotropica* **16**: 319–323.

Hulshof, J. and I. Vänninen. 2002. Western flower thrips feeding on pollen, and its implications for control. In *Thrips and Tospoviruses*, Proc. 7th Int. Symp. *Thysanoptera*, Canberra, pp. 173–179.

Huxley, C. R. 1980. Symbiosis between ants and epiphytes. *Biological Reviews* **55**: 321–340.

Ingold, C. T. 1971. *Fungal Spores: Their Liberation and Dispersal.* Oxford, UK: Clarendon Press.

Inouye, D. W. 1983. The ecology of nectar robbing. In B. Bentley and T. Elias (eds.) *The Biology of Nectaries*. New York: Columbia University Press, pp. 153–173.

Inouye, D. W. and O. R. Taylor. 1979. A temperate region plant–ant–seed predator system: consequences of extrafloral nectar secretion by *Helianthella quinquenervis*. *Ecology* **60**: 1–7.

Inouye, D. W. and G. D. Waller. 1984. Responses of honey bees (*Apis mellifera*) to amino acid solutions mimicking floral nectars. *Ecology* **65**: 618–625.

Izzo, T. J. and H. L. Vasconcelos. 2002. Cheating the cheater: domatia loss minimizes the effects of ant castration in an Amazonian ant-plant. *Oecologia* **133**: 200–205.

Jackson, S. and S. W. Nicolson. 2002. Xylose as a nectar sugar: from biochemistry to ecology. *Comparative Biochemistry and Physiology B* **131**: 613–620.

Jander, R. 1998. Olfactory learning of fruit odors in the eastern yellow jacket, *Vespula maculifrons* (Hymenoptera: Vespidae). *Journal of Insect Behavior* **11**: 879–888.

Janzen, D. H. 1966. Coevolution of mutualism between ants and acacias in Central America. *Evolution* **20**: 249–275.

1977. Why don't ants visit flowers? *Biotropica* **9**: 252.

Jervis, M. A. 1990. Predation of *Lissonota coracinus* (Gmelin) (Hymenoptera: Ichneumonidae) by *Dolichonabis limbatus* (Dahlborn) (Hemiptera: Nabidae). *Entomologist's Gazette* **41**: 231–233.

1998. Functional and evolutionary aspects of mouthpart stucture in parasitoid wasps. *Biological Journal of the Linnean Society* **63**: 461–493.

Jervis, M. A., N. A. C. Kidd, M. G. Fitton, T. Huddleston, and H. A. Dawah. 1993. Flower-visiting by hymenopteran parasitoids. *Journal of Natural History* **27**: 67–105.

Jervis, M. A., N. A. C. Kidd, and G. E. Heimpel. 1996. Parasitoid adult feeding behaviour and biocontrol: a review. *Biocontrol News and Information* **17**: 11N–26N.

Johnson, M. W., V. P. Jones, and N. C. Toscano. 1987. Diel activity patterns of tobacco budworm, *Heliothis virescens* (F.), and cabbage looper, *Trichoplusia ni* (Hübner) larvae. *Environmental Entomology* **16**: 25–29.

Jolivet, P. 1998. *Myrmecophily and Ant-Plants*. Boca Raton, FL: CRC Press.

Josens, R. B., W. M. Farina, and F. Roces. 1998. Nectar feeding by the ant *Camponotus mus* as a function of sucrose concentration. *Journal of Insect Physiology* **44**: 579–585.

Keeler, K. H. 1981. Function of *Mentzelia nuda* (Loasaceae) postfloral nectaries in seed defense. *American Journal of Botany* **68**: 295–299.

Kevan, P. G. 1972. Floral colors in the high arctic with reference to insect–flower relations and pollination. *Canadian Journal of Botany* **50**: 2289–2316.

1973. Parasitoid wasps as flower visitors in the Canadian high arctic. *Anzeiger für Schädlingskunde, Pflanzenschutz und Umweltschutz* **46**: 3–7.

Kevan, P. G. and H. G. Baker. 1983. Insects as flower visitors and pollinators. *Annual Review of Entomology* **28**: 407–453.

1998. Insects on flowers. In C. B. Huffaker and A. P. Gutierrez (eds.) *Ecological Entomology*. New York: John Wiley, pp. 553–583.

Kikuchi, T. 1963. Studies on the coaction among insects visiting flowers. III. Dominance relationship among flower-visiting flies, bees and butterflies. *Scientific Reports of the Tohoku University* **29**: 1–8.

Kirk, W. D. J. 1984. Pollen-feeding in thrips (Insecta: Thysanoptera). *Journal of Zoology* **204**: 107–117.

1985. Pollen-feeding and the host specificity and fecundity of flower thrips. *Ecological Entomology* **10**: 281–289.

Kiss, A. 1981. Melizitose, aphids and ants. *Oikos* **37**: 382.

Kleber, E. 1935. Hat das Zeitgedächtnis der Bienen biologische Bedeutung? *Journal of Comparative Physiology* **22**: 221–262.

Knoll, F. 1930. Uber Pollenkitt und Bestäubungsart. *Zeitschrift für Botanie* **23**: 609–675.

Knox, R. B., J. Kenrick, P. Bernhardt, *et al.* 1985. Extra-floral nectaries as adaptations for bird pollination in *Acacia terminalis*. *American Journal of Botany* **72**: 1185–1196.

Knudsen, J. T., L. Tollsten, and L. G. Bergström. 1993. Floral scents: a checklist of volatile compounds isolated by head-space techniques. *Phytochemistry* **33**: 253–280.

Koptur, S. 1989. Is extrafloral nectar production an inducible defence? In J. Bock and Y. Linhart (eds.) *Evolutionary Ecology of Plants*. Boulder, CO: Westview Press, pp. 323–339.

1992. Extrafloral nectary-mediated interactions between insects and plants. In E. Bernays (ed.) *Insect–Plant Interactions*. Boca Raton, FL: CRC Press, pp. 81–129.

1994. Floral and extrafloral nectars of Costa Rican *Inga* trees: a comparison of their constituents and composition. *Biotropica* **26**: 276–284.

Koptur, S. and N. Truong. 1998. Facultative ant–plant interactions: nectar sugar preferences of introduced pest ant species in South Florida. *Biotropica* **30**: 179–189.

Kretschmar, J. A. and T. W. Baumann. 1999. Caffeine in *Citrus* flowers. *Phytochemistry* **52**: 19–23.

Kretz, R. 1979. A behavioural analysis of colour vision in the ant *Cataglyphis bicolor* (Formicidae, Hymenoptera). *Journal of Comparative Physiology A* **131**: 217–233.

Krivan, V. and E. Sirot. 1997. Searching for food or hosts: the influence of parasitoids behavior on host–parasitoid dynamics. *Theoretical Population Biology* **51**: 201–209.

Kugler, H. 1970. *Blütenökologie*. Stuttgart, Germany: Fischer Verlag.

Kunkel, H. and W. J. Kloft (eds.). 1985. *Waldtracht und Waldhonig in der Imkerei*. Munich, Germany: Ehrenwirth Verlag.

Kunze, H. 1999. Pollination ecology in two species of *Gonolobus* (Asclepiadaceae). *Flora* **194**: 309–316.

Labandeira, C. C. 1997. Permian pollen eating. *Science* **277**: 1422–1423.

Lanza, J. 1988. Ant preferences for *Passiflora* nectar mimics that contain amino acids. *Biotropica* **20**: 341–344.

Lanza, J. and B. R. Krauss. 1984. Detection of amino acids in artificial nectars by two tropical ants, *Leptothorax* and *Monomorium*. *Oecologia* **63**: 423–425.

Larochelle, A. 1990. The food of carabid beetles (Coleoptera: Carabidae, including Cicindelinae). *Fabreries* (suppl.) **5**: 1–132.

Leatemia, J. A., J. E. Laing, and J. E. Corrigan. 1995. Effects of adult nutrition on longevity, fecundity and offspring sex ratio of *Trichogramma minutum* Riley (Hymenoptera: Trichogrammatidae). *Canadian Entomologist* **127**: 245–254.

Leduc, N., G. C. Douglas, M. Monnier, and V. Connolly. 1990. Pollination *in vitro*: effects on the growth of pollen tubes, seed set and gametophytic self-incompatibility in *Trifolium pratense* L. and *Trifolium repens* L. *Theoretical and Applied Genetics* **80**: 657–664.

Lee, J. C. and G. E. Heimpel. 2003. Sugar feeding by parasitoids in cabbage fields and the consequences for pest control. Proc. 1st Int. Symp. *Biological Control of Arthropods*, Honolulu, pp. 220–225.

Leius, K. 1960. Attractiveness of different foods and flowers to the adults of some hymenopterous parasites. *Canadian Entomologist* **92**: 369–376.

1961. Influence of various foods on fecundity and longevity of adults of *Scambus buolianae* (Htg.) (Hymenoptera: Ichneumonidae). *Canadian Entomologist* **93**: 1079–1084.

Letourneau, D. K. 1990. Code of ant–plant mutualism broken by a parasite. *Science* **248**: 215–217.

Leveau, J. H. J. and S. E. Lindow. 2001. Appetite of an epiphyte: quantitative monitoring of bacterial sugar consumption in the phyllosphere. *Proceedings of the National Academy of Sciences of the USA* **98**: 3446–3453.

Lewis, A. C. 1986. Memory constraints and flower choice in *Pieris rapae*. *Science* **232**: 863–865.

Limburg, D. D. and J. A. Rosenheim. 2001. Extrafloral nectar consumption and its influence on survival and development of an omnivorous predator, larval *Chrysoperla plorabunda* (Neuroptera: Chrysopidae). *Environmental Entomology* **30**: 595–604.

Lingren, P. D. and M. J. Lukefahr. 1977. Effects of nectariless cotton on caged populations of *Campoletis sonorensis*. *Environmental Entomology* **6**: 586–588.

Lunau, K. 2000. The ecology and evolution of visual pollen signals. In A. Dafni, M. Hesse and E. Pacini (eds.) *Pollen and Pollination*. Vienna: Springer-Verlag, pp. 89–113.

Lunau, K. and S. Wacht. 1994. Optical releasers of the innate proboscis extension in the hoverfly *Eristalis tenax* L. (Syrphidae, Diptera). *Journal of Comparative Physiology A* **174**: 575–579.

Malcolm, S. B. 1990. Chemical defenses in chewing and sucking insect herbivores: plant-derived cardenolides in the monarch butterfly and oleander aphid. *Chemoecology* **1**: 12–21.

Markin, G. P. 1970. Food distribution within laboratory colonies of the Argentine ant, *Iridomyrmex humilis* (Mayr). *Insectes Sociaux* **17**: 127–158.

May, P. G. 1985. Nectar uptake rates and optimal nectar concentrations of two butterfly species. *Oecologia* **66**: 381–386.

Mendel, Z., D. Blumberg, A. Zehavi, and M. Weissenberg. 1992. Some polyphagous Homoptera gain protection from their natural enemies by feeding on the toxic plants *Spartium junceum* and *Erythrina corallodendrum* (Leguminosa). *Chemoecology* **3**: 118–124.

Mercier, J. and S. E. Lindow. 2000. Role of leaf surface sugars in colonization of plants by bacterial epiphytes. *Applied and Environmental Microbiology* **66**: 369–374.

Meurer, B., V. Wray, R. Wiermann, and D. Starck. 1988. Hydroxy cinnamic acid-spermidine amides from pollen of *Alnus glutinosa*, *Betula verrucosa* and *Pterocarya fraxinifolia*. *Phytochemistry* **27**: 839–843.

Milewski, A. V. and W. J. Bond. 1982. Convergence of myrmecochory in mediterranean Australia and South Africa. In R. C. Buckley (ed.) *Ant–Plant Interactions in Australia*. The Hague, the Netherlands: Junk, pp. 89–98.

Milton, K. 1999. Nutritional characteristics of wild primate foods: do the diets of our closest living relatives have lessons for us? *Nutrition* **15**: 488–498.

Mittler, T. E. and T. Meikle. 1991. Effects of dietary sucrose concentration on aphid honeydew carbohydrate levels and rates of excretion. *Entomologia Experimentalis et Applicata* **59**: 1–7.

Morse, D. H. 1986. Predation risk to insect foraging at flowers. *Oikos* **46**: 223–228.

Ne'eman, G. and P. G. Kevan. 2001. The effect of shape parameters on maximal detection distance of model targets by honeybee workers. *Journal of Comparative Physiology A* **187**: 653–660.

Nepi, M. and G. G. Franchi. 2000. Cytochemistry of mature angiosperm pollen. In A. Dafni, M. Hesse, and E. Pacini (eds.) *Pollen and Pollination*. Vienna: Springer-Verlag, pp. 45–62.

Nettles, W. C. and M. L. Burks. 1971. Absorption and metabolism of galactose and galactitol in *Anthonomus grandis*. *Journal of Insect Physiology* **17**: 1615–1623.

O'Dowd, D. J. and E. A. Catchpole. 1983. Ants and extrafloral nectaries: no evidence for plant protection in *Helichrysum* spp.–ant interactions. *Oecologia* **59**: 191–200.

Oliveira, P. S. 1997. The ecological function of extrafloral nectaries: herbivore deterrence by visiting ants and reproductive output in *Caryocar brasiliense* (Caryocaraceae). *Functional Ecology* **11**: 323–330.

Olson, D. L. and J. R. Nechols. 1995. Effects of squash leaf trichome exudates and honey on adult feeding, survival, and fecundity of the squash bug (Heteroptera: Coreidae) egg parasitoid *Gryon pennsylvanicum* (Hymenoptera: Scelionidae). *Environmental Entomology* **24**: 454–458.

Orivel, J. and A. Dejean. 2002. Ant activity rhythms in a pioneer vegetal formation of French Guiana (Hymenoptera: Formicidae). *Sociobiology* **39**: 65–76.

Osche, G. 1983. Optische Signale in der Coevolution von Pflanze und Tier. *Berichte der deutschen botanischen Gesellschaft* **96**: 1–27.

Pascal, L. and M. Belin-Depoux. 1991. La correlation entre les rhythmes biologiques de l'association plante–fourmis: le cas des nectaires extra-floraux de Malpighiaceae américaines. *Comptes Rendus de l'Académie des Sciences Paris Series 3* **312**: 49–54.

Pate, J. S., M. B. Peoples, P. J. Storer, and C. A. Atkins. 1985. The extrafloral nectaries of cowpea (*Vigna unguiculata* (L.) Walp.). II. Nectar composition, origin of nectar solutes, and nectary functioning. *Plant* **166**: 28–38.

Patt, J. M., G. C. Hamilton, and J. H. Lashomb. 1997. Foraging success of parasitoid wasps on flowers: interplay of insect morphology, floral architecture and searching behavior. *Entomologia Experimentalis et Applicata* **83**: 21–30.

Pemberton, R. W. and N. J. Vandenberg. 1993. Extrafloral nectar feeding by ladybird beetles (Coleoptera, Coccinellidae). *Proceedings of the Entomological Society of Washington* **95**: 139–151.

Peng, Z. and P. W. Miles. 1991. Oxidases in the gut of an aphid, *Macrosiphum rosae* (L.) and their relation to dietary phenolics. *Journal of Insect Physiology* **37**: 779–787.

Percival, M. S. 1955. The presentation of pollen in certain angiosperms and its collection by *Apis mellifera*. *New Phytologist* **54**: 353–368.

1961. Types of nectar in angiosperms. *New Phytologist* **60**: 235–281.

1965. *Floral Biology*. Oxford, UK: Pergamon Press.

Petanidou, T. and D. Vokou. 1990. Pollination and pollen energetics in Mediterranean ecosystems. *American Journal of Botany* **77**: 986–992.

Peumans, W. J., K. Smeets, K. van Nerum, F. van Leuven, and E. J. M. van Damme. 1997. Lectin and alliinase are the predominant proteins in nectar from leek (*Allium porrum* L.) flowers. *Planta* **201**: 298–302.

Pfannenstiel, R. S. and K. V. Yeargan. 2002. Identification and diel activity patterns of predators attacking *Helicoverpa zea* (Lepidoptera: Noctuidae) eggs in soybean and sweet corn. *Environmental Entomology* **31**: 232–241.

Pickett, C. H. and W. D. Clark. 1979. The function of extrafloral nectaries in *Opuntia acanthocarpa* (Cactacea). *American Journal of Botany* **66**: 618–625.

Porter, S. D. 1989. Effects of diet on the growth of laboratory fire ant colonies (Hymenoptera: Formicidae). *Journal of the Kansas Entomological Society* **62**: 288–291.

Potter, C. F. and R. I. Bertin. 1988. Amino acids in artificial nectar: feeding preferences of the flesh fly *Sarcophaga bullata*. *American Midland Naturalist* **120**: 156–162.

Putman, W. L. 1958. Mortality of the European red mite (Acarina: Tetranychidae) from secretions of peach leaf nectaries. *Canadian Entomologist* **90**: 720–721.

Rahbé, Y., N. Sauvion, G. Febvay, W. J. Peumans, and A. M. R. Gatehouse. 1995. Toxicity of lectins and processing of ingested proteins in the pea aphid *Acyrthosiphon pisum*. *Entomologia Experimentalis et Applicata* **76**: 143–155.

Retana, J., J. Bosch, A. Alsina, and X. Cerda. 1987. Foraging ecology of the nectarivorous ant *Camponotus foreli* (Hymenoptera: Formicidae) in a savannah-like grassland. *Miscellania Zoologica* **11**: 187–193.

Ricks, B. L. and S. B. Vinson. 1970. Feeding acceptability of certain insects and various water-soluble compounds to varieties of the imported fire ant. *Journal of Economic Entomology* **63**: 145–148.

Rickson, F. R. 1980. Developmental anatomy and ultrastructure of the ant-food bodies (Beccarian bodies) of *Macaranga triloba* and *M. hypoleuca* (Euphorbiaceae). *American Journal of Botany* **67**: 285–292.

Rico-Gray, V. and L. B. Thien. 1989. Effect of different ant species on reproductive fitness of *Schomburgkia tibicinis* (Orchidaceae). *Oecologia* **81**: 487–489.

Rico-Gray, V., M. Palacios-Rios, and J. G. Garcia-Franco. 1998. Richness and seasonal variation of ant–plant associations as mediated by plant-derived food resources in the semiarid Zapotitlan Valley, Mexico. *American Midland Naturalist* **140**: 21–26.

Risch, S. J. and F. R. Rickson. 1981. Mutualism in which ants must be present before plants produce food bodies. *Nature* **291**: 149–150.

Rogers, C. E. 1985. Extrafloral nectar: entomological implications. *Bulletin of the Entomological Society of America* **31**: 15–20.

Romeis, J. and F. L. Wäckers. 2000. Feeding responses by female *Pieris brassicae* butterflies to carbohydrates and amino acids. *Physiological Entomology* **25**: 247–253.

2002. Nutritional suitability of individual carbohydrates and amino acids for adult *Pieris brassicae*. *Physiological Entomology* **27**: 148–156.

Romeis, J. and C. P. W. Zebitz. 1997. Searching behaviour of *Encarsia formosa* as mediated by colour and honeydew. *Entomologia Experimentalis et Applicata* **82**: 299–309.

Romeis, J., D. Babendreier and F. L. Wäckers. 2003. Consumption of snowdrop lectin (*Galanthus nivalis*) agglutinin causes direct effects on adult parasitic wasps. *Oecologia* **34**: 528–536.

Rosenheim, J. A. 1998. Higher order predators and the regulation of insect herbivore populations. *Annual Review of Entomology* **43**: 421–447.

Roulston, T. H. and S. L. Buchmann. 2000. A phylogenetic reconsideration of the pollen starch–pollination correlation. *Evolutionary Ecology Research* **2**: 627–643.

Roulston, T. H. and J. H. Cane. 2000. Pollen nutritional content and digestibility for animals. In A. Dafni, M. Hesse, and E. Pacini (eds.) *Pollen and Pollination*. Vienna: Springer-Verlag, pp. 187–211.

Roy, B. A. 1993. Floral mimicry by a plant pathogen. *Nature* **362**: 56–58.

Ruhren, S. and S. Handel. 1999. Jumping spiders (Salticidae) enhance the seed production of a plant with extrafloral nectaries. *Oecologia* **119**: 227–230.

Runions, C. J., K. H. Rensing, T. Takaso and J. Owens. 1999. Pollination of *Picea orientalis* (Pinaceae): saccus morphology governs pollen buoyancy. *American Journal of Botany* **86**: 190–197.

Rusterholz, H. P. and A. Erhardt. 1997. Preferences for nectar sugars in the peacock butterfly *Inachis io*. *Ecological Entomology* **22**: 220–224.

1998. Effects of elevated CO_2 on flowering phenology and nectar production of nectar plants important for butterflies of calcareous grasslands. *Oecologia* **113**: 341–349.

Salvucci, M. E., R. C. Rosell, and J. K. Brown. 1998. Uptake and metabolism of leaf proteins by the silverleaf whitefly. *Archives of Insect Biochemistry and Physiology* **39**: 155–165.

Sandström, J. P. and N. Moran. 1999. How nutritionally imbalanced is phloem sap for aphids? *Entomologia Experimentalis et Applicata* **91**: 203–210.

2001. Amino acid budgets in three aphid species using the same host plant. *Physiological Entomology* **26**: 202–211.

Sasaki, T. and H. Ishikawa. 1991. Amino acids and their metabolism in symbiotic and aposymbiotic pea aphids. *Miscellaneous Publications of the Agricultural Experimental Station of Oklahoma State University* **288**.

Sasaki, T., T. Aoki, H. Hayashi, and H. Ishikawa. 1990. Amino acid composition of the honeydew of symbiotic and aposymbiotic pea aphids *Acyrthosiphon pisum*. *Journal of Insect Physiology* **36**: 35–40.

Schneider, P. 1972. Versuche zur Frage der individuellen Futterverteilung bei der kleinen roten Waldameise (*Formica polyctena*). *Insectes Sociaux* **19**: 279–299.

Schoonhoven, L. M. 1972. Secondary plant substances and insects. *Recent Advances in Phytochemistry* **5**: 197–224.

Schubert, A. 1972. Bienenfeindliche Pflanzen. *Bienenpflege* **8**: 171–172.

Schwarz, H. H. and K. Huck. 1997. Phoretic mites use flowers to transfer between foraging bumblebees. *Insectes Sociaux* **44**: 303–310.

Schwörer, U. and W. Völkl. 2001. Foraging behavior of *Aphidius ervi* (Haliday) (Hymenoptera: Braconidae: Aphidiinae) at different spatial scales: resource utilization and suboptimal weather conditions. *Biological Control* **21**: 111–119.

Scott, H. J. and C. T. Stojanovich. 1963. Digestion of juniper pollen by Collembola. *Florida Entomologist* **46**: 189–191.

Seeman, O. D. 1996. Flower mites and phoresy: the biology of *Hattena panopla* Domrow and *Hattena cometis* Domrow (Acari: Mesostigmata: Ameroseiidae). *Australian Journal of Zoology* **44**: 193–203.

Seibert, T. F. 1993. A nectar-secreting gall wasp and ant mutualism: selection and counter-selection shaping gall wasp phenology, fecundity and persistence. *Ecological Entomology* **18**: 247–253.

Sheldon, J. K. and E. G. MacLeod. 1971. Studies on the biology of Chrysopidae. II. The feeding behavior of the adult *Chrysopa carnea* (Neuroptera). *Psyche* **78**: 107–121.

Shiraishi, A. and M. Kuwabra. 1970. The effects of amino acids on the labellar hair chemosensory cells of the fly. *Journal of Genetic Physiology* **56**: 768–782.

Shivanna, K. R., H. F. Linskens, and M. Cresti. 1991. Responses of tobacco pollen to high humidity and heat-stress: viability and germinability *in vitro* and *in vivo*. *Sexual Plant Reproduction* **4**: 104–109.

Shykoff, J. A. and E. Bucheli. 1995. Pollinator visitation patterns, floral rewards and the probability of transmission of *Microbotryum violaceum*, a venereal disease of plants. *Journal of Ecology* **83**: 189–198.

Siekmann, G., B. Tenhumberg, and M. A. Keller. 2001. Feeding and survival in parasitic wasps: sugar concentration and timing matter. *Oikos* **95**: 425–430.

Simpson, S. J. and D. Raubenheimer. 2001. The geometric analysis of nutrient–allelochemical interactions: a case study using locusts. *Ecology* **82**: 422–439.

Smith, L. L., J. Lanza, and G. C. Smith. 1990. Amino acid concentrations in extrafloral nectar of *Impatiens sultani* increase after simulated herbivory. *Ecology* **71**: 107–115.

Solberg, Y. and G. Remedios. 1980. Chemical composition of pure and bee-collected pollen. *Medlinger fra Norges Landbrukshoegskole* **59**: 2–12.

Sols, A., E. Cadenas, and F. Alvarado. 1960. Enzymatic basis of mannose toxicity in honey bees. *Science* **131**: 297–298.

Stadler, B. and A. F. G. Dixon. 1999. Ant attendance in aphids: why different degrees of myrmecophily? *Ecological Entomology* **24**: 363–369.

Stamp, N. E. and M. D. Bowers. 1988. Direct and indirect effects of predatory wasps on gregarious larvae of the buckmoth, *Hemileuca lucina* (Saturnidae). *Oecologia* **75**: 619–624.

Stanley, R. G. and H. F. Linskens. 1974. *Pollen: Biology, Biochemistry, Management.* Heidelberg, Germany: Springer-Verlag.

Stanton, M. L. 2003. Interacting guilds: moving beyond the pairwise perspective on mutualisms. *American Naturalist* **162**: S10–S23.

Stapel, J. O., A. M. Cortesero, C. M. De Moraes, J. H. Tumlinson, and W. J. Lewis. 1997. Extrafloral nectar, honeydew and sucrose effects on searching behavior and efficiency of *Microplitis croceipes* (Hymenoptera: Braconidae) in cotton. *Environmental Entomology* **26**: 617–623.

Steinbauer, M. J. 1996. A note on manna feeding by ants (Hymenoptera: Formicidae). *Journal of Natural History* **30**: 1185–1192.

Stephens, M. J., C. M. France, S. D. Wratten, *et al.* 1998. Enhancing biological control of leafrollers (Lepidoptera: Tortricidae) by sowing buckwheat (*Fagopyrum esculentum*) in an orchard. *Biocontrol Science and Technology* **8**: 547–558.

Stephenson, A. G. 1982a. Iridoid glycosides in the nectar of *Catalpa speciosa* are unpalatable to nectar thieves. *Journal of Chemical Ecology* **8**: 1025–1034.

1982b. The role of the extrafloral nectaries of *Catalpa speciosa* in limiting herbivory and increasing fruit production. *Ecology* **63**: 663–669.

Stoffolano, J. G. 1995. Regulation of a carbohydrate meal in the adult Diptera, Lepidoptera, and Hymenoptera. In R. F. Chapman and G. de Boer (eds.) *Regulatory Mechanisms in Insect Feeding*. New York: Chapman and Hall, pp. 210–247.

Sudd, J. H. and N. R. Franks. 1987. *The Behavioural Ecology of Ants*. New York: Chapman and Hall.

Swirski, E., Y. Izhar, M. Wysoki, E. Gurevitz, and S. Greenberg. 1980. Integrated control of the long-tailed mealybug, *Pseudococcus longispinus* (Hom., Pseudococcidae), in avocado plantations in Israel. *Entomophaga* **25**: 415–426.

Takasu, K. and W. J. Lewis. 1993. Host- and food-foraging of the parasitoid *Microplitis croceipes*: learning and physiological state effects. *Biological Control* **3**: 70–74.

1995. Importance of adult food sources to host searching of the larval parasitoid *Microplitis croceipes*. *Biological Control* **5**: 25–30.

1996. The role of learning in adult food location by the larval parasitoid, *Microplitis croceipes*. *Journal of Insect Behavior* **9**: 265–281.

Takeda, S., K. Kinomura and H. Sakurai. 1982. Effects of ant-attendance on the honeydew excretion and larviposition of the cowpea aphid, *Aphis craccivora* Koch. *Applied Entomology and Zoology* **17**: 133–135.

Tanowitz, B. D. and D. L. Koehler. 1986. Carbohydrate analysis of floral and extrafloral nectars in selected taxa of *Sansivieria* (Agavaceae). *Annals of Botany* **58**: 541.

Thien, L. B., P. Bernhardt, G. W. Gibbs, *et al.* 1985. The pollination of *Zygogynum* (Winteraceae) by a moth, *Sabatinca* (Micropterigidae): an ancient association. *Science* **227**: 540–543.

Tilman, D. 1978. Cherries, ants, and tent caterpillars: timing of nectar production in relation to susceptibility of caterpillars to ant-predation. *Ecology* **59**: 686–692.

Tobin, J. E. 1994. Ants as primary consumers: diet and abundance in the Formicidae. In J. H. Hunt and C. A. Nalepa (eds.) *Nourishment and Evolution in Insect Societies*. Boulder, CO: Westview Press, pp. 279–307.

Triltsch, H. 1997. Contents in field sampled adults of *Coccinella septempunctata* (Col.: Coccinellidae). *Entomophaga* **42**: 125–131.

Tuckey, H. B. 1971. Leaching of substances from plants. In T. F. Peece and C. H. Dickinson (eds.) *Ecology of Leaf Surface Organisms*. New York: Academic Press, pp. 67–80.

Turlings, T. C. J. and F. L. Wäckers. 2004. Recruitment of predators and parasitoids by herbivore-injured plants. In R. T. Cardé and J. Millar (eds.) *Advances in Insect Chemical Ecology*. Cambridge, UK: Cambridge University Press, pp. 21–75.

Turlings, T. C. J., M. Bernasconi, R. Bertossa, *et al.* 1998. The induction of volatile emissions in maize by three herbivore species with different feeding habits: possible consequences for their natural enemies. *Biological Control* **11**: 122–129.

Vaissière, B. E. and S. B. Vinson. 1994. Pollen morphology and its effect on pollen collection by honey-bees, *Apis mellifera* L. (Hymenoptera, Apidae), with special reference to upland cotton, *Gossypium hirsutum* L. (Malvaceae). *Grana* **33**: 128–138.

Van Baalen, M., V. Krivan, P. C. J. van Rijn, and M. W. Sabelis. 2001. Alternative food, switching predators, and the persistence of predator–prey systems. *American Naturalist* **157**: 512–524.

Van Dam, N. M., J. A. Harvey, F. L. Wäckers, *et al.* 2003. Interactions between aboveground and belowground induced responses against phytophages. *Basic and Applied Ecology* **4**: 63–77.

Van der Pijl, L. 1951. On the morphology of some tropical plants: *Gloriosa, Bougainvillea, Honckenya* and *Rottboellia*. *Phytomorphology* **1**: 185–188.

Van Rijn, P. C. J. and L. K. Tanigoshi. 1999a. The contribution of extrafloral nectar to survival and reproduction of the predatory mite *Iphiseius degenerans* on *Ricinus communis*. *Experimental and Applied Acarology* **23**: 281–296.

1999b. Pollen as food for the predatory mites *Iphiseius degenerans* and *Neoseiulus cucumeris* (Acari: Phytoseiidae): dietary range and life history. *Experimental and Applied Acarology* **23**: 785–802.

Van Rijn, P. C. J., Y. M. van Houten, and M. W. Sabelis. 2002. How plants benefit from providing food to predators even when it is also edible to herbivores. *Ecology* **83**: 2664–2679.

Vet, L. E. M., F. L. Wäckers, and M. Dicke. 1991. How to hunt for hiding hosts: the reliability-detectability problem in foraging parasitoids. *Netherlands Journal of Zoology* **41**: 202–213.

Vinson, S. B. 1968. The distribution of an oil, carbohydrate and protein food source to members of the imported fire ant colony. *Journal of Economic Entomology* **61**: 712–714.

Vitzthum, H. G. 1930. Investigations of the causes of May sickness. *Bee World* **11**: 14–15.

Völkl, W. 1992. Aphids or their parasitoids: who actually benefits from ant-attendance? *Journal of Animal Ecology* **61**: 273–281.

2001. Parasitoid learning during interactions with ants: how to deal with an aggressive antagonist. *Behavioral Ecology and Sociobiology* **49**: 135–144.

Völkl, W. and W. Kraus. 1996. Foraging behaviour and resource utilization of the aphid parasitoid *Pauesia unilachni*: adaptation to host distribution and mortality risks. *Entomologia Experimentalis et Applicata* **79**: 101–109.

Völkl, W. and A. S. Kroupa. 1997. Effects of adult mortality risks on parasitoid foraging tactics. *Animal Behaviour* **54**: 349–359.

Völkl, W. and M. Mackauer. 1993. Interactions between ants attending *Aphis fabae* ssp. *cirsiiacanthoidis* on thistles and foraging parasitoid wasps. *Journal of Insect Behavior* **6**: 301–312.

Völkl, W., J. Woodring, M. Fischer, M. W. Lorenz, and K. H. Hoffmann. 1999. Ant-aphid mutualisms: the impact of honeydew production and honeydew sugar composition on ant preferences. *Oecologia* **118**: 483–491.

Von Frisch, K. 1934. Über den Geschmackssinn der Biene: Ein Beitrag zur vergleichenden Physiologie des Geschmacks. *Zeitschrift für vergleichenden Physiologie* **21**: 1–45.

Vrieling, K., W. Smit, and E. van der Meijden. 1991. Tritrophic interactions between aphids (*Aphis jacobaeae* Schrank.), ant species, *Tyria jacobaeae* L., and *Senecio jacobaea* L. lead to maintenance of genetic variation in pyrrolizidine alkaloid concentration. *Oecologia* **86**: 177–182.

Wäckers, F. L. 1994. The effect of food deprivation on the innate visual and olfactory preferences in the parasitoid *Cotesia rubecula*. *Journal of Insect Physiology* **40**: 641–649.

1999. Gustatory response by the hymenopteran parasitoid *Cotesia glomerata* to a range of nectar and honeydew sugars. *Journal of Chemical Ecology* **25**: 2863–2877.

2000. Do oligosaccharides reduce the suitability of honeydew for predators and parasitoids? A further facet to the function of insect-synthesized honeydew sugars. *Oikos* **90**: 197–201.

2001. A comparison of nectar and honeydew sugars with respect to their utilization by the hymenopteran parasitoid *Cotesia glomerata*. *Journal of Insect Physiology* **47**: 1077–1084.

2003a. The effect of food supplements on parasitoid–host dynamics. Proc. 1st Int. Symp. *Biological Control of Arthropods*, Honolulu, pp. 226–231.

2003b. The parasitoids' need for sweets: sugars in mass rearing and biological control. In J. C. van Lenteren (ed.) *Quality Control of Natural Enemies*. Wallingford, UK: CAB International, pp. 59–72.

2004. Assessing the suitability of flowering herbs as parasitoid food sources: flower attractiveness and nectar accessibility. *Biological Control* **29**: 307–314.

Wäckers, F. L. and T. M. Bezemer. 2003. Root herbivory induces an above-ground indirect defence. *Ecology Letters* **6**: 9–12.

Wäckers, F. L. and C. Bonifay. 2004. How to be sweet? Extrafloral nectar allocation in *Gossypium hirsutum* fits optimal defense theory predictions. *Ecology* **85**: 1512–1518.

Wäckers, F. L. and A. Steppuhn. 2003. Characterizing nutritional state and food source use of parasitoids collected in fields with high and low nectar availability. *IOBC/WPRS Bulletin* **26**: 203–208.

Wäckers, F. L. and C. P. M. Swaans. 1993. Finding floral nectar and honeydew in *Cotesia rubecula*: random or directed? *Proceedings of the Section Experimental and Applied Entomology of the Netherlands Entomological Society* **4**: 67–72.

Wäckers, F. L. and R. Wunderlin. 1999. Induction of cotton extrafloral nectar production in response to herbivory does not require a herbivore-specific elicitor. *Entomologia Experimentalis et Applicata* **91**: 149–154.

Wäckers, F. L., A. Björnsen and S. Dorn. 1996. A comparison of flowering herbs with respect to their nectar accessibility for the parasitoid *Pimpla turionellae*. *Proceedings of the Section Experimental and Applied Entomology of the Netherlands Entomological Society* **7**: 177–182.

Wäckers, F. L., C. Bonifay, and W. J. Lewis. 2002. Conditioning of appetitive behavior in the hymenopteran parasitoid *Microplitis croceipes*. *Entomologia Experimentalis et Applicata* **103**: 135–138.

Wäckers, F. L., D. Zuber, R. Wunderlin, and F. Keller. 2001. The effect of herbivory on temporal and spatial dynamics of extrafloral nectar production in cotton and castor. *Annals of Botany* **87**: 365–370.

Wada, A., Y. Isobe, S. Yamaguchi, R. Yamaoka, and M. Ozaki. 2001. Taste-enhancing effects of glycine on the sweetness of glucose: a gustatory aspect of symbiosis between the ant *Camponotus japonicus* and the larvae of the lycaenid butterfly *Niphanda fusca*. *Chemical Senses* **26**: 983–992.

Wagner, D. 1997. The influence of ant nests on *Acacia* seed production, herbivory and soil nutrients. *Journal of Ecology* **85**: 83–93.

Wagner, D. and A. Kay. 2002. Do extrafloral nectaries distract ants from visiting flowers? An experimental test of an overlooked hypothesis. *Evolutionary Ecology Research* **4**: 293–305.

Waller, G. D. 1972. Evaluating responses of honey bees to sugar solutions using an artificial-flower feeder. *Annals of the Entomological Society of America* **65**: 857–862.

Washburn, J. O. 1984. Mutualism between a cynipid gall wasp and ants. *Ecology* **65**: 654–656.

Wasserthal, L. T. 1993. Swing-hovering combined with long tongue in hovermoths, an antipredator adaptation during flower visits. In W. Barthlott, C. M. Naumann, K. Schmidt-Loske, and K. L. Schuchmann (eds.) *Animal–Plant Interactions in Tropical Environments*. Bonn, Germany: Zoologische Forschungsinstitut und Museum Alexander Koenig, pp. 77–87.

Watt, W. B., P. C. Hoch, and S. G. Mills. 1974. Nectar resource use by *Colias* butterflies: chemical and visual experiments. *Oecologia* **14**: 353–374.

Way, M. J. 1963. Mutualism between ants and honeydew producing Homoptera. *Annual Review of Entomology* **8**: 307–344.

Weber, G., S. Oswald, and U. Zöllner. 1986. Die Wirtseignung von Rapssorten unterschiedlichen Glucosinolatsgehaltes für *Brevicoryne brassicae* (L.) und *Myzus persicae*. (Sulzer) (Hemiptera, Aphididae). *Zeitschrift für Pflanzenkrankheiten und Pflanzenschutz* **93**: 113–124.

Weevers, T. 1952. Flower colours and their frequency. *Acta Botanica Neerlandica* **1**: 81–92.

Wehner, R. and M. V. Srinivasan. 1989. The world as the insect sees it. In T. Lewis (ed.) *Insect Communication*. London: Academic Press, pp. 29–47.

Weisser, W. W., A. I. Houston, and W. Völkl. 1994. Foraging strategies in solitary parasitoids: the trade-off between female and offspring mortality risks. *Evolutionary Ecology* **8**: 587–597.

Wellenstein, G. 1952. Zur Ernährungsbiologie der Roten Waldameise. *Zeitschrift für Pflanzenkrankheiten und Pflanzenschutz*: 430–451.

Whitman, D. W. 1996. Plant bodyguards: mutualistic interactions between plants and the third trophic level. In T. N. Ananthakrishnan (ed.) *Functional Dynamics of Phytophagous Insects*. New Delhi: Oxford University Press / IBH Publishing, pp. 207–248.

Wilkinson, T. L., D. A. Ashford, J. Pritchard, and A. E. Douglas. 1997. Honeydew sugars and osmoregulation in the pea aphid *Acyrthosiphon pisum*. *Journal of Experimental Biology* **200**: 2137–2143.

Willmer, P. G. 1980. The effects of insect visitors on nectar constituents in temperate plants. *Oecologia* **47**: 270–277.

Willmer, P. G. and G. N. Stone. 1997. How aggressive ant-guards assist seed-set in *Acacia* flowers. *Nature* **388**: 165–167.

Wink, M. and P. Römer. 1986. Acquired toxicity: the advantages of specializing on alkaloid-rich lupins to *Macrosiphon albifrons* (Aphidae). *Naturwissenschaften* **73**: 210–212.

Wink, M. and L. Witte. 1991. Storage of quinolizidine alkaloids in *Macrosiphum albifrons* and *Aphis genistae* (Homoptera: Aphididae). *Entomologia Generalis* **15**: 237–254.

Winkler, K., F. L. Wäckers, L. V. Valdivia, V. Larraz, and J. C. van Lenteren. 2003. Strategic use of nectar sources to boost biological control. *IOBC/WPRS Bulletin* **26**: 209–214.

Wirth, R. and I. R. Leal. 2001. Does rainfall affect temporal variability of ant protection in *Passiflora coccinea*? *Ecoscience* **8**: 450–453.

Wootton, J. T. and I. F. Sun. 1990. Bract liquid as a herbivore defense mechanism for *Heliconia wagneriana* inflorescences. *Biotropica* **22**: 155–159.

Wratten, S. D., A. J. White, M. H. Bowie, N. A. Berry, and U. Weigmann. 1995. Phenology and ecology of hoverflies (Diptera, Syrphidae) in New Zealand. *Environmental Entomology* **24**: 595–600.

Wunnachit, W., C. F. Jenner, and M. Sedgley. 1992. Floral and extrafloral nectar production in *Anacardium occidentale* L (Anacardiaceae): an andromonoecious species. *International Journal of Plant Sciences* **153**: 413–420.

Yano, S. 1994. Flower nectar of an autogamous perennial *Rorippa indica* as an indirect defense mechanism against herbivorous insects. *Researches in Population Ecology* **36**: 63–71.

Yao, I. and S. Akimoto. 2001. Ant attendance changes the sugar composition of the honeydew of the drepanosiphid aphid *Tuberculatus quercicola*. *Oecologia* **128**: 36–43.

2002. Flexibility in the composition and concentration of amino acids in honeydew of the drepanosiphid aphid *Tuberculatus quercicola*. *Ecological Entomology* **27**: 745–752.

Yue, B. S. and J. H. Tsai. 1996. Development, survivorship, and reproduction of *Amblyseius largoensis* (Acari: Phytoseiidae) on selected plant pollens and temperatures. *Environmental Entomology* **25**: 488–494.

Yue, B. S., C. C. Childers, and A. H. Fouly. 1994. A comparison of selected plant pollens for rearing *Euseius mesembrinus* (Acari: Phytoseiidae). *International Journal of Acarology* **20**: 103–108.

Zimmerman, M. 1932. Über die extra floralen Nectarien der Angiospermen. *Botanisches Zentralblatt Beihefte* **49**: 99–196.

Zimmerman, M. H. and H. Ziegler. 1975. List of sugars and sugar alcohols in sieve-tube exudates. In M. H. Zimmerman and J. A. Milburn (eds.) *Encyclopedia of Plant Physiology* vol. 1. New York: Springer-Verlag, pp. 480–503.

Zoebelein, G. 1956. Der Honigtau als Nahrung der Insekten. *Zeitschrift für angewandte Entomologie* **38**: 369–416.

3

Nectar as fuel for plant protectors

SUZANNE KOPTUR

Introduction

Nectar is a sweet liquid produced by plants on various parts of the plant body. Most people are familiar with nectar in flowers, collected by bees to make honey, and utilized by a variety of floral visitors, some of whom serve as pollinators for the plant. Less familiar is extrafloral nectar, produced outside the flowers in extrafloral nectaries and usually not associated with pollination. Plants produce nectar in various ways (Elias 1983; Koptur 1992a), and whether they do it purposefully (secretion) or passively (excretion) has been the subject of debate between physiologists and evolutionary ecologists for many years (reviewed in Bentley 1977; see also Sabelis *et al.*, Chapter 4). Over evolutionary time, myriad selective forces have shaped not only the morphology and function of nectaries, but also the composition of the substances secreted and whether or not the structures secrete under different circumstances. Thompson's (1994) synthetic theory of the "co-evolutionary mosaic", in which different populations of a given species experience different interactions over space and time, helps to explain the variable findings researchers encounter in studying interactions between plants and predatory insects, especially those mediated by nectar (or other direct or indirect food rewards from plants). Carnivorous organisms, which can benefit plants as protectors, may rely on nectar as an energy source. If ants, wasps, other predators, and parasitoids are more likely to encounter their herbivore prey if they utilize a plant's nectar, mutualisms are thus promoted.

In this chapter I examine the many interactions promoted by nectar, with particular attention to its role in support of predatory insects on plants. Parallels will be drawn between systems involving nectar (primarily extrafloral, but also

Plant-Provided Food for Carnivorous Insects, ed. F. L. Wäckers, P. C. J. van Rijn, and J. Bruin.
Published by Cambridge University Press.

floral nectar) and those that involve insect analogs of extrafloral nectaries (honeydew-producing insects, alternative resources for predators), since the literatures are intertwined and discoveries in one area help answer questions in the other (see also Wäckers, Chapter 2). In all areas examined, I highlight an arbitrary selection – acknowledging a particular bias towards ants – out of many studies chosen for their relevance, but my review of the subjects covered here is by no means complete.

Floral and extrafloral nectaries

Floral nectar is one of the primary rewards for animals visiting flowers, and is produced in flowers in a variety of locations. Most commonly, floral nectaries are located between the reproductive and sterile whorls of a flower, outside the androecium (stamens) of the flower. In monocots, septal nectaries are the norm, located on the surface of the ovary where the carpels (its component parts) come together to create the septae (or divisions) of the ovary. Nectar may be secreted and left exposed to the environment and visitors, open to all who might come, or pouring out of flowers if it is not collected. Many plants are economical in their production of nectar, and limit access to the nectar to visitors that are the right size, shape, or behavior to effect pollination. In some flowers nectar is produced by one flower part and collects in another, e.g., in *Viola* spp. where the nectar is produced from the lower staminal filaments and collects in a spur formed by the petal bases.

Extrafloral nectaries are known from ferns and flowering plants (Bentley 1977; Elias 1983; Koptur 1992b) but not yet recorded in gymnosperms. Though scientists have long been interested in extrafloral nectaries, their occurrence in many floras has not been carefully studied, and important contributions to our knowledge of the distribution and basic biology of extrafloral nectaries are still being made (Galetto and Bernardello 1992; Fiala *et al.* 1994; Hunter 1994; Morellato and Oliveira 1994; Valenzuela-Gonzalez *et al.* 1994; Fiala and Linsenmair 1995; O'Brien 1995; Oliveira and Pie 1998; Zachariades and Midgley 1999; Dejean *et al.* 2000b; Junqueira *et al.* 2001; Blüthgen and Reifenrath 2003) and could certainly continue for many years. Extrafloral nectaries are simply nectaries located outside the flowers, and may more specifically be named with reference to their position (e.g., foliar nectaries, petiolar nectaries, bracteal nectaries; or simply "nectaries" in ferns, which do not have flowers!). The position of extrafloral nectaries and the timing of their functioning suggest vulnerabilities in a plant that may benefit from protection by organisms attracted to the nectar. For example, the inflorescence bracts and stem tips of

Caryocar (Oliveira 1997) attract ants that reduce herbivores that attack its young leaves and reproductive parts; the nectaries on unfurling fronds of *Polypodium plebeium* (Koptur *et al.* 1998) attract ants that deter herbivory during leaf expansion. Nectar-imbibing organisms visit the nectaries to fuel their own energetic and nutritional needs, and in doing so are placed in proximity to plant parts that are vulnerable to enemies.

Possessing floral nectar is not a prerequisite for producing extrafloral nectar; not only do ferns not have flowers, but many flowering plant species with no floral nectaries have prominent extrafloral nectaries. Sometimes the extrafloral nectaries function as do most floral nectaries, attracting visitors to the vicinity of the flowers to inadvertently pollinate them. *Acacia terminalis* has large reddish foliar nectaries that are visited by passerine birds in Australia that pollinate the flowers located on branches with leaves (Knox *et al.* 1985). Australian *Acacia* species lack floral nectaries and most are entomophilous, mostly pollinated by bees collecting pollen and extrafloral nectar (Kenrick *et al.* 1987; Thorp and Sugden 1990). *Marcgravia* and *Norantea* (Marcgraviaceae) inflorescences have large nectaries at the base of each pedicel, and their self-incompatible flowers are pollinated by bats (*Marcgravia*), hummingbirds, and passerine birds (*Norantea* spp.) foraging on the nectar (Sazima and Sazima 1980; Sazima *et al.* 1993; Pinheiro *et al.* 1995).

Some floral nectars may also have a protective function, attracting non-pollinating visitors that protect the plant against detrimental herbivory (Dominguez *et al.* 1989; Yano 1994; Altshuler 1999). In lowland tropical dry forests of coastal Veracruz and in the Zapotitlan Valley (Mexico), ants are more commonly associated with floral nectar and nectar on plant reproductive structures than they are with extrafloral nectar on vegetative plant parts (Rico-Gray 1993; Rico-Gray *et al.* 1998).

Floral and extrafloral nectar

Nectar is an aqueous solution of sugars, amino acids, and other components including lipids, vitamins, and other compounds (Baker and Baker 1973, 1983; Baker 1977). Nectar is consumed by a variety of organisms that interact with plants in a variety of ways, and there has been much research conducted to illuminate patterns of nectar use by visitors and the resultant effects on plants that produce the nectar.

Looking for patterns in nectar composition and its utilization, Herbert and Irene Baker collected hundreds of nectars from many locations. Their extensive analyses revealed that the sucrose/hexose ratios of the nectar correspond to the main type of visitor utilizing floral nectar (Baker and Baker 1983, 1990), enabling prediction of probable pollinators by analysis of nectar sugars (an additional floral character to include in "pollination syndromes"). There is also a statistically significant tendency for amino acids to be correlated with pollinator types (Baker and Baker 1986). Some researchers have objected to these generalizations, concerned that the age and condition of flowers, as well as pollen contamination, may affect apparent nectar composition (Gottsberger *et al.* 1990). Secreted nectar certainly changes and the chemical composition of what has accumulated on the nectaries (and what therefore is used for chemical analysis) is not identical with what has been secreted by the glands. However, as correlations made from large sets of laboratory measurements and field observations, the findings of the Bakers are robust.

The Bakers also observed that nectars open to the air are hexose-dominated, whereas concealed nectars may be sucrose-dominated, and sucrose-dominated nectars open to the air for some time degrade to hexose-dominated nectars, presumably through the actions of microorganisms degrading sucrose to its hexose components (Baker and Baker 1983). They extended their sugar-ratio observations to fruit (Martinez del Rio *et al.* 1992) and studies on various birds found that those which prefer hexose-dominated foods lack sucrase, the enzyme needed to digest sucrose, in their intestinal tract (Martinez del Rio *et al.* 1988, 1989; Martinez del Rio and Stevens 1989; Martinez del Rio and Karasov 1990).

Many organisms can distinguish sugars by taste, and some show distinct preferences for sucrose or hexose sugars (Martinez del Rio 1990; Rusterholz and Erhardt 1997; Koptur and Truong 1998; Blüthgen and Fiedler 2003). In a quest to discover any potential ant-attracting compounds from natural nectar sources, the sugar preferences of a variety of pest ant species were investigated (Koptur and Truong 1998). Most pest ant species preferred fructose over the other sugars, when a preference was demonstrated (interpreted as shorter time to discovery of the bait and more total ants at the bait). Fire ants (*Solenopsis invicta*) can even distinguish between the diastereomers D- and L-glucose (Vandermeer *et al.* 1995). In all species there is potential for individual variation in taste, but only the dominant species can exercise their ability to discriminate (Blüthgen and Fiedler 2003).

Nectars range from dilute (10% sugar or less) to highly concentrated (70% sugar or more), and the viscosity of nectar corresponds to sugar concentration. Viscosity may affect the ability of visitors to drink nectars. If the nectar is concealed in a floral spur or tube, and must be obtained through a long proboscis, it must be dilute enough to travel up by capillary action; in contrast, nectars produced in sunny, dry environments may be viscous and collected in open mandibles, especially extrafloral nectars (Corbet *et al.* 1979; Willmer and Corbet 1981; Wäckers, Chapter 2). The greater the sugar content of nectar, the greater its energetic reward for the visitor that can harvest it. Many studies have shown floral visitor preferences for larger nectar quantities over smaller quantities, and for nectars with higher sugar concentrations (Ricks and Vinson 1970; Bennett and Breed 1985; Lanza 1988, 1991; Galetto and Bernadello 1992; Burd 1995). However, there is likely an optimal nectar concentration curve for most visitors due to slow intake of highly concentrated nectar (e.g., Roces *et al.* 1993).

In plants that possess both floral and extrafloral nectars, the sugar concentration and composition of the nectars can differ dramatically. *Inga* floral nectars are sucrose-dominated and much more dilute compared to extrafloral nectars, which are hexose-dominated and much more concentrated (Koptur 1994). *Inga* flowers are visited by a variety of flying animals, and pollinated primarily by hawkmoths (Sphingidae) and hummingbirds (Trochilidae) (Koptur 1983, 1984a); their extrafloral nectaries are visited by ants, flies, and wasps (Koptur 1984b, 1985). Cashew (*Anacardium occidentale*) is romonoecious and has nectar in both hermaphrodite and male flowers, and extrafloral nectar on the panicle (Wunnachit *et al.* 1992). All three of the nectars have very high sugar concentrations (greater than 70%), and the panicle nectar had a significantly higher sugar content than the floral nectars. All of these cashew nectars were hexose-dominated, with similar sucrose/hexose ratios.

The amino acid complements of the different nectars of the same plant species may also differ (Baker *et al.* 1978), though nectars cluster more in accordance with taxonomy than function (Koptur 1994). In many plants, the more concentrated extrafloral nectars tend to have greater amino acid contents (Blüthgen *et al.* 2004). However, extrafloral nectar of cashew had more amino acids than its floral nectar, but the overall amino acid content of floral nectar was about three times that of the extrafloral nectar (Wunnachit *et al.* 1992). Galetto and Bernardello (1992) analyzed extrafloral nectars of Argentinian Bromeliaceae and found them to have very high sugar concentrations and to be sucrose-dominated, with amino acids present in all species studied. Amino acids are important constituents of extrafloral nectars of pitcher plants (Dress *et al.* 1997). Amino acids have been shown to be important in the attraction of ants to extrafloral nectars and artificial nectar solutions (Lanza 1988, 1991; Wagner and Kay 2002), and Lanza *et al.* (1993) demonstrated that the amino acid content in *Passiflora* extrafloral nectar increases when plants are defoliated, and that it is preferred by ants over nectar from non-defoliated plants.

(Non-pollinating) visitors to nectar

Visitors to plant nectar encompass a wide array of animals, whose effects on the plant secreting the nectar range from positive to negative (Rico-Gray 2001). The effect of the plants on the visitors is usually positive (providing liquid nourishment), and plants may actually manipulate their visitors via nectar secretion patterns (see below). Many visitors to floral nectar pollinate the flowers they visit, a benefit to the plant and essential for maintaining genetic diversity via sexual reproduction. The extent to which genetic diversity is enhanced is related to the mobility of the visitor: crawling insects (such as ants or beetles) tend to move pollen within individual flowers and plants more than do flying insects such as butterflies, bees, and moths. Non-pollinating visitors fall into a number of categories: nectar robbers, herbivores, plant prey (for carnivorous plants), and plant protectors.

Nectar robbers

All sequestered nectar may be subjected to robbing by visitors who manage to collect the nectar without touching floral reproductive parts, apparently taking advantage of the plant. Note that "robbing" is used in a broad sense – sometimes a distinction is made between robbers and thieves, based on whether or not the flowers are physically damaged (see, e.g., Romeis *et al.*, Chapter 7 – thieves leave the flower undamaged), but here the actual mechanism of theft is left in the middle.

By definition, nectar robbers utilize nectar with no direct benefit to the plant (Maloof and Inouye 2000). Robbing is, however, not always detrimental to the plant; robbers may enhance pollination in some indirect ways, e.g., they may decrease self-pollination and increase genetic diversity by making pollinators move farther. Some nectar robbers decrease nectar volume and may cause legitimate pollinators to visit more flowers (Colwell *et al.* 1974). Flower mites, for example, live inside the flowers and can decrease nectar volume to such an extent that flowers are visited less by hummingbirds (Colwell 1995; Lara and Ornelas 2001); however, the flower mites can also act as secondary pollinators. In certain systems, nectar robbery may be an integral part of the mutualism between plants and their pollinators (Morris 1996): bumblebees rob nectar from mature flowers and buzz-pollinate earlier stage flowers on the same plants; the nectar they rob may entice them to legitimately pollinate other flowers. *Mertensia paniculata* stems with robbers excluded donate less pollen and set fewer fruit than those with robbing. Robbed flowers of *Anthyllis vulneraria* in Spain set more seed than unrobbed flowers, because the robbers' bodies effect pollination (Navarro 2000). Extrafloral nectar can be robbed as well, if visitors imbibe nectar with no benefit to the plant, as the Japanese white-eye bird (*Zosterops japonica*) taking extrafloral nectar of the tropical almond (*Terminalia catappa*) in the Bonin Islands (Pemberton 1993), or opportunistic ants that do not provide benefit to their host plant, such as *Cataulacus* on *Leonardoxa africana* (Gaume and McKey 1998). The function of calyx and bud nectaries of *Thunbergia grandiflora* in Malaysia was hypothesized to be for ant protection against nectar-robbing carpenter bees, but ants apparently protect the inflorescences against herbivores (Fiala *et al.* 1996).

Herbivores

Certain herbivores may be attracted to plants by nectar, or by other insects that are attracted to nectar. Among those that are attracted to plants by extrafloral nectar, and that are more likely to oviposit on the plant than if it did not have nectaries, are Lepidoptera with ant-tended caterpillars. *Saraca thaipingensis*, a legume tree with extrafloral nectaries of peninsular Malaysia, is host to 10 species of lycaenid butterflies, and females of some species oviposit exclusively on trees occupied by their specific host ants (Seufert and Fiedler 1996). Wagner and Kurina (1997) found that lycaenid butterflies lay substantially more eggs on *Acacia constricta* plants with ants than plants without ants, using ants as oviposition cues. Ant-tended butterfly caterpillars may even drink extrafloral nectar (DeVries and Baker 1989), adding "insult to herbivory"! Some ant-tended herbivores may also exhibit inducible defenses, secreting more honeydew to elicit ant protection when they are threatened (Agrawal and Fordyce 2000). Homoptera may also be attracted to plants that secrete nectar. The gregarious, polyphagous tettigometrid *Hilda patruelis* is frequently found on fig-bearing branches of *Ficus sur* in South Africa, where ants tend fig secretions as well as the homopterans (Compton and Robertson 1991).

Plant prey

A special case of nectar use exists in some carnivorous plants. Pitcher plants (Sarraceniaceae, Nepenthaceae, Cephalotaceae, Bromeliaceae) secrete extrafloral nectar on their pitcher leaves to lure insects that may fall into the pitchers to be digested as prey (Joel 1988). Ants are frequent visitors and sometimes become food for the pitcher plants, but probably not as often as various flying nectar-drinking insects.

Plant protectors

Plant-protecting predators and parasitoids are more likely to discover their prey/hosts after having been recruited to the plant by nectar. Ants are common on plant foliage foraging for food, and often affect the herbivores of the plants on which they forage (Oliveira *et al.* 2002). Fiala *et al.* (1996) found that ant-free inflorescences of *Thunbergia grandiflora* in Malaysia were attacked and destroyed by moths (Pyralidae) whereas buds were normally protected from these herbivores by the presence of ants. Ant–plant protection studies have contributed much to theory of mutualism and to our understanding of species interactions (Bronstein 1998). Many interactions between ants (and other protectors) and plants are facultative: the plant and the ant can survive without the other, but with diminished fitness. These facultative associations tend to rely solely on the plant providing extrafloral nectar (e.g., Torres-Hernandez *et al.* 2000) – see Box 3.1 for some examples of facultative ant–plant relationships in the legume family. In some cases the ants may build their nests on the surface of the plants bearing nectaries, as seen in *Clerodendrum fragrans* in Southeast Asia (Jolivet 1983) or *Inga sapindoides* in Panama (S. Koptur, personal observation). Ants may even hide the nectaries from competitors by constructing shelters of debris over the nectaries (Eskildsen *et al.* 2001), though this may alternatively be interpreted as protecting the nectar from rain.

Enhanced plant resources make plants bigger, and larger plants provide more resources for herbivores and potentially for mutualistic predators in bottom-up trophic cascades. Studies by Letourneau and Dyer yielded no evidence for bottom–up control of predators, but robust evidence for top–down indirect effects on host plant biomass by predators (Letourneau and Dyer 1998; Dyer and Letourneau 1999a, 1999b). A study of wood ants and birch herbivores came to similar conclusions: predation by ants on herbivores created the "green island" effects much more than their soil amelioration could explain (Karhu and Neuvonen 1998). Interestingly, wood ants are the sole ant that demonstrated some level of anti-herbivore defense on bracken fern (Heads and Lawton 1985). *Acacia constricta* plants with ant nests below them were associated with greater fruit set than plants without ant nests; ant numbers were higher on these plants (which bear extrafloral nectaries), but ants did not appear to reduce damage from leaf herbivores; fruit set enhancement may be due to nutrient enhancement (Wagner 1997). It would be interesting to measure extrafloral nectar production in both sets of plants.

The presence of extrafloral nectaries increases the number of predators (especially ants) on a plant, and ants on plants frequently reduce herbivory or enhance plant reproduction. Though ants are the main predator in most of the examples in this section, it is important to recognize that other arthropods, such as spiders (Ruhren and Handel 1999) and wasps (Cuatle and Rico-Gray 2003; see also Olson *et al.*, Chapter 5 and Eubanks and Styrsky, Chapter 6) that are attracted to extrafloral nectar also can act as effective defenders of plants.

Box 3.1 Extrafloral nectaries: examples of facultative mutualisms in the legume family (Fabaceae)

Many families of plants are characterized by extrafloral nectaries, but the Fabaceae has many diverse genera bearing extrafloral nectaries on various parts of the plant body. I discuss one example from each of the three subfamilies, but this is not entirely representative of this enormous family.

Vicia

Extrafloral nectaries have been shown to serve as a generalized defense, wherever the plants might live. In their native habitat the nectaries may have co-evolved with ants and other biotic protective agents over long periods of time, and their protective agents may repel new herbivorous species on the plant in question. However, there may be many specialized herbivores that have outsmarted the defense system. An example is the common vetch, *Vicia sativa*, a weed in natural areas of northern California (USA) that has a truly facultative mutualism with ants (in many locations, the non-native Argentine ant, *Iridomyrmex humilis*). Eliminating ants by excising the nectar-bearing stipules resulted in plants losing more leaf area to externally feeding herbivores than control plants with intact nectaries and attendant ants (Koptur 1979), which also decreased fruit and seed set in plants with nectaries and ants removed. When common vetch was studied in its native land (UK: Koptur and Lawton 1988), we found specialized herbivores (most important were caterpillars that silked leaves together and fed inside their shelters, and others that fed inside developing fruit) that actually benefited from the presence of ants.

Lysiloma

Some plants may be long-lived, but have relatively short-lived leaves that are quite palatable to herbivores and benefit from ant protection during their development. *Lysiloma bahamensis* is a native tree of pine rocklands and hardwood hammocks of south Florida (USA). *Lysiloma* is dry-season deciduous, with leaves lasting less than a year. Leaves have extrafloral nectaries that attract four species of ants (*Pheidole dentata*, *Pseudomyrmex elongata*, *Pseudomyrmex simplex*, and *Solenopsis geminata*) in Everglades National Park, and exclusion experiments during the period of leaf expansion showed that ants reduce leaf damage from herbivores (S. Koptur, unpublished data). Caterpillars were more numerous and herbivory was greater on trees in pinelands than trees on hammock edges (Rodriguez

Figure 3.1 *Pheidole megacephala* ant at *Cassia bahamensis* foliar nectaries in Miami.

1995), and though ant activity (as well as several other plant species with extrafloral nectaries) was greater in pinelands than in hammocks (Koptur 1992b), ants were more common on hammock *Lysiloma* trees than on pineland counterparts.

Cassia

Unlike the species of the Fabaceae described above, *Cassia* have no floral nectaries (their floral reward is categorized as "pollen only"), but most species have petiolar nectaries (Fig. 3.1). The morphology of these nectaries varies among the species: in some the nectaries are flat pads, in others nectaries are elongated into points. The genus *Cassia* (*sensu lato*, including *Senna* and *Chamaecrista*) has been studied in many locations, and is an important group of plants in the formulation of ant–plant interaction theory. Schimper (1903) excised *Cassia* nectaries to learn their function, and finding no change, concluded they did not have a physiological role. Boecklen (1984) found no support for the "protectionist hypothesis"; in his exclusion experiments with *Cassia fasiculata* (he used both nectary excision and tanglefoot banding at two sites) treatment plants produced as many fruit as control plants. Kelly (1986) used small fences ringed with tanglefoot to exclude ants from *Cassia fasiculata* and found that ant protection and plant fecundity varied geographically; Barton (1986) demonstrated striking spatial variation in ant–plant interactions of this species. Using clear nail polish to eliminate nectar in *Chamaecrista nictitans*, Ruhren and Handel (1999) found that spider predators chose plants with active extrafloral nectaries over those with nectaries blocked, spending most of their time on these plants and enhancing seed set of the plants despite the presence of a bruchid predator immune to spider defense. Large-scale ant-exclusion experiments conducted with both potted and naturally occurring *Senna occidentalis* in east Texas (USA) demonstrated that introduced fire ants (*Solenopsis invicta*) greatly reduce foliar damage to plants by caterpillars and enhance fruit and seed set (Fleet and Young 2000). In southern Florida several species of *Cassia* (*sensu lato*) occur. All serve as host plants for sulfur butterflies (Pieridae) of the genus *Phoebis*. *Phoebis sennae*, the cloudless sulfur, is a native species; *P. philea*, the orange-barred sulfur, is naturalized in Florida, colonizing south Florida from Mexico in the 1920s (Glassberg 1999). Caterpillars of both species are much more common on *Cassia bahamensis* than on *C. ligustrina* (S. Koptur, unpublished data); nectar secretion patterns and perhaps nectar constitution of *C. ligustrina* may be more attractive to ants, and their presence may deter oviposition and deter young caterpillar presence on *C. ligustrina* (S. Koptur, personal observation). On most of the *Cassia* species, caterpillars feed on new growth, often eating flowers and sometimes preventing fruit development (their damage has a large impact on plant fitness). Plants in natural habitats appear to have greater reproductive success than plants in gardens and urban/suburban landscape settings where ants and other protective agents are often less abundant (S. Koptur, unpublished data).

Abiotic and biotic influences on nectar production

Nectar production is influenced by a number of factors, physical and biotic. Sunlight, soil moisture, and humidity are factors that can affect nectar secretion in plants. Sunlight promotes photosynthesis, which produces sugars, also involved in nectar production. Well-watered plants have adequate turgor and extra water available for nectar production. Many succulent plants of xeric habitats use their valuable water to make nectar, attracting both pollinators and protective agents. Relative humidity affects the concentration of nectar (Corbet *et al.* 1979), with lower humidity promoting evaporation of water from nectar, leaving exposed nectars more concentrated than when conditions are humid.

Some defensive food supplements are affected by nutrient availability. Field experiments demonstrated that food body production by *Macaranga triloba* is limited by soil nutrient supply at the plant's natural growing site (Heil *et al.* 2001a, 2002). In myrmecophytic *Cecropia* species (Folgarait and Davidson 1995), glycogen-rich Müllerian bodies increased with greater carbon and nitrogen levels in greenhouse studies, whereas lipid- and amino acid-rich pearl bodies increased under conditions of high nutrient levels and low light (which contribute to an excess of nitrogen); carbon-based defenses (tannins and phenolics) reached higher concentrations in lower nutrient conditions. *Cecropia* do not produce nectar, but predictions for plants that do, suggest that nectar (carbon-based) should not be affected by nutrient limitation as food bodies might be. Studies are needed examining the effects of nutrient enhancement on nectar production in plants with extrafloral nectaries.

Removal of nectar has been shown to increase total nectar output in flowers of some plants (Cruden and Hermann 1983; Koptur 1983; Gill 1988; Pyke 1991), providing the plants with a way to respond to increased visitation by producing more nectar. Some plants with extrafloral nectaries produce more nectar when extrafloral nectar is removed, as demonstrated in careful studies with *Macaranga tanarius*. In this species, when nectar was removed at 3-hour intervals, nectaries produced on average 2.5 times more nectar than if it was collected only once a day (Heil *et al.* 2000). Some plants will maintain constant levels in active nectaries when nectar is not removed (*Inga densiflora*, *I. punctata*, *Vicia sativa*), while others will secrete continuously whether nectar is removed or not (*Polypodium plebeium*, *Turnera ulmifolia*, *C. ligustrina*). Careful quantification and field observations of these differences may reveal differing strategies in attracting protective agents, and in avoiding colonization by fungi.

Extrafloral nectar production may be an inducible anti-herbivore defense in some plants (Koptur 1989). Both artificial defoliation and insect herbivory have been shown to increase nectar secretion in a number of plants with extrafloral nectaries (Stephenson 1982; Koptur 1989; Stevens 1990; Wäckers and Wunderlin 1999; Heil *et al.* 2000; Wäckers *et al.* 2001). Moderate artificial damage to leaves of *V. sativa* increased extrafloral nectar volume secreted for 2 days following the damage (Koptur 1989). Leaf damage to *V. faba* caused a dramatic increase in the number of extrafloral nectaries on stipules (Mondor and Addicott 2003), interpreted as an adaptive plastic response. Agrawal and Rutter (1998) postulated that physical damage to leaves or other plant parts may trigger plant signals such as plant sap release or green leaf volatiles that attract ant defenders. Ants are recruited to a number of cues associated with herbivory (leaf damage, caterpillar presence, plant sap, and hexanal) in *Cecropia* (Agrawal 1998). Kawano *et al.* (1999) found that volatile substances (that serve as a generalized parasitoid attractant) were emitted after leaf damage on extrafloral nectary-bearing *Fallopia* spp. (Polygonaceae). Brouat *et al.* (2000) demonstrated that leaf volatiles influence ant patrolling by attracting ants to the younger leaves where their protection is needed. Induced defenses may be favored over constitutive defenses in plants for many reasons (Karban and Baldwin 1997; Agrawal and Karban 1999), including economy, reduced host-finding by specialists, reduced susceptibility to pathogens, increased variability in food quality for herbivores, increased herbivore movement and dispersion of herbivory, and reduced pollinator deterrence. Heil *et al.* (2001b) showed that extrafloral nectar production is elicited by jasmonic acid, as an induced defensive response to herbivory and artificial damage in *Macaranga tanarius*.

Hypotheses on the function of extrafloral nectar production

Why have extrafloral nectaries arisen in such a wide variety of plants? Apparently they impart some selective advantage to plants that possess them. The historical development of the "protectionist" and "exploitationist" views of extrafloral nectar production are reviewed by Bentley (1977). For the sake of completeness, I discuss these competing hypotheses here, prior to discussing two other alternatives that have received recent attention.

Hypothesis 1 – exploitation of plant physiological waste products

Nectar secretion has been envisioned a passive process, the nectaries functioning as "sap valves", excreting excess carbohydrates from growing points of the plant when they shift from photosynthetic sink to source (reviewed by Bentley 1977). The concept of extrafloral nectar as secretion of excess carbohydrates to achieve a more balanced carbohydrate/nutrient level has not been supported by experimental evidence (Baker *et al.* 1978).

Hypothesis 2 – to attract protective agents

The plant protection hypothesis has been repeatedly supported in investigations conducted to decide between this and the preceding hypothesis (Janzen 1966, 1967; many studies reviewed in Bentley 1977; Buckley 1982; Jolivet 1986; Koptur 1992a; and many newer examples discussed in this chapter), notwithstanding that interactions vary spatially and temporally, and several studies have found no evidence of protection.

Hypothesis 3 – to prevent flower plundering

Kerner (1878) proposed that the role of extrafloral nectaries is to distract ants from flowers. Plants potentially lose energy invested in floral rewards if these rewards were taken by non-pollinating visitors like ants, so if ants could be occupied outside of flowers, the floral rewards could be "better spent" on worthy pollinators. Though some floral nectars appear to be repellent to ants (Feinsinger and Swarm 1978; Guerrant and Fiedler 1981; Prys-Jones and Willmer 1992), many are readily eaten when accessible (Frankie *et al.* 1981; Koptur and Truong 1998; Rico-Gray *et al.* 1998). Due to their ubiquity and abundance it is not surprising that ants are known to pollinate certain species (Hickman 1974; Peakall *et al.* 1991; Gomez and Zamora 1992, 2000; Gomez *et al.* 1996; Bosch *et al.* 1997; Puterbaugh 1998; Schuerch *et al.* 2000) despite apparent drawbacks, such as limited pollen movement and antibiotic substances on ant-bodies which negatively affect pollen germination and pollen tube growth (Beattie *et al.* 1984; Wagner 2000). Though ants assist seed set in east African acacias by protective patrolling during flowering and fruit development, they are repelled from flowers by a volatile chemical signal during the time they might interfere with pollination (Willmer and Stone 1997). Recent experimental work shows such floral repellents to be widespread among plants, to be effective against most ants, and to potentially prevent ants from parasitizing plant–pollinator mutualisms (Ghazoul 2001).

An innovative approach to testing Kerner's hypothesis was taken by Wagner and Kay (2002). These researchers conducted experiments with artificial plants bearing different arrays of nectars of different qualities to simulate floral nectar and extrafloral nectar. They found that fewer ants (of both species tested) visited floral nectaries when extrafloral nectar sources were present, evidence that extrafloral nectaries might indeed distract ants from flowers. They found that workers of one ant species preferred sugar solutions with amino acids over sugar alone, as has been demonstrated for other species (Lanza 1988, 1991; Blüthgen and Fiedler 2003), corroborating the idea that extrafloral nectars' relatively high concentrations of amino acids have evolved under selection pressure from ants.

Hypothesis 4 – to lure ants away from tending honeydew-producing insects

Becerra and Venable (1989) published a thought-provoking idea that extrafloral nectar production is selected to distract ants from homopterans, reducing recruitment of ant-tended insects and concomitant damage to plants. Many researchers responded with observations and experiments from their own systems (e.g. Del-Claro and Oliveira 1993), mostly refuting their hypothesis.

Fiala (1990) reported that more ants visit homopterans than extrafloral nectaries on six species of *Macaranga*. She reasoned that nectar secretion may be less constant than honeydew, and this may be why ants prefer scale insects. Blüthgen *et al.* (2000) observed that honeydew collection from Homoptera in the Amazonian rainforest canopy was usually monopolized by a single species of dominant ant, whereas extrafloral nectaries, in contrast, attracted a wider array of ants, with more co-occurrence of ant species on plants with nectaries. Dejean *et al.* (2000b) came to similar conclusions for the ant mosaic in a Cameroonian rainforest: dominant ants prefer Homoptera, the rest depend on extrafloral nectaries. In South Africa, *Anoplolepis steingroeveri* ants prefer homopterans to extrafloral nectar of *Mimetes fimbriifolius* (Proteaceae), and two species of *Crematogaster* commonly tended scale insects and nested in Proteaceae plants with extrafloral nectaries (Zachariades and Midgley 1999). Ants in Mexico have been observed to switch between extrafloral nectar and honeydew, preferring extrafloral nectar in the driest conditions (Moya-Raygoza and Larsen 2001). Also, Rico-Gray (1993) reported that two-thirds of ant species in lowland dry forest of Veracruz (Mexico) foraging for floral nectar switched to homopteran honeydew, with an alternating pattern through the year. However, another recent study (Engel *et al.* 2001) found that ants visited the more concentrated extrafloral nectar even in the presence of the higher quality (containing melezitose) honeydew from aphids. See Box 3.2 for additional examples of the relative attractiveness of extrafloral nectar and honeydew to ants.

All hypotheses considered

These four hypotheses are not mutually exclusive, and they all have received some support. Undoubtedly they all provide valid explanations in specific cases. However, the most generally applicable seems to be the second. Extrafloral nectaries have most likely been selected because of the selective advantage imparted to individuals that possess them, and predatory insects visiting the nectaries and providing the plants with some protection is the most likely evolutionary scenario. There are certainly examples of plants with both nectaries and ant-tended Homoptera, but in most cases the ants prefer the honeydew to extrafloral nectar (see also Box 3.2). And ants do like floral nectar, when they can get it, though they may often prefer extrafloral nectar due to its higher concentrations of sugars and amino acids. So extrafloral nectaries may serve myriad functions, but may primarily have evolved to promote protection of plant parts vulnerable to herbivores.

Box 3.2 Ants, extrafloral nectar, and honeydew

According to some authors, phloem sap can be extracted in two main forms: extrafloral nectar and honeydew (Blüthgen *et al.* 2000; Blüthgen and Fiedler 2002). (We will allow them to overlook floral nectar and plant sap exuded through fissures in the plant body.) Extrafloral nectaries attract a wider array of ants than honeydew, with more co-occurrence of

Figure 3.2 *Paratrechina longicornis* ants and coccid scale on cultivated *Annona* sp. In Miami.

ant species utilizing it than honeydew sources (which are more often monopolized by dominant ant species). Extrafloral nectaries are perhaps more important for general ant nourishment. High-performance liquid chromatography (HPLC) analyses of honeydew and nectar reveal that honeydew sources have a higher number and higher concentrations of amino acids than most nectar sources (Blüthgen *et al.* 2004; Wäckers, Chapter 2). Ants tending Homoptera often protect their Homoptera from enemies (Del-Claro and Oliveira 1999, 2000) and may protect plants hosting the Homoptera in ways similar to ants visiting extrafloral nectaries (Messina 1981; and others, see below) (Fig. 3.2). However, the majority of honeydew-producing herbivores are not tended by ants (Bristow 1991; Sakata and Hashimoto 2000), and ants do not always benefit honeydew-producing herbivores.

Homopterans may appear at first to be insect analogs of extrafloral nectaries (Dansa and Rocha 1992; Koptur 1992a), and interactions between ants and Homoptera are considered mutualistic (Way 1963; Bach 1991; Del-Claro and Olivera 1999, 2000), but unlike the secretion of extrafloral nectar, the plant has no control over how much sap the honeydew-producing insects ingest (Becerra and Venable 1989). Some amount of control may be effected by the tending ants, and indeed the resident ants of many myrmecophytes tend honeydew-producing insects in or on the plant body. In turn, Homoptera may attract ants by honeydew flicking: *Guayaquila xiphias* treehoppers flick honeydew to attract ground-dwelling ants to tend the treehoppers on their host plants (Del-Claro and Oliveira 1996; Oliveira *et al.* 2002). Lepidoptera and Homoptera may interact in unusual ways:

a Neotropical riodinid butterfly, *Alesa amesis*, was discovered to be entomophagous (previously unknown in New World Lepidoptera). It oviposits preferentially on a variety of host plants upon which one ant species tends Homoptera (DeVries and Penz 2000), not for their honeydew, but for the caterpillars to eat the homopteran nymphs.

Cushman (1991) postulated that host plants could mediate (ant-tended) herbivore–ant interactions via differences in plant host quality and effects on tended-herbivore fitness. Indeed, some ants switch their roles as mutualists and predators, tending insects for honeydew but also eating the same insects at certain times. In some circumstances, plants may benefit from ants tending Homoptera if ants deter herbivory by other insects (Messina 1981; Compton and Robertson 1988, 1991; Dansa and Rocha 1992; Rashbrook *et al.* 1992; Figueiredo 1997; Moog *et al.* 1998).

Some aphids may attract ants, while others may repel ants (Sakata and Hashimoto 2000); ants preferred tending *Aphis craccivora* to extrafloral nectaries of *Vicia faba* and avoided tending *Megoura crassicauda* in microcosm experiments in Japan. Ants tending *A. craccivora* were more likely to consume *Megoura*. Offenberg (2000) maintains that the evolution of extrafloral nectaries and the evolution of ant–aphid associations may be correlated, and that aphid species (in two families) associated with ants tend to utilize host plants with extrafloral nectaries, and that aphids on host plants with extrafloral nectaries are more likely to evolve associations with ants.

Ant–plant mutualism

Obligate relationships between ants and plants are found in many myrmecophytes, i.e., plants with inhabitant ants, nesting in special structures or hollow stems (Janzen 1966, 1967; Jolivet 1986, 1998). Ants may even induce domatia formation in some species (Blüthgen and Wesenberg 2001). Many myrmecophytes provide food for their ants via nectar and/or food bodies. *Acacia cornigera* supplies its resident *Azteca* ants not only with extrafloral nectar, but also with lipid-rich Beltian bodies produced on the tips of leaflets (Janzen 1966, 1967). Interestingly, myrmecophytic *Acacia* species in Africa apparently produce only extrafloral nectar, while those in central America produce both nectar and food bodies. In the genus *Leonardoxa* there is a match between nectary size (and number) in plants and tending of homopterans by associated ants (McKey 1991); nectar and homopteran secretions are alternative food for ants. Thus on many myrmecophytes nectaries are lacking and ants may tend honeydew-producing insects on the plants.

The presence of ants can also directly influence the production of food rewards produced by the plant, as demonstrated experimentally for food body production in *Macaranga triloba* (Heil *et al.* 1997). The production of food bodies inside the stems of *Piper coenocladum* takes place in response to occupation of their hollow stems by ants (Risch and Rickson 1981; Letourneau 1983). *Cecropia* feed the ants that nest in their hollow stems via Müllerian bodies produced on specialized areas of the leaf bases (Janzen 1969; Schupp 1986). Some plants provide only pearl bodies, produced on the surface of the leaves, e.g., *Ochroma* and other species (O'Dowd 1982) and *Maieta guianensis* (Vasconcelos 1991).

In their study of Panamanian forest plants, Schupp and Feener (1991) found that plants with ant rewards (extrafloral nectaries and/or food bodies) are over-represented in secondary habitats. Ants are much more abundant on *Caryocar brasiliense* shrubs than on nearby plants without extrafloral nectaries, and are much more likely to attack termite baits and presumably insect herbivores on *Caryocar* plants (Oliveira 1997). Though this protection did not translate into greater fruit set on ant-tended individuals of *Caryocar*, another cerrado plant (*Qualea multiflora*) experienced less damage to buds and flowers when ants were not excluded from plants, and plants with ants had greater fruit set (Del-Claro *et al.* 1996). Ant visitors to the extrafloral nectaries of *Opuntia stricta* substantially reduced herbivore damage and increased fruit set by 50% on ant-tended branches over ant-excluded branches (Oliveira *et al.* 1999). Exclusion experiments on *Stryphnodendron microstachyum* saplings showed that ants visiting extrafloral nectaries benefit plants, not by killing herbivores but by bothering them, so that they damage plants less, and ant-visited saplings grew taller than ant-excluded individuals (De la Fuente and Marquis 1999). Some myrmecophytes benefit from the presence of ants not only in reduced folivory, but may hold their leaves longer and gain higher stature due to ant occupation (Fonseca 1994).

In most ants, only workers collect the nectar, sometimes transferring it to other workers to eat or to feed the rest of the colony. Some ant species show behavioral specialization, with some workers serving as prey-hunters, and others as nectar-collectors (Passera *et al.* 1994). Such task partitioning (Ratnieks and Anderson 1999) may make for more efficient patrolling and plant protection, where dedicated foragers may be more likely to protect the resources important to their colony.

Many plants with extrafloral nectaries have been shown to benefit from associations with more than one species of ant, though ant species often differ in their protective ability (Schemske 1982; Koptur 1984a; Mody and Linsenmair 2004). In Cameroon, some ants (arboreal-nesters with diurnal activity) visiting the extrafloral nectaries of the pioneer tree *Alchornea cordifolia* provided substantial protection against the variegated locust, *Zonocerus variegatus*, while other species (ground-nesting, nocturnally active arboreal foragers) did not, except during outbreaks of the locust (Dejean *et al.* 2000a). Daily turnover of ant species has been documented in a number of systems, such as *Ouratea hexasperma* in Brazilian cerrado (Oliveira *et al.* 1995) and *Opuntia stricta* (Oliveira *et al.* 1999). Sympatric species with extrafloral nectaries may host different ant assemblages that vary between day and night (Hossaert-McKey *et al.* 2000). Labeyrie *et al.* (2001) found that *Passiflora glandulosa* benefits from visits from two sympatric ant species in French Guiana, one of which is active diurnally, the other nocturnally. Sympatric *Passiflora* species studied in successional neotropical forests of Costa Rica were not associated with particular ant species, or vice versa (Apple and Feener 2001).

Even plants with demonstrated mechanical defenses may benefit from ants visiting extrafloral nectaries in some circumstances: spinescent *Acacia drepanolobium* are protected from (young) giraffe herbivory by ants visiting extrafloral nectaries on shoot tips, acting aggressively against the vertebrates (Madden and Young 1992). Other impressive examples of a mechanically defended plant that benefits from ants are seen in species of *Opuntia* (Pickett and Clark 1979; Oliveira *et al.* 1999).

Non-protective predators and counter-adapted herbivores

Although ants and other predators feed at extrafloral nectaries of South African Proteaceae, they do not reduce herbivory (Zachariades and Midgley 1999). Similar conclusions have been drawn from other southern hemisphere systems (O'Dowd and Catchpole 1983; Rashbrook *et al.* 1992) and elsewhere in the world (Barton 1986; Boecklen 1984; Tempel 1983; Heads and Lawton 1984; Lawton and Heads 1985).

A variety of herbivores are immune to ant predation, protected by their behavior or feeding mode (Heads and Lawton 1985; Eubanks *et al.* 1997). *Eunica bechina* caterpillars can find refuge from predatory ants on *Caryocar* in the Brazilian cerrado by climbing to the end of stick-like frass chains they build at leaf margins (Oliveira *et al.* 2002); *Smyrna blomfildia* caterpillars exhibit similar behavior on *Urera* plants (Machado and Freitas 2001). See Sabelis *et al.* (Chapter 4) for additional examples of what can be viewed as cheaters and thieves in the food-for-protection mutualism.

Biological control using extrafloral nectar

Biological control can be promoted by co-planting crop species with species possessing extrafloral nectaries (see Wilkinson and Landis, Chapter 10 and Gurr *et al.*, Chapter 11). Both floral and extrafloral nectar resources can be important in supporting insect parasitoids in agroecosystems (Stapel *et al.*

1997; Baggen *et al.* 1999; Olson *et al.*, Chapter 5; Heimpel and Jervis, Chapter 9). Stettmer (1993) suggests that extrafloral nectar of the cornflower *Centaurea cyanus* is an important food source for many beneficial insects and may be used to stabilize their population densities in agriculture when growing as a weed in crop borders. Floral nectar is generally a much shorter-lived potential resource for beneficial insects than extrafloral nectar. Ants can be effective biological control agents in neotropical agroecosystems: after experimentally reducing ant numbers with insecticides, pest pressure increased in maize monocultures and maize–bean bicultures (Perfecto and Sediles 1992).

The nectaries of *Theobroma cacao* can support *Pachycondyla* ants (predators of a variety of phytophagous insects) for biological control in cacao plantations (Valenzuela-Gonzalez *et al.* 1994), where there is a large arboreal ant community (Majer *et al.* 1994). Cashew has leaves covered with extrafloral nectaries, and inflorescences develop lying on the surface of leaves; the inflorescences also have nectaries on floral bracts (Rickson and Rickson 1998). Cashew crops may benefit from protection by ants and other predators if plantings are made with appropriate considerations for encouraging ants in countries where ants are normally regarded as undesirable in agroecosystems (Rickson and Rickson 1998). Even homopterans (extrafloral nectary analogs) have been suggested as a way to attract ants onto plants to protect them from gypsy moth attack (Weseloh 1995).

Co-adaptation and co-speciation

Some myrmecophytes produce extrafloral nectar (e.g., swollen-thorn *Acacia* spp.), but in many genera the ant-attracting trait of extrafloral nectar production is lost in favor of food bodies or hosting honeydew-producing insects. Food rewards offered by *Macaranga* in Malaysia affect ant colony size (Itino *et al.* 2001a), particularly in species where ants tending Homoptera regulate their own biomass by regulating the biomass of their honeydew-producers; in *Macaranga* species where ants feed on food bodies produced by the plants, ant colony size is larger. The authors postulate that energy transfer directly to ant from plant may be more efficient than transfer through another trophic level. In the paleotropical genus *Macaranga*, only the glands of non-myrmecophytic species function as nectaries; liquids secreted by glands of myrmecophytic species do not contain sugar (Fiala and Maschwitz 1991). Young plants of *Macaranga hosei* secrete extrafloral nectar until they are colonized by their mutualist ant partner. McKey (1988) described the same phenomenon for *Barteria*, where young plants have extrafloral nectaries, but larger plants, which have developed their domatia, lack extrafloral nectaries.

Some myrmecophytes provide compelling examples of co-accommodation (*Leonardoxa* spp.: McKey 1991) and co-speciation (*Macaranga* spp.: Itino *et al.*

2001b). *Leonardoxa africana* T3 feeds its resident ants via both extrafloral nectaries and Homoptera (Gaume and McKey 1998), and *Aphomomyrmex afer* eat small herbivore larvae, and disturb the feeding of larger larvae, protecting the plants against their microlepidopteran herbivores. Even small ants are very important in protecting plants against smaller chewing, and especially sucking, insect herbivores (Gaume *et al.* 1997); *L. africana* leaves with *Petalomyrmex* ants excluded did not expand as much as ant-tended leaves, likely due to damage from sucking insects. *Leonardoxa africana* plants benefit more from larger ant colonies (Rocha and Bergallo 1992; Gaume *et al.* 1998), and colonies are bigger when resident ants tend pseudococcids rather than coccids (pseudococcids are less expensive for the plant to support).

Myrmecophytic African ant acacias have large nectaries and hold their leaves longer than non-myrmecophytic congeners, and one species supports at least nine ant species, with widely varying relationships with the host plants (Young *et al.* 1997). All of these mutually exclusive resident ants collect nectar, some tend scale insects inside the plants, some eat the nectaries, and one species eats the axillary shoots rendering the host tree sterile but inciting more terminal shoots with healthier leaves and more extrafloral nectar.

Future directions

Interactions between plants and their protective agents fueled by nectar have repeatedly been suggested to have great potential for biological control. Ants especially may have drawbacks and present health hazards to agricultural workers, so ant protection is not always a simple solution, especially in areas where fire ants are abundant. This direct form of plant protection is not the only interaction that could benefit crops, however, and using nectar to encourage other predators and parasitoids of crop herbivores should certainly be pursued (see, e.g., Van Rijn and Sabelis, Chapter 8 and Heimpel and Jervis, Chapter 9).

We need to know much more about the ways that plants control the insect visitors to their nectar. Much recent work has examined how herbivory and other damage to plants can elicit greater production of nectar, and the mechanisms that may underlie this evolutionarily advantageous plant response. Inducible nectar production appears to be a widespread phenomenon, but many more systems warrant investigation. We have not yet adequately investigated all the environmental influences (such as nutrients, water, and sunlight) on nectar secretion (amounts and composition). More work is also needed in determining what makes some plants more attractive than others to various ants, and other nectar collectors.

Acknowledgments

I thank the editors for providing me with the opportunity to reflect on what has been accomplished with nectar as fuel for plant protectors, as well as contemplating potentially fruitful areas for future research. I also thank those colleagues who sent pre-prints and discussed their unpublished results, and the editors and anonymous reviewers whose critiques improved this paper.

References

Agrawal, A. A. 1998. Leaf damage and associated cues induce aggressive ant recruitment in a neotropical ant-plant. *Ecology* **79**: 2100–2112.

Agrawal, A. A. and J. A. Fordyce. 2000. Induced indirect defence in a lycaenid-ant association: the regulation of a resource in a mutualism. *Proceedings of the Royal Society of London Series B* **1455**: 1857–1861.

Agrawal, A. A. and R. Karban. 1999. Why induced defenses may be favored over constitutive strategies in plants. In R. Tollrian and C. D. Harvell (eds.) *The Ecology and Evolution of Inducible Defenses*. Princeton, NJ: Princeton University Press, pp. 45–61.

Agrawal, A. A. and M. T. Rutter. 1998. Dynamic anti-herbivore defense in ant-plants: the role of induced responses. *Oikos* **83**: 227–236.

Altshuler, D. L. 1999. Novel interactions of non-pollinating ants with pollinators and fruit consumers in a tropical forest. *Oecologia* **119**: 600–606.

Apple, J. and D. J. Feener. 2001. Ant visitation of extrafloral nectaries of *Passiflora*: the effects of nectary attributes and ant behavior on patterns in facultative ant-plant mutualisms. *Oecologia* **127**: 409–416.

Bach, C. E. 1991. Direct and indirect interactions between ants (*Pheidole megacephala*), scales (*Coccus viridis*) and plants (*Pluchea indica*). *Oecologia* **87**: 233–239.

Baggen, L. R., G. M. Gurr, and A. Meats. 1999. Flowers in tri-trophic systems: mechanisms allowing selective exploitation by insect natural enemies for conservation biological control. *Entomologia Experimentalis et Applicata* **91**: 155–161.

Baker, H. G. 1977. Non-sugar chemical constituents of nectar. *Apidologie* **8**: 349–356.

Baker, H. G. and I. Baker. 1973. Some anthecological aspects of the evolution of nectar-producing flowers, particularly amino acid production in nectar. In V. H. Heywood (ed.) *Taxonomy and Ecology*. London: Academic Press, pp. 243–264.

1983. Floral nectar sugar constituents in relation to pollinator type. In C. E. Jones and R. J. Little (eds.) *Handbook of Experimental Pollination Biology*. New York: Van Nostrand Reinhold, pp. 117–141.

1986. The occurrence and significance of amino acids in floral nectar. *Plant Systematics and Evolution* **151**: 175–186.

1990. The predictive value of nectar chemistry to the recognition of pollinator types. *Israel Journal of Botany* **39**: 157–166.

Baker, H. G., P. A. Opler, and I. Baker. 1978. A comparison of the amino acid complements of floral and extrafloral nectars. *Botanical Gazette* **139**: 322–332.

Barton, A. M. 1986. Spatial variation in the effect of ants on an extrafloral nectary plant. *Ecology* **67**: 495–504.

Beattie, A. J., C. Turnbull, R. B. Knox, and E. G. Williams. 1984. Ant inhibition of pollen function: a possible reason why ant pollination is rare. *American Journal of Botany* **71**: 421–426.

Becerra, J. X. and D. L. Venable. 1989. Extrafloral nectaries: a defense against ant–homopteran mutualism? *Oikos* **55**: 276–280.

Bennett, B. and M. D. Breed. 1985. The association between *Pentaclethra macroloba* (Mimosaceae) and *Paraponera clavata* (Hymenoptera: Formicidae) colonies. *Biotropica* **17**: 253–255.

Bentley, B. 1977. Extrafloral nectaries and protection by pugnacious bodyguards. *Annual Review of Ecology and Systematics* **8**: 407–427.

Blüthgen, N. and K. Fiedler. 2002. Interactions between weaver ants (*Oecophylla smaragdina*), homopterans, trees and lianas in an Australian rainforest canopy. *Journal of Animal Ecology* **71**: 793–801.

2003. Preferences for sugars and amino acids and their conditionality in a diverse nectar-feeding ant community. *Journal of Animal Ecology* **73**: 155–166.

Blüthgen, N. and K. Reifenrath. 2003. Extrafloral nectaries in an Australian rain forest: structure and distribution. *Australian Journal of Botany* **51**: 515–527.

Blüthgen, N. and J. Wesenberg. 2001. Ants induce domatia in a rain forest tree (*Vochysia vismiaefolia*). *Biotropica* **33**: 637–642.

Blüthgen, N., G. Gebauer, and K. Fiedler. 2003. Disentagling a rainforest food web using stable isotopes: dietary diversity in a species-rich ant community. *Oecologia* **137**: 426–435.

Blüthgen, N., G. Gottsberger, and K. Fiedler. 2004. Sugar and amino acid composition of ant-attended nectar and honeydew sources from an Australian rainforest. *Austral Ecology* **29**: 418–429.

Blüthgen, N., M. Verhaagh, W. Goitia, *et al.* 2000. How plants shape the ant community in the Amazonian rainforest canopy: the key role of extrafloral nectaries and homopteran honeydew. *Oecologia* **125**: 229–240.

Boecklen, W. J. 1984. The role of extrafloral nectaries in the herbivore defence of *Cassia fasiculata*. *Ecological Entomology* **9**: 243–249.

Bosch, J., J. Retana, and X. Cerda. 1997. Flowering phenology, floral traits and pollinator composition in a herbaceous Mediterranean plant community. *Oecologia* **109**: 583–591.

Bristow, C. M. 1991. Why are so few aphids ant-tended? In C. R. Huxley and D. F. Cutler (eds.) *Ant–Plant Interactions*. Oxford, UK: Oxford University Press, pp. 104–119.

Bronstein, J. L. 1998. The contribution of ant–plant protection studies to our understanding of mutualism. *Biotropica* **30**: 150–161.

Brouat, C., D. McKey, J. M. Bessiere, L. Pascal, and M. Hossaert-McKey. 2000. Leaf volatile compounds and the distribution of ant patrolling in an ant–plant protection mutualism: preliminary results on *Leonardoxa* (Fabaceae: Caesalpinioideae) and *Petalomyrmex* (Formicidae: Formicinae). *Acta Oecologica* **21**: 349–357.

Buckley, R. 1982. *Ant–Plant Interactions in Australia*. The Hague, the Netherlands: Junk.

Burd, M. 1995. Pollinator behavioural responses to reward size in *Lobelia deckenii*: no escape from pollen limitation of seed set. *Journal of Ecology* **83**: 865–872.

Colwell, R. K. 1995. Effects of nectar consumption by the hummingbird flower mite *Proctolaelaps kirmsei* on nectar availability in *Hamelia patens*. *Biotropica* **27**: 206–217.

Colwell, R. K., B. J. Betts, P. Bunnel, F. L. Carpenter, and P. Feinsinger. 1974. Competition for nectar of *Centropogon velerii* by the hummingbird *Colibri thalassinus* and the flower-piercer *Diglossa plumbea*, and its ecological implications. *Condor* **76**: 447–452.

Compton, S. G. and H. G. Robertson. 1988. Complex interactious between mutualists: ants tending homopterans protect fig seeds and pollinators. *Ecology* **69**: 1302–1305.

1991. Effects of ant–homopteran systems on fig-figwasp interactions. In C. R. Huxley and D. F. Cutler (eds.) *Ant–Plant Interactions*. Oxford, UK: Oxford University Press, pp. 120–130

Corbet, S. A., D. M. Unwin, and O. E. Prys-Jones. 1979. Humidity, nectar and insect visits to flowers, with special reference to *Crataegus, Tilia* and *Echium*. *Ecological Entomology* **4**: 9–22.

Cruden, R. W. and S. M. Hermann. 1983. Studying nectar? Some observations on the art. In B. L. Bentley and T. S. Elias (eds.) *The Biology of Nectaries*. New York: Columbia University Press, pp. 223–241.

Cuatle, M. and V. Rico-Gray. 2003. The effect of wasps and ants on the reproductive success of the extrafloral nectaried plant *Turnera ulmifolia* (Turneraceae). *Functional Ecology* **17**: 417–423.

Cushman, J. H. 1991. Host-plant mediation of insect mutualisms: variable outcomes in herbivore-ant interactions. *Oikos* **61**: 138–144.

Dansa, C. V. and C. F. D. Rocha. 1992. An ant–membracid–plant interaction in a cerrado area of Brazil. *Journal of Tropical Ecology* **8**: 339–348.

Dejean, A., M. Gibernau, J. L. Durand, D. Abehassera, and J. Orivel. 2000a. Pioneer plant protection against herbivory: impact of different ant species (Hymenoptera: Formicidae) on a proliferation of the variegated locust. *Sociobiology* **36**: 227–236.

Dejean, A., D. McKey, M. Gibernau, and M. Belin. 2000b. The arboreal ant mosaic in a Cameroonian rainforest (Hymenoptera: Formicidae). *Sociobiology* **35**: 403–424.

De La Fuente, M. A. S. and R. J. Marquis. 1999. The role of ant-tended extrafloral nectaries in the protection and benefit of a Neotropical rainforest tree. *Oecologia* **118**: 192–202.

Del-Claro, K. and P. S. Oliveira. 1993. Ant–Homoptera interaction: do alternative sugar sources distract tending ants? *Oikos* **68**: 202–206.

1996. Honeydew flicking by treehoppers provides cues to potential tending ants. *Animal Behavior* **51**: 1071–1075.

1999. Ant–Homoptera interactions in a neotropical savanna: the honeydew-producing treehopper, *Guayaquila xiphias* (Membracidae), and its associated ant fauna on *Didymopanax vinosum* (Araliaceae). *Biotropica* **31**: 135–144.

2000. Conditional outcomes in a neotropical treehopper–ant association: temporal and species-specific variation in ant protection and homopteran fecundity. *Oecologia* **124**: 156–165.

Del-Claro, K., V. Berto, and W. Reu. 1996. Effect of herbivore deterrence by ants on the fruit set of an extrafloral nectary plant, *Qualea multiflora* (Vochysiaceae). *Journal of Tropical Ecology* **12**: 887–892.

DeVries, P. J. and I. Baker. 1989. Butterfly exploitation of an ant–plant mutualism: adding insult to herbivory. *Journal of the New York Entomological Society* **97**: 332–340.

DeVries, P. J. and C. Penz. 2000. Entomophagy, behavior, and elongated thoracic legs in the myrmecophilous Neotropical butterfly *Ales amesis* (Riodinidae). *Biotropica* **32**: 712–721.

Dominguez, C. A., R. Dirzo, and S. H. Bullock. 1989. On the function of floral nectar in *Croton suberosus* (Euphorbiaceae). *Oikos* **56**: 109–114.

Dress, W. J., S. J. Newell, A. J. Nastase, and J. C. Ford. 1997. Analysis of amino acids in nectar from pitchers of *Sarracenia purpurea* (Sarraceniaceae). *American Journal of Botany* **84**: 1701–1706.

Dyer, L. A. and D. K. Letourneau. 1999a. Relative strengths of top-down and bottom-up forces in a tropical forest community. *Oecologia* **119**: 265–274.

1999b. Trophic cascades in a complex terrestrial community. *Proceedings of the National Academy of Sciences of the USA* **96**: 5072–5076.

Elias, T. S. 1983. Extrafloral nectaries: their structure and distribution. In B. Bentley and T. S. Elias (eds.) *The Biology of Nectaries*. New York: Columbia University Press, pp. 174–203

Engel, V., M. I. Fischer, F. L. Wäckers, and W. Voelkl. 2001. Interactions between extrafloral nectaries, aphids and ants: are there competition effects between plant and homopteran sugar sources? *Oecologia* **129**: 577–584.

Eskildsen, L. I., A. B. Lindberg, and J. M. Olesen. 2001. Ants monopolise plant resources by shelter-construction. *Acta Amazonica* **31**: 155–157.

Eubanks, M. D., K. A. Nesci, M. K. Petersen, Z. Liu, and H. B. Sanchez. 1997. The exploitation of an ant-defended host plant by a shelter-building herbivore. *Oecologia* **109**: 454–460.

Feinsinger, P. and L. A. Swarm. 1978. How common are ant-repellent nectars? *Biotropica* **10**: 238–239.

Fiala, B. 1990. Extrafloral nectaries vs. ant–Homoptera mutualisms: a comment on Becerra and Venable. *Oikos* **59**: 281–282.

Fiala, B. and K. E. Linsenmair. 1995. Distribution and abundance of plants with extrafloral nectaries in the woody flora of a lowland primary forest in Malaysia. *Biodiversity and Conservation* **4**: 165–182.

Fiala, B. and U. Maschwitz. 1991. Extrafloral nectaries in the genus *Macaranga* (Euphorbiaceae) in Malaysia: comparative studies of their possible significance as predispositions for myrmecophytes. *Biological Journal of the Linnean Society* **44**: 287–305.

Fiala, B., H. Grunsky, U. Maschwitz, and K. E. Linsenmair. 1994. Diversity of ant–plant interactions: protective efficacy in *Macaranga* species with different degrees of ant association. *Oecologia* **97**: 186–192.

Fiala, B., S. A. Krebs, H. S. Barlow, and U. Maschwitz. 1996. Interactions between the climber *Thunbergia grandiflora*, its pollinator *Xylocopa latipes* and the ant *Dolichoderus thoracicus*: the "nectar-thief hypothesis" refuted? *Malayan Nature Journal* **50**: 1–14.

Figueiredo, R. A. 1997. Interactions between stingless meliponine bees, honeydew-producing homopterans, ants and figs in a cerrado area. *Naturalia (São Paulo)* **22**: 159–164.

Fleet, R. R. and B. L. Young. 2000. Facultative mutualism between imported fire ants (*Solenopsis invicta*) and a legume (*Senna occidentalis*). *Southwestern Naturalist* **45**: 289–298.

Folgarait, P. J. and D. W. Davidson. 1995. Myrmecophytic *Cecropia*: antiherbivore defenses under different nutrient treatments. *Oecologia* **104**: 198–206.

Fonseca, C. R. 1994. Herbivory and the long-lived leaves of an Amazonian ant-tree. *Journal of Ecology* **82**: 833–842.

Frankie, G. W., W. A. Haber, H. G. Baker, I. Baker, and S. Koptur. 1981. Ants like flower nectar. *Biotropica* **13**: 211–214.

Galetto, L. and L. M. Bernardello. 1992. Extrafloral nectaries that attract ants in Bromeliaceae: structure and nectar composition. *Canadian Journal of Botany* **70**: 1101–1105.

Gaume, L. and D. McKey. 1998. Protection against herbivores of the myrmecophyte *Leonardoxa africana* (Baill.) Aubrev. T3 by its principal ant inhabitant *Aphomomyrmex afer* Emery. *Comptes Rendus de l'Académie des Sciences Pari's Series 3* **321**: 593–601.

Gaume, L., D. McKey, and M. C. Anstett. 1997. Benefits conferred by "timid" ants: active anti-herbivore protection of the rainforest tree *Leonardoxa africana* by the minute ant *Petalomyrmex phylax*. *Oecologia* **113**: 209–216.

Gaume, L., D. McKey, and S. Terrin. 1998. Ant–plant–homopteran mutualism: how the third partner affects the interaction between a plant-specialist ant and its myrmecophyte host. *Proceedings of the Royal Society of London Series B* **265**: 569–575.

Ghazoul, J. 2001. Can floral repellents pre-empt potential ant–plant conflicts? *Ecology Letters* **4**: 295–299.

Gill, F. B. 1988. Effects of nectar removal on nectar accumulation in flowers of *Heliconia imbricata* (Heliconiaceae). *Biotropica* **20**: 169–171.

Glassberg, J. 1999. *Butterflies through Binoculars: The East.* New York: Oxford University Press.

Gomez, J. M. and R. Zamora. 1992. Pollination by ants: consequences of the quantitative effects on a mutualistic system. *Oecologia* **91**: 410–418.

2000. Spatial variation in the selective scenarios of *Hormathophylla spinosa* (Cruciferae). *American Naturalist* **155**: 657–668.

Gomez, J. M., R. Zamora, J. A. Hodar, and D. Garcia. 1996. Experimental study of pollination by ants in Mediterranean high mountain and arid habitats. *Oecologia* **105**: 236–242.

Gottsberger, G., T. Arnold, and H. F. Linskens. 1990. Variation in floral nectar amino acids with aging flowers, pollen contamination, and flower damage. *Israel Journal of Botany* **39**: 167–176.

Guerrant, E. O. and P. L. Fiedler. 1981. Flower defenses against nectar-pilferage by ants. *Biotropica* **13** (suppl): 25–33.

Heads, P. A. and J. H. Lawton. 1984. Bracken, ants, and extrafloral nectaries. II. The effect of ants on the insect herbivores of bracken. *Journal of Animal Ecology* **53**: 1015–1032.

1985. Bracken, ants, and extrafloral nectaries. III. How insect herbivores avoid ant predation. *Ecological Entomology* **10**: 29–42.

Heil, M., B. Fiala, B. Baumann, and K. E. Linsenmair. 2000. Temporal, spatial and biotic variations in extrafloral nectar secretion by *Macaranga tanarius*. *Functional Ecology* **14**: 749–757.

Heil, M., B. Fiala, K. E. Linsenmair, *et al.* 1997. Food body production in *Macaranga triloba* (Euphorbiaceae): a plant investment in anti-herbivore defence via symbiotic ant partners. *Journal of Ecology* **85**: 847–861.

Heil, M., B. Fiala, U. Maschwitz, and K. E. Linsenmair. 2001a. On benefits of indirect defence: short- and long-term studies of antiherbivore protection via mutualistic ants. *Oecologia* **126**: 395–403.

Heil, M., T. Koch, A. Hilpert, *et al.* 2001b. Extrafloral nectar production of the ant-associated plant, *Macaranga tanarius*, is an induced, indirect, defensive response elicited by jasmonic acid. *Proceedings of the National Academy of Sciences of the USA* **98**: 1083–1088.

Heil, M., A. Hilpert, B. Fiala, *et al.* 2002. Nutrient allocation of *Macaranga triloba* ant plants to growth, photosynthesis and indirect defense. *Functional Ecology* **16**: 475–483.

Hickman, J. C. 1974. Pollination by ants: a low-energy system. *Science* **184**: 1290–1292.

Hossaert-McKey, M., J. Orivel, E. Labeyrie, *et al.* 2000. Differential associations with ants of three co-occurring extrafloral nectary-bearing plants. *Ecoscience* **8**: 325–335.

Hunter, J. C. 1994. Extrafloral nectaries on *Arbutus menziesii* (Madrone). *Madrono* **412**: 127.

Itino, T., T. Itioka, A. Hatada, and A. A. Hamid. 2001a. Effects of food rewards offered by ant-plant *Macaranga* on the colony size of ants. *Ecological Research* **16**: 775–786.

Itino, T., S. J. Davies, H. Tada, *et al.* 2001b. Cospeciation of ants and plants. *Ecological Research* **16**: 787–793.

Janzen, D. H. 1966. Coevolution between ants and acacias in Central America. *Evolution* **20**: 249–275.

1967. Interaction of the bull's horn acacia (*Acacia cornigera*) with an ant inhabitant (*Pseudomyrmex ferruginea*) in East Mexico. *Kansas University Science Bulletin* **47**: 315–558.

1969. Allelopathy by myrmecophytes: the ant *Azteca* as an allelopathic agent of *Cecropia*. *Ecology* **50**: 147–153.
Joel, D. M. 1988. Mimicry and mutualism in carnivorous pitcher plants (Sarraceniaceae, Nepenthaceae, Cephalotaceae, Bromeliaceae). *Biological Journal of the Linnean Society* **35**: 185–197.
Jolivet, P. 1983. A hemimyrmecophyte with Chrysomelidae Coleoptera of Southeast Asia, *Clerodendrum fragrans* (Verbenaceae). *Bulletin de la Société Linnéenne (Lyon)* **52**: 242–261.
1986. *Les Fourmis et Les Plantes.* Paris: Société Nouvelle des Editions Boubée.
1998. *Interrelationship between Insects and Plants.* Boca Raton, FL: CRC Press.
Junqueira, L. K., E. Diehl, and E. Diehl-Fleig. 2001. Visitor ants (Hymenoptera: Formicidae) of *Ilex paraguariensis* (Aquifoliaceae). *Neotropical Entomology* **30**: 161–164.
Karban, R. and I. T. Baldwin. 1997. *Induced Responses to Herbivory*. Chicago, IL: University of Chicago Press.
Karhu, K. J. and S. Neuvonen. 1998. Wood ants and a geometrid defoliator of birch: predation outweighs beneficial effects through the host plant. *Oecologia* **113**: 509–516.
Kawano, S., H. Azuma, M. Ito, and K. Suzuki. 1999. Extrafloral nectaries and chemical signals of *Fallopia japonica* and *Fallopia sachalinensis* (Polygonaceae), and their roles as defense systems against insect herbivory. *Plant Species Biology* **14**: 167–178.
Kelly, C. A. 1986. Extrafloral nectaries: ants, herbivores and fecundity in *Cassia fasciculata*. *Oecologia* **69**: 600–605.
Kenrick, J., P. Bernhardt, R. Marginson, *et al.* 1987. Pollination-related characteristics in the mimosoid legume *Acacia terminalis* (Leguminosae). *Plant Systematics and Evolution* **157**: 49–62.
Kerner, A. 1878. *Flowers and their Unbidden Guests*. London: C. Kegan Paul and Co.
Knox, R. B., J. Kenrick, P. Bernhardt, *et al.* 1985. Extra-floral nectaries as adaptations for bird pollination in *Aacia terminalis*. *American Journal of Botany* **72**: 1185–1196.
Koptur, S. 1979. Facultative mutualism between weedy vetches bearing extrafloral nectaries and weedy plants in California. *American Journal of Botany* **66**: 1016–1020.
1983. Flowering phenology and floral biology of *Inga*. *Systematic Botany* **8**: 354–368.
1984a. Experimental evidence for defense of *Inga* (Mimosoideae) saplings by ants. *Ecology* **65**: 1787–1793.
1984b. Outcrossing and pollinator limitation of fruit set: breeding systems of Neotropical *Inga* trees (Fabaceae: Mimosoideae). *Evolution* **38**: 1130–1143.
1985. Alternative defenses against herbivores in *Inga* (Fabaceae: Mimosoideae) over an elevation gradient. *Ecology* **66**: 1639–1650.
1989. Is extrafloral nectar production an inducible defense? In J. Bock and Y. Linhart (eds.) *Evolutionary Ecology of Plants*. Boulder, CO: Westview Press, pp. 323–339.
1992a. Extrafloral nectary-mediated interactions between insects and plants. In E. Bernays (ed.) *Insect–Plant Interactions*, vol 4. Boca Raton, FL: CRC Press, pp. 81–129.
1992b. Plants with extrafloral nectaries and ants in Everglades habitats. *Florida Entomologist* **75**: 38–50.

1994. Floral and extrafloral nectars of neotropical *Inga* trees: a comparison of their constituents and composition. *Biotropica* **26**: 276–284.

Koptur, S. and J. H. Lawton. 1988. Interactions among vetches bearing extrafloral nectaries, their biotic protective agents, and herbivores. *Ecology* **69**: 278–293.

Koptur, S. and N. Truong. 1998. Facultative ant/plant interactions: nectar sugar preferences of introduced pest ant species in South Florida. *Biotropica* **30**: 179–189.

Koptur, S., V. Rico-Gray, and M. Palacios-Rios. 1998. Ant protection in neotropical ferns bearing foliar nectaries. *American Journal of Botany* **85**: 736–739.

Labeyrie, E., L. Pascal, J. Delabie, *et al.* 2001. Protection of *Passiflora glandulosa* (Passifloraceae) against herbivory: impact of ants exploiting extrafloral nectaries. *Sociobiology* **38**: 317–322.

Lanza, J. 1988. Ant preferences for *Passiflora* nectar mimics that contain amino acids. *Biotropica* **20**: 341–344.

1991. Response of fire ants (Formicidae: *Solenopsis invicta* and *S. geminata*) to artificial nectars with amino acids. *Ecological Entomology* **16**: 203–210.

Lanza, J., E. L. Vargo, S. Pulim, and Y. Z. Chang. 1993. Preference of the fire ants *Solenopsis invicta* and *S. geminata* (Hymenoptera: Formicidae) for amino acid and sugar components of extrafloral nectars. *Environmental Entomology* **22**: 411–417.

Lara, C. and J. F. Ornelas. 2001. Nectar "theft" by hummingbird flower mites and its consequences for seed set in *Moussonia deppeana*. *Functional Ecology* **15**: 78–84.

Lawton, J. H. and P. A. Heads. 1985. Bracken, ants, and extrafloral nectaries. I. The components of the system. *Journal of Animal Ecology* **53**: 995–1015.

Letourneau, D. K. 1983. Passive aggression: an alternative hypothesis for the *Piper–Pheidole* association. *Oecologia* **60**: 122–126.

Letourneau, D. K. and L. A. Dyer. 1998. Experimental test in lowland tropical forest shows top-down effects through four trophic levels. *Ecology* **79**: 1678–1687.

Machado, G. and V. L. Freitas. 2001. Larval defence against ant predation in the butterfly *Smyrna blomfildia*. *Ecological Entomology* **26**: 436–439.

Madden, D. and T. P. Young. 1992. Symbiotic ants as an alternative defense against giraffe herbivory in spinescent *Acacia drepanolobium*. *Oecologia* **91**: 235–238.

Majer, J. D., J. H. C. Delabie, and M. R. B. Smith. 1994. Arboreal ant community patterns in Brazilian cocoa farms. *Biotropica* **26**: 73–83.

Maloof, J. E. and D. W. Inouye. 2000. Are nectar robbers cheaters or mutualists? *Ecology* **81**: 2651–2661.

Martinez del Rio, C. 1990. Sugar preferences in hummingbirds: the influence of subtle chemical differences on food choice. *Condor* **92**: 1022–1030.

Martinez del Rio, C. and W. H. Karasov. 1990. Digestion strategies in nectar-eating and fruit-eating birds and the sugar composition of plant rewards. *American Naturalist* **136**: 618–637.

Martinez del Rio, C. and B. R. Stevens. 1989. Physiological constraint on feeding behavior: intestinal membrane disaccharidases of the starling. *Science* **243**: 794–796.

Martinez del Rio, C., H. G. Baker, and I. Baker. 1992. Ecological and evolutionary implications of digestive processes, bird preferences, and the sugar constituents of floral nectar and fruit pulp. *Experientia* **48**: 544–551.
Martinez del Rio, C., W. H. Karasov, and D. J. Levey. 1989. Physiological basis and ecological consequences of sugar preferences in cedar waxwings. *Auk* **106**: 64–71.
Martinez del Rio, C., B. R. Stevens, D. E. Daneke, and P. T. Andreadis. 1988. Physiological correlates of preference and aversion for sugars in three species of birds. *Physiological Zoology* **61**: 222–229.
McKey, D. 1988. Promising new directions in the study of ant–plant mutualisms. Proc. 14th Int. Botanical Congr. Koeltz, Konigstein, Germany, pp. 335–355.
1991. Phylogenetic analysis of the evolution of a mutualism: *Leonardoxa* (Caesalpiniaceae) and its associated ants. In C. R. Huxley and D. F. Cutler (eds.) *Ant–Plant Interactions*. Oxford, UK: Oxford University Press, pp. 310–334 .
Messina, F. J. 1981. Plant protection as a consequence of an ant–membracid mutualism: interactions on golden rod (*Solidago* sp.). *Ecology* **62**: 1433–1460.
Mody, K. and K. E. Linsenmair. 2004. Plant-attracted ants affect arthropod community structure but not necessarily herbivory. *Ecological Entomology* **29**: 217–225.
Mondor, E. B. and J. F. Addicott. 2003. Conspicuous extra-floral nectaries are inducible in *Vicia faba*. *Ecology Letters* **6**: 495–497.
Moog, J., T. Drude, and U. Maschwitz. 1998. Protective function of the plant-ant *Cladomyrma maschwitzi* to its host, *Crypteronia griffithii*, and the dissolution of the mutualism (Hymenoptera: Formicidae). *Sociobiology* **31**: 105–130.
Morellato, L. P. C. and P. S. Oliveira. 1994. Extrafloral nectaries in the tropical tree *Guarea macrophylla* (Meliaceae). *Canadian Journal of Botany* **72**: 157–160.
Morris, W. F. 1996. Mutualism denied? Nectar-robbing bumblebees do not reduce female or male success of bluebells. *Ecology* **77**: 1451–1462.
Moya-Raygoza, G. and K. J. Larsen. 2001. Temporal resource switching by ants between honeydew produced by the fivespotted gama grass leafhopper (*Dalbulus quinquenotatus*) and nectar produced by plants with extrafloral nectaries. *American Midland Naturalist* **146**: 311–320.
Navarro, L. 2000. Pollination ecology of *Anthyllis vulneraria* subsp. *vulgaris* (Fabaceae): nectar robbers as pollinators. *American Journal of Botany* **87**: 980–985.
O'Brien, S. P. 1995. Extrafloral nectaries in *Chamelaucium uncinatum*: a first record in the Myrtaceae. *Australian Journal of Botany* **43**: 407–413.
O'Dowd, D. J. 1982. Pearl bodies as ant food: an ecological role for some leaf emergences of tropical plants. *Biotropica* **14**: 40–49.
O'Dowd, D. J. and E. A. Catchpole. 1983. Ants and extrafloral nectaries: no evidence for plant protection in *Helichrysum* spp.–ant interactions. *Oecologia* **59**: 191–200.
Offenberg, J. 2000. Correlated evolution of the association between aphids and ants and the associations between aphids and plants with extrafloral nectaries. *Oikos* **91**: 146–152.
Oliveira, P. S. 1997. The ecological function of extrafloral nectaries: herbivore deterrence by visiting ants and reproductive output in *Caryocar brasiliense* (Caryocaraceae). *Functional Ecology* **11**: 323–330.

Oliveira, P. S. and M. R. Pie. 1998. Interaction between ants and plants bearing extrafloral nectaries in cerrado vegetation. *Anais de Sociedade Entomologica do Brasil* **27**: 161–176.

Oliveira, P. S., A. V. L. Freitas, and K. Del-Claro. 2002. Ant foraging on plant foliage: contrasting effects on the behavioral ecology of insect herbivores. In P. S. Oliveira and R. J. Marquis (eds.) *The Cerrados of Brazil: Ecology and Natural History of a Neotropical Savanna*. New York: Columbia University Press, pp. 287–305.

Oliveira, P. S., C. Klitzke, and E. Vieira. 1995. The ant fauna associated with the extrafloral nectaries of *Ouratea hexasperma* (Ochnaceae) in an area of cerrado vegetation in Central Brazil. *Entomologist's Monthly Magazine* **131**: 77–82.

Oliveira, P. S., V. Rico-Gray, C. Diaz-Castelazo, and C. Castillo-Guevara. 1999. Interaction between ants, extrafloral nectaries and insect herbivores in Neotropical coastal sand dunes: herbivore deterrence by visiting ants increases fruit set in *Opuntia stricta*. *Functional Ecology* **13**: 623–631.

Passera, L., J.-P. Lachaud, and L. Gomel. 1994. Individual food source fidelity in the neotropical ponerine ant *Ectatomma ruidum* Roger (Hymenoptera Formicidae). *Ethology, Ecology, and Evolution* **6**: 13–21.

Peakall, R., S. N. Handel, and A. J. Beattie. 1991. The evidence for, and importance of, ant pollination. In C. R. Huxley and D. F. Cutler (eds.) *Ant–Plant Interactions*. Oxford, UK: Oxford University Press, pp. 421–429.

Pemberton, R. W. 1993. Extrafloral nectar feeding by the Japanese white-eye. *Tropics* **2**: 183–186.

Perfecto, I. and A. Sediles. 1992. Vegetational diversity, ants (Hymenoptera: Formicidae), and herbivorous pests in a Neotropical agroecosystem. *Environmental Entomology* **21**: 61–67.

Pickett, C. H. and W. D. Clark. 1979. The function of extrafloral nectaries in *Opuntia acanthocarpa* (Cactaceae). *American Journal of Botany* **66**: 618–625.

Pinheiro, M. C. B., W. T. Ormond, H. A. De Lima, and M. C. R. Correia. 1995. Biology of reproduction of *Norantea brasiliensis* Choisy (Marcgraviaceae). *Revista Brasileira de Biologia* **55** (suppl. 1): 79–88.

Prys-Jones, O. E. and P. G. Willmer. 1992. The biology of alkaline nectar in the purple toothwort (*Lathraea clandestina*): ground level defences. *Biological Journal of the Linnean Society* **45**: 373–388.

Puterbaugh, M. N. 1998. The roles of ants as flower visitors: experimental analysis in three alpine plant species. *Oikos* **83**: 36–46.

Pyke, G. H. 1991. What does it cost a plant to produce floral nectar? *Nature* **350**: 58–59.

Rashbrook, V. K., S. G. Compton, and J. H. Lawton. 1992. Ant–herbivore interactions: reasons for the absence of benefits to a fern with foliar nectaries. *Ecology* **73**: 2167–2174.

Ratnieks, F. L. W. and C. Anderson. 1999. Task partitioning in insect societies. *Insectes Sociaux* **46**: 95–108.

Ricks, B. L. and S. B. Vinson. 1970. Feeding acceptibility of certain insects and various water soluble compounds to two varieties of the imported fire ant. *Journal of Economic Entomology* **63**: 145–148.

Rickson, F. R. and M. M. Rickson. 1998. The cashew nut, *Anacardium occidentale* (Anacardiaceae), and its perennial association with ants: extrafloral nectary location and the potential for ant defense. *American Journal of Botany* **85**: 835–849.

Rico-Gray, V. 1993. Use of plant-derived food resources by ants in the dry tropical lowlands of coastal Veracruz, Mexico. *Biotropica* **25**: 301–315.

2001. Interspecific interaction. In *Encyclopedia of Life Sciences*. London: Macmillan; available at http://www.els.net.

Rico-Gray, V., M. Palacios-Rios, J. G. Garcia-Franco, and W. P. MacKay. 1998. Richness and seasonal variation of ant–plant associations mediated by plant-derived food resources in the semiarid Zapotitlan Valley, Mexico. *American Midland Naturalist* **140**: 21–26.

Risch, S.J. and F. R. Rickson. 1981. Mutualism in which ants must be present before plants produce food bodies. *Nature* **291**: 149–150.

Roces, F., Y. Winter, and O. V. Helversen. 1993. Nectar concentration preference and water balance in a flower visiting bat, *Glossophaga soricina antillarum*. In W. Barthlott, C. Naumann, C. Schmidt-Loske, and K. Schuhmann (eds.) *Animal–Plant Interactions in Tropical Environments*. Bonn, Germany: Museum Alexander Koenig, pp. 159–165.

Rocha, C. R. and H. G. Bergallo. 1992. Bigger ant colonies reduce herbivory and herbivore residence time on leaves of an ant-plant: *Azteca muelleri* vs. *Coelomera ruficornis* on *Cecropia pachystachya*. *Oecologia* **91**: 249–252.

Rodriguez, M. C. 1995. Interactions between *Lysiloma bahamensis* Benth. (Fabaceae: Mimosoideae) and herbivorous insects in the Everglades. M.S. thesis, Florida International University, Miami.

Ruhren, S. and S. N. Handel. 1999. Jumping spiders (Salticidae) enhance the seed production of a plant with extrafloral nectaries. *Oecologia* **119**: 227–230.

Rusterholz, H. P. and A. Erhardt. 1997. Preferences for nectar sugars in the peacock butterfly, *Inachis io*. *Ecological Entomology* **22**: 220–224.

Sakata, H. and Y. Hashimoto. 2000. Should aphids attract or repel ants? Effect of rival aphids and extrafloral nectaries on ant–aphid interactions. *Population Ecology* **42**: 171–178.

Sazima, I., S. Buzato, and M. Sazima. 1993. The bizarre inflorescence of *Norantea brasiliensis* (Marcgraviaceae): visits of hovering and perching birds. *Botanica Acta* **106**: 507–513.

Sazima, M. and I. Sazima. 1980. Bat visits to *Marcgravia myriostigma* Tr. et Planch. (Marcgraviaceae) in Southeastern Brazil. *Flora* **169**: 84–88.

Schemske, D. W. 1982. Ecological correlates of a neotropical mutualism: ant assemblages at *Costus* extrafloral nectaries. *Ecology* **63**: 932–941.

Schimper, A. F. W. 1903. *Plant-Geography upon a Physiological Basis*. Oxford, UK: Clarendon Press.

Schuerch, S., M. Pfunder, and B. A. Roy. 2000. Effects of ants on the reproductive success of *Euphorbia cyparissias* and associated pathogenic rust fungi. *Oikos* **88**: 6–12.

Schupp, E. W. 1986. *Azteca* protection of *Cecropia*: ant occupation benefits juvenile trees. *Oecologia* **70**: 379–385.

Schupp, E. W. and D. H. Feener. 1991. Phylogeny, lifeform, and habitat dependence of ant-defended plants in a Panamanian forest. In C. R. Huxley and D. F. Cutler (eds.) *Ant–Plant Interactions*. Oxford, UK: Oxford University Press, pp. 175–197.

Seufert, P. and K. Fiedler. 1996. The influence of ants on patterns of colonization and establishment within a set of coexisting lycaenid butterflies in a south-east Asian tropical rain forest. *Oecologia* **106**: 127–136.

Stapel, J. O., A. M. Cortesero, C. M. De Moraes, J. H. Tumlinson, and W. J. Lewis. 1997. Extrafloral nectar, honeydew, and sucrose effects on searching behavior and efficiency of *Microplitis croceipes* (Hymenoptera: Braconidae) in cotton. *Environmental Entomology* **26**: 617–623.

Stephenson, A. G. 1982. The role of the extrafloral nectaries of *Catalpa speciosa* in limiting herbivory and increasing fruit production. *Ecology* **63**: 663–669.

Stettmer, C. 1993. [Flower-visiting beneficial insects on extrafloral nectaries of the cornflower *Centaurea cyanus* (Asteraceae).] *Mitteilungen der schweizischen entomologischen Gesellschaft* **66**: 1–8. (In German)

Stevens, J. A. 1990. Response of *Campsis radicans* (Bignoniaceae) to simulated herbivory and ant visitation. M.S. thesis, University of Missouri, St. Louis.

Tempel, A. S. 1983. Bracken fern (*Pteridium aquilinum*) and nectar-feeding ants: a nonmutualistic interaction. *Ecology* **64**: 1411–1422.

Thompson, J. N. 1994. *The Coevolutionary Process*. Chicago, IL: University of Chicago Press.

Thorp, R. W. and E. A. Sugden. 1990. Extrafloral nectaries producing rewards for pollinator attraction in *Acacia longifolia* (Andr.) Willd. *Israel Journal of Botany* **39**: 177–186.

Torres-Hernandez, L., V. Rico-Gray, C. Castillo-Guevara, and J. A. Vergara. 2000. Effect of nectar-foraging ants and wasps on the reproductive fitness of *Turnera ulmifolia* (Turneraceae) in a coastal sand dune in Mexico. *Acta Zoologica Mexicana* **81**: 13–21.

Valenzuela-Gonzalez, J., A. Lopez-Mendez, and A. Garcia-Ballinas. 1994. [Activity patterns and foraging habits of *Pachycondyla villosa* (Hymenoptera, Formicidae) in cacao plantations of Soconusco, Chiapas, Mexico.] *Folia Entomologica Mexicana* **91**: 9–21. (In Spanish)

Vandermeer, R. K., C. S. Lofgren, and J. A. Seawright. 1995. Specificity of the red imported fire ant (Hymenoptera: Formicidae) phagostimulant response to carbohydrates. *Florida Entomologist* **78**: 144–154.

Vasconcelos, H. L. 1991. Mutualism between *Maieta guianensis* Aubl., a myrmecophytic melastome, and one of its ant inhabitants: ant protection against insect herbivores. *Oecologia* **87**: 295–298.

Wäckers, F. L. and R. Wunderlin. 1999. Induction of cotton extrafloral nectar production in response to herbivory does not require a herbivore-specific elicitor. *Entomologia Experimentalis et Applicata* **91**: 149–154.

Wäckers, F. L., D. Zuber, R. Wunderlin, and F. Keller. 2001. The effect of herbivory on temporal and spatial dynamics of foliar nectar production in cotton and castor. *Annals of Botany* **87**: 365–370.

Wagner, D. 1997. The influence of ant nests on *Acacia* seed production, herbivory and soil nutrients. *Journal of Ecology* **85**: 83–93.

2000. Pollen viability reduction as a potential cost of ant association for *Acacia constricta* (Fabaceae). *American Journal of Botany* **87**: 711–715.

Wagner, D. and A. Kay. 2002. Do extrafloral nectaries distract ants from visiting flowers? An experimental test of an overlooked hypothesis. *Evolutionary Ecology Research* **4**: 293–305.

Wagner, D. and L. Kurina. 1997. The influence of ants and water availability on oviposition behaviour and survivorship of a facultatively ant-tended herbivore. *Ecological Entomology* **22**: 352–360.

Way, M. J. 1963. Mutualism between ants and honeydew producing Homoptera. *Annual Review of Entomology* **8**: 307–344.

Weseloh, R. M. 1995. Ant traffic on different tree species in Connecticut. *Canadian Entomologist* **127**: 569–575.

Willmer, P. G. and S. A. Corbet. 1981. Temporal and microclimatic partitioning of the floral resources of *Justicea aurea* amongst a concourse of pollen vectors and nectar robbers. *Oecologia* **51**: 67–78.

Willmer, P. G. and G. N. Stone. 1997. How aggressive ant-guards assist seed-set in *Acacia* flowers. *Nature* **388**: 165–167.

Wunnachit, W., C. F. Jenner, and M. Sedgley. 1992. Floral and extrafloral nectar production in *Anacardium occidentale* L. (Anacardiaceae): an andromonoecious species. *International Journal of Plant Sciences* **153**: 413–420.

Yano, S. 1994. Flower nectar of an autogamous perennial *Rorippa indica* as an indirect defense mechanism against herbivorous insects. *Researches on Population Ecology* **36**: 63–71.

Young, T. P., C. H. Stubblefield, and L. A. Isbell. 1997. Ants on swollen-thorn acacias: species coexistence in a simple system. *Oecologia* **109**: 98–107.

Zachariades, C. and J. J. Midgley. 1999. Extrafloral nectaries of South African Proteaceae attract insects but do not reduce herbivory. *African Entomology* **7**: 67–76.

4

Fitness consequences of food-for-protection strategies in plants

MAURICE W. SABELIS, PAUL C. J. VAN RIJN, AND ARNE JANSSEN

Behind the idle and quiet appearance of plants, warfare is an everyday issue. Herbivorous arthropods, below- and aboveground, continue to threaten a plant's existence, whereas their attack is countered by the plant in many ways. Plants defend themselves directly by modifying plant structure (e.g., cuticle thickness, leaf hairiness), lowering nutritional quality, decreasing digestibility and increasing toxicity, but also indirectly by promoting the effectiveness of enemies of the herbivores (Price *et al.* 1980). This indirect plant defense implies that plants provide chemical lures, shelter and/or food, whereas they gain protection in exchange (Sabelis *et al.* 1999a,b,c,d, 2002). Central American *Acacia* trees stand out as a landmark example (Janzen 1966). They have stipular thorns that are expanded and hollow and provide nesting sites for certain ants. In addition, they secrete nectar from large foliar nectaries and produce nutritive organs called Beltian bodies on the leaf pinnules. These food bodies are eagerly harvested by foraging ants and fed to their larvae. The ants in turn kill insect herbivores, repel mammalian herbivores, and destroy plants interfering with the *Acacia* tree. In this chapter, we focus on food provisioning as a strategy of the plant to boost the third trophic level and we discuss the conditions under which this particular mode of defense is favored by natural selection.

The argument that plants benefit from consumers of the foods they provide dates back to Thomas Belt in his book published in 1874. It was subject to intense debate for many decades, and repeatedly discarded. Then Janzen (1966) revived the idea, almost a century later. From thereon, a suite of experiments has been published, showing that consumers of plant-provided foods could in fact (though not always) benefit plants (Beattie 1985; Jolivet 1996). The foods provided may consist of sugars, lipids, and amino acids, either alone or in combination (Hagen 1986; Wäckers, Chapter 2). Since these nutrients are

Plant-Provided Food for Carnivorous Insects, ed. F. L. Wäckers, P. C. J. van Rijn, and J. Bruin. Published by Cambridge University Press.

vital for plant survival, growth, and reproduction, it seems reasonable to hypothesize that producing this food goes at the expense of investment in other fitness-related plant traits. Such trade-off relations are expected to constrain investment in "food-for-protection", especially because food provisions rank among the *relatively* more energy-demanding modes of indirect defense (Dicke and Sabelis 1992; Heil *et al.* 1997; Fischer *et al.* 2002). Together with investment in other indirect and direct defenses, these relations may set limits to plant defense strategies.

That plants gain by investing in food-for-protection is not at all self-evident. This is not simply because the benefits may not outweigh the costs. The much more fundamental reason is that once such food is provided, the plant has no – or at best limited – control over which organisms are going to utilize it. If organisms at the second or fourth trophic level, rather than those at the third, reap the benefits of plant foods, the plant may not gain anything or even incur more damage from herbivory. For example, pollens are important alternative foods for predatory arthropods, but also for folivorous thrips, and extrafloral nectaries may be frequented not only by predators and parasitoids of herbivorous arthropods, but also by hyperpredators and hyperparasitoids. Therefore the core problem is to unravel why many plants evidently invest in providing foods, despite the danger of "robbery" (Sabelis *et al.* 1999a,b,c,d, 2002).

Which combination of defenses proves best depends not only on the level of herbivory and the effectiveness of carnivores, but also on the strategies of other plants in the population or community. Selection within populations acts on between-individual differences in plant defendedness, not on the absolute level of defense per se. Plants may compete for acquiring protection by the enemies of the herbivores, and this promotes investment in defenses. Plant competition is not population-wide, however. It is bound to prevail in the direct neighborhood of a plant. Thus, to understand defense strategies in plants it is paramount to consider how a plant's investment in defenses will affect local competition with other plants.

The aim of this chapter is to review assumptions and hypotheses on selective advantages of food provision by plants and to derive predictions that are testable by experiments. Published experimental results will be discussed to illustrate some points, but they will not be reviewed exhaustively. In the sections to follow we first argue against the skeptic view that plant-provided food is no more than a by-product of plant physiological processes, then we discuss the evidence for plants investing in products that are not directly vital to their growth and reproduction, yet are food for organisms at higher trophic levels. Next, we review evidence that plant-provided foods attract enemies of herbivores and result in plant protection. Then, we consider how plants tune investments in space and time to imminent herbivore threat. The subsequent sections are about avoiding conflicts with organisms that do not provide protection, eat plant-provided foods, yet benefit the plant (e.g., pollinators), and about preventing (mis-)use by organisms that do not benefit the plant. The discussion centers around recent theoretical insights as to how mutualisms persist in the face of robbers and cheaters.

Plants provide foods as a by-product of their physiology or reproduction

The simplest explanation for plant-provided foods is to ascribe a purely physiological function to them. Guttation, exudation, and secretion may well help to regulate the salt, sugar, and water balance of a plant. Just as leaf stomata help maintain a physiological balance in leaves, glandular hairs may do so in leaves and stems, and extrafloral nectaries in phloem tissue. Feeding on those plant secretions by arthropods may then be considered as a mere by-product of plant physiological processes. If these secretions promote the visitation rate by arthropods, there will be selection to increase or decrease secretory activity depending on the predator/herbivore ratio among the visitors. Indeed, this may well have been the ancestral state of food-for-protection mutualisms (Beattie 1985). The important point to note is that once plants secrete nutritious fluids and this affects overall herbivore damage, natural selection will inevitably act on the secretory process. Hence, even if food provisioning arose as a by-product from other processes, it is not necessarily selectively neutral. Any inference of a purely plant-physiological function of exudation and secretion requires demonstrating that the trait is selectively neutral.

Secretory sites may also help to reduce the concentration of harmful compounds in the plant. For example, herbicides and pesticides have been found in plant secretions such as extrafloral nectars (Rogers 1985), but an active process whereby such undesired compounds are selectively transported to secretory sites has never been demonstrated. If toxin removal were the main function, then it is not clear why arthropods observed feeding on plant secretions are much more often predators than herbivores, whereas predators tend to develop toxin-resistance more slowly than herbivores (e.g., Jansen and Sabelis 1995). Secretory toxin removal remains a possible physiological function of extrafloral nectaries, but it does not match the common observation of predatory arthropods feeding on the nectar. Moreover, there are other secretory sites, such as glandular hairs on leaves and stems, that are thought to function in direct defense against herbivores, and these seem better suited for toxin removal as well.

Whereas toxins are well known from plant structures that are supposed to function in direct defense of plant tissue vital for growth and reproduction (glandular hairs), they sometimes also occur in plant structures that are thought to increase attractiveness. For example, some plant species produce toxic floral nectar (Adler 2000). This is puzzling: is it a by-product of direct plant defense or does it restrict the use of such secretions by robbers and give access to more specialized pollinators? These are open questions with respect to nectaries in flowers, and their relevance to extrafloral nectaries needs further critical assessment.

Another example of a plant-provided food is pollen. Although its primary function is undoubtedly linked to plant reproduction, it can also be utilized as food by pollinators and many herbivorous or predaceous arthropods. Pollens from different plant species differ widely in edibility for these consumers (e.g., Roulston *et al.* 2000). If edibility were negative to the plant (e.g., lower chances to pollinate), reduced edibility would be generally favored by selection. It is striking to see, however, how frequently predatory arthropods can utilize pollen as food for growth and reproduction (Sabelis and Van Rijn 1997; Van Rijn and Tanigoshi 1999a; Beckman and Hurd 2003; Wäckers, Chapter 2). Hence, one may wonder whether edible pollen has evolved to attract predators and thereby promote plant protection. Such an effect is unlikely to jeopardize its function in plant reproduction because predatory arthropods feed usually on "waste" pollen, i.e., pollen kernels that drop from the stamina onto the flower and lower leaves of the same plant or drift on wind currents onto the surface of other plants. An unresolved question is whether pollen production rates evolved solely to maximize the number of plant offspring (taking pollen wastage for granted), or evolved to even higher levels because pollen wastage is to some degree adaptive to the plant (Van Rijn and Tanigoshi 1999b; Van Rijn *et al.* 2002).

Thus, plant-provided foods, such as nectar and pollen, may well have evolved for their function in plant physiology and reproduction, but their edibility and food value are unlikely to be simply by-products of these functions. Edibility and food value may have evolved in response to the benefits from attracting predatory arthropods, thereby acquiring protection against herbivores.

Investment in plant-provided foods diverts energy from other vital processes

In contrast to pollen and nectar, there are also plant-provided foods that have no obvious alternative function and that are bound to be produced at a cost to the plant. This is especially true for highly specialized protection mutualisms (e.g., so-called ant-plants or myrmecophytes), as opposed to more generalized protection mutualisms involving relatively variable costs of herbivory and low investment in rewards (e.g., plants that produce extrafloral nectar). Striking are the food bodies – specialized epidermal structures on leaves of myrmecophytes

(Beattie 1985). In *Macaranga* trees, so-called "pearl bodies" contain more protein in trees that are myrmecophytes than in trees that are not (Heil *et al.* 1998; Hatada *et al.* 2002). Treating *Macaranga* trees with fertilizer leads to the production of more food bodies, larger populations of ant workers, and less herbivory (Heil *et al.* 2001a). Thus, these trees make scarce nutrients available for consumption by other organisms, whereas they gain protection against herbivores (and other benefits) from the ants consuming the plant-provided food (Beattie 1985; Jolivet 1996).

Investment in the production of food bodies is very low if expressed as a percentage of total plant dry matter (*c.*0.01% in *Piper* plants: Fischer *et al.* 2002). However, food bodies contain nitrogen, stored mainly in the form of proteins (*c.*20% of food body dry matter in *Piper* plants). Soluble protein and especially lipids are the main constituents of food bodies. They are therefore a high-energy food source (*c.*23 kJ/g in *Piper* plants: Fischer *et al.* 2002). In terms of total assimilation products, this investment is considerable (up to 5% in *Macaranga triloba*: Heil *et al.* 1997, 1998). Such an investment in limiting nutrients and assimilates may well go at the expense of other processes vital for the plant. The resulting trade-off may then act as a constraint, causing plants to optimize their allocation to indirect defense (Linsenmair *et al.* 2001). This trade-off with direct defense is supported by observations showing increased herbivore growth on obligate (as compared to facultative) myrmecophytes that had their associated ants removed (Heil *et al.* 2002). Although it is known for some plant species which secondary plant compounds (e.g., chitinases, tannins, amides) are negatively affected by investment in food bodies (Heil *et al.* 1999, 2000a; Dyer *et al.* 2001; Eck *et al.* 2001), the overall picture is one of much variability in presence and strength of the trade-off, and great difficulty in pinpointing the defensive compounds subject to a trade-off (Heil *et al.* 2002). Other plant-provided foods, such as extrafloral nectar, are less energy-demanding than food bodies. For example, the amount of sugar excreted through extrafloral nectaries in damaged castor bean leaves represented 0.1% of the leaf's daily assimilate production (Wäckers *et al.* 2001). Like food bodies, extrafloral nectars also contain nitrogen-based compounds, not as proteins but as amino acids (Lanza 1991; Wäckers, Chapter 2). Therefore, we tentatively conclude that investment in plant-provided foods is low, but not negligible, because it involves nutrients that limit plant growth and that are thus channeled away from other vital processes (such as direct defense, growth, and reproduction). It is actually no surprise that trade-off relations are hard to detect given that investments in plant-provided foods are low, and given that alternative allocation channels are numerous. If the question is whether plant-provided foods trade off against other components of plant fitness, the answer is bound to be affirmative, but if the question is how to detect and represent their multidimensional nature, there are some obvious research challenges for the future.

Modified plant-food supply alters degree of plant protection

If we assume that plant investments are not negligible, then what do plants receive in return? There is proof that many carnivorous arthropods feed on plant-provided foods and in many (but not all) cases it is shown that plants benefit from the increased visitation by carnivorous arthropods. One line of evidence comes from carnivore exclusion experiments in which the consequences for herbivore abundance are measured, as well as for plant damage and seed set. Ant removal from plants with extrafloral nectaries (Koptur 1992; De La Fuente and Marquis 1999; Di Giusto *et al.* 2001; Labeyrie *et al.* 2001 and/or food bodies (Heil *et al.* 2001a, b; Itino and Itioka 2001) leads to increased herbivory, reduced plant growth, and reduced seed set (Beattie 1985; Wagner 1997; Oliveira *et al.* 1999; Schmitz *et al.* 2000). Removal of ants may also cause herbivores to spend more time feeding and to switch feeding at sites that are more vital to the plant (Rudgers *et al.* 2003). Thus, carnivore removal need not necessarily affect herbivore abundance, it may also affect herbivore behavior. If carnivore removal were to affect neither of the two, this might be due to low herbivore abundance, herbivore resistance to predation, or herbivore-induced plant resistance/tolerance, acting instead of carnivory.

The second line of evidence for plants benefiting from food provision comes from manipulation experiments. One mode of manipulation is to make plant-provided foods inaccessible. Covering extrafloral nectaries with clear nail polish decreased visitation by jumping spiders (Araneae: Salticidae) and increased seed set of *Chamaecrista nictitans*, an annual legume native to the eastern USA (Ruhren and Handel 1999). Another mode of manipulation is to supply alternative foods to plants that would otherwise not provide them. Spraying sugar-rich supplementary foods on plants tends to increase populations of predatory arthropods, such as chrysopids and coccinellids (Hagen 1986; Evans and Swallow 1993). Adding sugar droplets to cassava plants that have had their phloem exudates secluded by plastic wrapping has a positive numerical impact on predatory mites (Acari: Phytoseiidae), and a negative impact on herbivorous mite numbers (Bakker and Klein 1992). Adding cattail (*Typha latifolia*) pollen to male-sterile cucumber (*Cucumis sativa*) plants causes populations of predatory mites to increase and those of herbivorous thrips to decrease (Van Rijn *et al.* 2002; see also Sabelis and Van Rijn 1997) (Fig. 4.1). Much the same results were obtained in a system with predatory mites and whiteflies as prey on male-sterile cucumber plants (Nomikou *et al.* 2002).

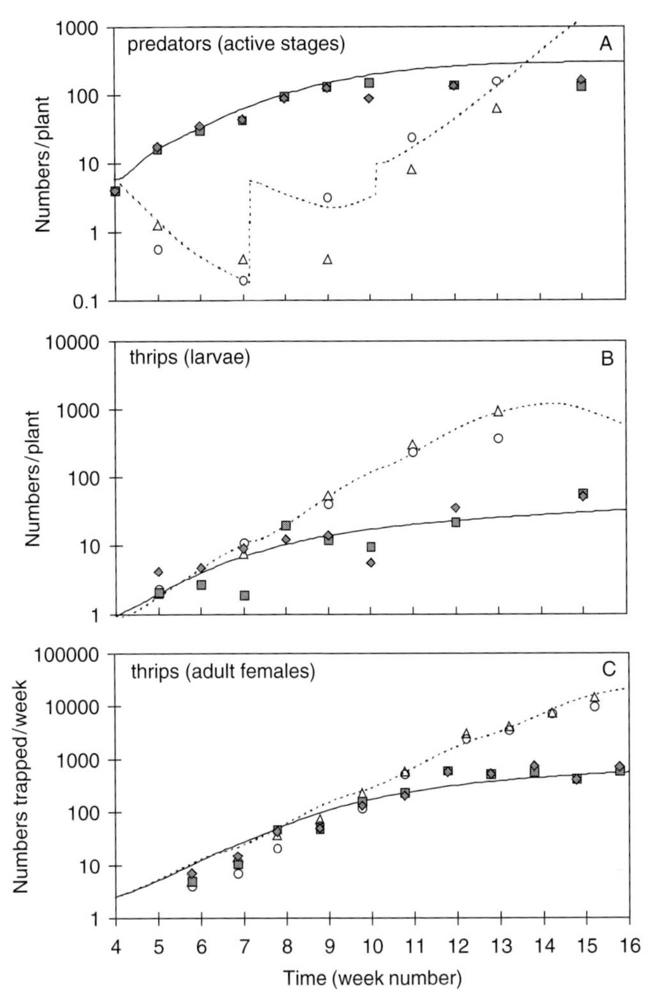

Figure 4.1 Population dynamics of the predatory mite *Iphiseius degenerans* (all mobile stages), and the western flower thrips, *Frankliniella occidentalis* (larvae and adult females), in the presence and absence of cattail pollen on male-sterile cucumber plants (Van Rijn *et al.* 2002). Experimental results are indicated by symbols (closed symbols for treatment and open symbols for control). Computer simulation results are indicated by lines (solid for treatment with pollen and dotted for control). Experiments and simulations concern numbers per plant. Adult thrips densities were converted into numbers trapped per week (bottom panel). The proportion of the area with pollen is fixed at 10%. Initial numbers were two thrips per plant (all stages in proportion to stable stage distribution) and six adult predators per plant (two males and four females, week 4). Predator introductions were repeated only in the control experiments in weeks 7 and 10, to prevent the predator population from going extinct. Note that the supply of pollen has a strong positive impact on the population growth of predatory mites and a strong negative impact on the population growth of the thrips larvae and the thrips adults (albeit with a delay corresponding to the time needed to develop from egg to maturity).

The final line of evidence comes from intra- and interspecific comparisons of food provision by plants and the associated degree of protection from predators. For example, *Macaranga* species differ in the extent to which they provide food bodies and these interspecific differences are reflected in the colony size of the attending ants (Itino and Itioka 2001; Itino *et al.* 2001; Hatada *et al.* 2002).

It can thus be concluded that differences in food supply, either through manipulation or as part of natural variation, have consequences for ant attendance and plant protection against herbivory.

Plants time, target, and tune food provision to reduce costs and robbery

If investment in foods is not negligible and returns in terms of protection are significant, then plants may be expected to save energy by allocating resources at the right site and at the right time. Usually, plants provide foods near to the most vital parts of the plant. Phloem exudation on cassava is most pronounced on petioles and veins of the first few leaves below the apex (Bakker and Klein 1992). These leaves are photosynthetically most active and are preferred by herbivores such as cassava green mites (Acari: Tetranychidae). Extrafloral nectaries also tend to be more active on young leaves (Heil *et al.* 2000b). Nectaries on young phyllodes of *Acacia pycnantha* are notably larger (Beattie 1985). Usually, extrafloral nectaries stop being productive and dry out on older leaves and during seed set (Jolivet 1996). All these observations on positioning near photosynthetically active parts of the plant, however, are compatible with the hypothesis on plant physiological by-products.

Plants provide food not only near the site where protection is most vital, but also at the time protection is most needed. Extrafloral nectaries in cherry trees become active around the time when tent caterpillars are most vulnerable to predation by ants (Tilman 1978). How the timing of nectar production is brought about is now beginning to become clear. Leaf wounding in some plant species appears to induce an increase in the number of extrafloral nectaries (Mondor and Addicott 2003), as well as in secretory activity (Smith *et al.* 1990; Heil *et al.* 2000b; Wäckers *et al.* 2001; Ness 2003a). Not only nectar volume, but also sugar and amino-acid concentrations increase, which attracts predatory arthropods such as ants (Lanza 1991; Agrawal and Rutter 1998). Such games of inducing nitrogen-rich foods are to be expected if it is true that predators require more nitrogen than herbivores and detritivores (Fagan *et al.* 2002). These induced responses, however, appear not to be herbivore-specific in castor bean and cotton, but they are rather generalized reactions to leaf tissue damage (Wäckers and Wunderlin 1999) and even to root damage (Wäckers and Bezemer 2003).

Artificial damage, as well as herbivore damage, up-regulates the octadecanoid signaling pathway in *Macaranga tanarius*, leading to increased concentrations of jasmonic acid and increased nectar production (Heil *et al.* 2001c; Linsenmair *et al.* 2001). Jasmonic acid probably mediates control, since its exogenous application also promotes nectar flow independent of plant damage.

Increased supply of food by plants can also be induced by carnivores. This is probably mediated simply by exploitation of the food. Artificial nectar removal and ant activity trigger increased nectar flow (Heil *et al.* 2000b). On *Piper cenocladum*, unicellular food bodies are 100 times more abundant on plants with the ant *Pheidole bicornis* than on ant-free plants (Risch and Rickson 1981). Moreover, congeneric ant species did not trigger food body production (Risch and Rickson 1981; Letourneau 1990). This may point at some degree of carnivore specificity, but the mechanism may be purely dependent on the behavior of ants that exploit the food bodies. As yet, there is no evidence for carnivore-specific induction mechanisms intrinsic to the plant.

Taken together, induced production of plant-provided foods does occur, but – as yet – there is at best weak evidence that the underlying mechanisms in the plant (recognition, signal transduction, and response) are specific with respect to herbivore or carnivore species involved. Specificity, however, may in principle arise from the behavior of the herbivores and carnivores, a possibility that has not been given sufficient attention. Whether specificity arises in the plant, in the carnivore, or both, is an important question. Its answer determines whether to interpret induction of plant-provided food as by-product or as intent. So far, the best evidence for intent comes from the induction of nectaries, nectar flow, and altered nectar composition following leaf wounding (Agrawal and Rutter 1998). Such induced responses are not compatible with the hypothesis on plant physiological by-products and they are compatible with the hypothesis based on trade-off relations between alternative allocation goals. However, compatibility does not imply proof, especially because there are several alternative explanations for the evolution of induced defenses and induced food rewards in particular (Agrawal and Karban 1999). Induced rewarding may minimize negative effects of carnivores on pollinators, minimize food theft, and increase heterogeneity in food supplied by plants, a feature that can be of critical importance in determining who is going to monopolize these rewards: herbivores, thieves, or predators (Van Rijn *et al.* 2002).

Plants compete for protection from herbivory

If food-for-protection strategies affect reproductive success and natural enemy abundance is a limiting factor, then one would expect plants to compete for protection from natural enemies of the herbivores. This may be expressed through raising the quantity and/or quality of the foods provided (e.g., Smith *et al.* 1990; Lanza 1991). In addition they may provide refuge or nesting space (domatia: e.g., Beattie 1985; Dicke and Sabelis 1988; Jolivet 1996) and betray the presence of herbivores to their enemies by releasing herbivory-induced volatile chemicals (e.g., Dicke and Sabelis 1988; Sabelis *et al.* 1999a,b,c,d; Agrawal 1999; Agrawal and Dubin-Thaler 1999). If this results in increased densities of the herbivore's enemies on a plant, herbivores will be subject to selection for avoiding these enemy-dense sites. Indeed, there is sound evidence that herbivores are deterred from enemy-occupied plants (e.g., Oliveira 1997; Oliveira *et al.* 1999; Pallini *et al.* 1999; Dyer *et al.* 2001; Itino and Itioka 2001; Nomikou *et al.* 2002; Werner and Peacor 2003). This in turn creates a selective disadvantage for plants that do not invest in attracting, arresting, and housing of the herbivore's enemies (unless they are standing close enough to plants that do invest).

If the arguments in favor of selection for indirect plant defenses seem so general, then why are the investments apparently so low? One major reason is that the fitness benefit of protection does not scale with the number of protector individuals. The net benefit of protection may either saturate at intermediate protector numbers, or else peak and then decline with increased protector numbers (Beattie 1985). Thus selection may optimize investment and come to halt at an intermediate level of investment where the benefit of protection is maximized. Another important reason for low investment may be that a plant has no direct control over the organisms that benefit from its investments. Once a plant has made refuge space, food, and/or herbivore-alarms available, it cannot control whether these facilities ("bed, breakfast and alarm clock") are used by their herbivores, their predators and parasitoids, or other organisms that may be fitness-neutral to the plant (e.g., Sabelis *et al.* 1999a,b,c,d, 2002). It also cannot determine which of the herbivore's enemies will utilize the facilities. As far as the evidence goes, however, dominance in competition for plant-provided facilities often seems to correlate with high efficiency in terms of protection against herbivory, as exemplified by experiments with plant-associated ants (e.g., Davidson *et al.* 1989; Lanza 1991; Gaume and McKey 1999; Stanton *et al.* 1999; Ness 2003b).

Another reason why the plant's investments in indirect defenses are low is that any benefit also directly accrues to neighboring plants, which are usually the plant's competitors for local resources and light (Sabelis *et al.* 1999a,b,c,d, 2002). This advantage to neighbors may arise not only because herbivores in the direct surroundings are killed or deterred, but also because there are enemies of herbivores nearby and they will go wherever they can find their potential victims, thereby also protecting the investor's neighbor against herbivory. As shown theoretically (Sabelis and De Jong 1988), there are broad conditions (patchy environment; the herbivore's enemies attracted by the investor protect all plants in the patch; non-investors saving energy to be spent on extra seed production) under which this leads to variability in the investments in indirect defenses and this variability can be maintained as a stable polymorphism in populations by frequency-dependent selection: low investors will gain if the frequency of higher (moderate) investors is sufficiently high, whereas the latter will gain if the frequency of low investors is sufficiently low. Thus, we predict low and variable investment levels for indirect plant defenses (Sabelis *et al.* 1999a,b,c,d, 2002). This prediction still awaits rigorous experimental testing.

Below, we discuss issues that influence how plant–predator mutualisms may or may not persist on an evolutionary timescale: predator-tolerant herbivores, measures to minimize negative effects on pollination, and cheaters, i.e., organisms that reap the benefits from plant-provided facilities, but give the plant nothing in return or even harm the plant's reproductive success.

Plants with enemies of herbivores provide enemy-free space for resistant/tolerant herbivores

Some herbivores have become resistant or tolerant to their plant-associated enemies, thereby finding a niche in which they are protected by their former enemies against competitors and current enemies. For example, *Polyhymno* caterpillars feed on ant-defended acacias in the tropical lowlands of Veracruz (Mexico) and can cause serious damage to this plant. They construct sealed shelters by silking together the pinnae of acacia leaves. These shelters serve as partial refuges from ants, thereby providing enemy-free space on ant-guarded plants (Eubanks *et al.* 1997). Another example is the leaf beetle *Pterocomma salicis* on willow trees; this herbivore is chemically defended against ants and gains protection from ants patrolling their shared host plant (Sipura 2002). There are many more ways in which herbivores can achieve resistance to ants and thereby protection against other enemies (e.g., Eubanks *et al.* 1997).

Homopterans are an example of enemy-tolerant herbivores. They offer ants honeydew as a reward and gain protection from their predators and parasitoids by ants that tend their colonies, but this protective effect decreases with homopteran population size per plant (Breton and Addicott 1992; Morales 2000; Larsen *et al.* 2001) and with increasing quality of plant-provided foods (Engel *et al.* 2001). It also depends on the homopteran species (Gaume *et al.* 1998). The ant–aphid symbiosis is full of conflicts (Bronstein 2001). It is not uncommon for ants to consume the aphids they are associated with. This occurs especially in periods of scarcity of protein relative to carbohydrates (Sakata 1994). Ants then adjust the carbohydrate–protein balance by consuming aphids. Thus, ants are thought to behave in a mutualistic or exploitative fashion towards aphids depending on their dietary requirements (Bronstein 2001). The crucial question to be answered is whether plants and aphids manipulate the quality of the foods they each provide so as to maximize their net gain; from the viewpoint of the plant this is the balance between homopteran feeding damage and plant protection by ants, whereas for the viewpoint of the homopterans this is the balance between protection and predation by ants (apart from quality and quantity of food in the plant). This game between plant and homopterans may result in a net benefit for both, as suggested by some experimental studies (Ito and Hagashi 1991; Breton and Addicott 1992; Rico-Gray and Castro 1996; Gaume *et al.* 1998; Morales 2000; Offenberg 2001). If, however, maintenance of aphids is an inefficient and potentially risky way of providing nectar to plant-protecting ants, the plant may gain by getting rid of the aphids in three possible ways: (1) they may develop direct resistance, (2) they may lure ants away from aphids by providing additional nectar sources (Becerra and Venable 1989; but see Fiala 1990; Del-Claro and Oliveira 1993), or (3) they may alter the nutrient composition in nectars such that ants become exploitative towards aphids. Which measure is taken by plants is largely an open question and one may wonder which countermeasures are available to the aphids (e.g., Sakata and Hashimoto 2000).

Plants suffer from conflicts between protection and pollination, but try to reduce them

Food-for-protection mutualism may entail a conflict between the host plants and the predators when the latter compete with pollinators for floral nectar or when they attack and/or deter pollinators. This may lead to reduced seed production through fewer and shorter floral visits by pollinating insects, as was shown for crab spiders in inflorescences of the invasive plant *Leucanthemum vulgare* in northern California (Suttle 2003).

This protection–pollination conflict of the plant can be resolved in two ways. One is to reduce ant–pollinator interactions by luring ants away from the flowers. This can be achieved through extrafloral nectaries that distract ants from foraging at floral nectaries. Indeed, extra nectar sources reduce flower visitation by *Forelius* ants and the inclusion of amino acids in the extra nectar reduces visits to primary nectaries even more, but flower visits of another ant species, *Formica perpilosa*, are unaffected (Wagner and Kay 2002). Thus, ant distraction by extrafloral nectaries may be a viable strategy for some, but not all ant species (see also Fiala 1990). The conditions for this to evolve seem rather special (Rosenzweig 2002): the cost of each extrafloral nectary divided by the cost of each flower must be less than the proportion of reproduction that is threatened by ant visits.

The other strategy of the plant is to produce flower deterrents. In acacia trees, odors from flowers or pollen appear to deter ants from young flowers at the crucial stage of dehiscence, allowing bees and other pollinators to visit and transfer pollen (Willmer and Stone 1997; Ghazoul 2001). Ants patrol the young buds prior to and after dehiscence, thus protecting the fertilized ovules and developing seeds. Because the result is an improved seed set, this may explain the production of a deterrent during dehiscence, but the ultimate reasons for ant deterrence are still not understood. The flower only sends a signal and - unless the ants are manipulated - they can "decide" whether to be attracted or deterred. Thus, the crucial question is whether the ants and the plant share an interest in the reproductive success of the plant. Since many seeds will germinate nearby the mother plant, this might increase the availability of suitable nesting sites for the ant.

Given that ant-repellents in flowers - as well as in floral nectar (Wäckers, Chapter 2) - require special conditions to be selectively advantageous for both the plant and the ants, one would expect the evolution of physical structures that prevent ants from getting access to flowers. Such ant barriers exist and they may be manifested near or in flowers as sticky belts (e.g., in *Viscaria vulgaris*: see Faegri and Van der Pijl 1979), glandular hairs, or waxy corollas (Feinsinger and Swarm 1978; Beattie 1985). Possibly, such barriers to enter the flowers did not only evolve to prevent ants from reducing visits of pollinators (attack, deterrence, competition for nectar: Faegri and Van der Pijl 1979), because ants may also castrate flowers and decrease pollen viability (Beattie 1985; Wagner 2000). There is no evidence, yet, for ants overcoming the barriers to enter flowers.

Food-for-protection mutualisms are vulnerable to cheating and thievery

Food-for-protection mutualisms are vulnerable to exploitation by organisms that consume the food provided, yet give no equivalent value or even nothing in return. One possibility is that predators of the predators that protect the plant consume the plant-provided food. In Costa Rica, a clerid beetle (*Phyllobaenus* sp.), known as a predator of ants, was sometimes found to thrive on *Piper* plants that contained no ant colonies or that had their *Pheidole* ant colonies destroyed by this predator. It appeared that this beetle was capable of stimulating the plant to produce food bodies and to consume them (Letourneau 1990).

Thus, the beetle did not only harm the plant by consuming the plant's bodyguards, but also by consuming plant-provided food bodies that would otherwise be eaten by the bodyguards.

Another possibility for cheating is that herbivores consume the plant-provided foods. Myrmecophilous *Thisbe* butterflies have caterpillars that do not only feed on leaf tissue, but also on extrafloral nectar of *Croton billbergianus* trees (DeVries and Baker 1989). Also, pollen is fed upon by herbivorous arthropods, such as thrips, lycaenid butterflies, and chrysomelid and curculionid beetles (Baggen *et al.* 1999; Romeis *et al.*, Chapter 7). Some of these herbivore species may pose a serious threat to the plant. A striking example, resulting from our own research (van Rijn *et al.* 2002), is given by the western flower thrips, whose mobile stages consume pollen in addition to the content of leaf parenchyme cells. Addition of pollen to a diet of leaf tissue promotes their growth and reproduction. However, phytoseiid mites that prey on young thrips larvae congregate on leaves around the sites with pollen, yet they also visit pollen-free leaves (Fig. 4.2), whereas thrips larvae seem to distribute themselves randomly over leaves with and without pollen. When visiting leaves with pollen, thrips larvae will be attacked by the mites, who in this way monopolize the pollen as a food source.

If, however, the predator itself causes harm to the plant, there may not be a fourth trophic level to rescue the plant. For example, flowers of the alpine skypilot (*Polemonium viscosum*) are attacked by floral-nectar-thieving ants (*Formica neorufibarbus*). Ants visiting these flowers interfere with seed production by dislodging the style from the ovary. This is thought to impose selection on floral form, resulting in less preferred short, narrow flowers below the timberline, as opposed to long, broadly flared flowers in the ant-poor alpine tundra (Galen 1999). This example illustrates that plants providing nectar – even if offered solely to attract pollinators – solicit unwanted visitors that can have negative side-effects on the plants. Similar risks may therefore arise when plants offer nectar extrafloraly. There is evidence that extrafloral nectar is robbed by parasitoids and predators, that in turn do not attack the herbivores on that plant (Pemberton and Lee 1996), or even protect members of the fourth trophic level (Novak 1994).

Decreased seed production or even complete plant castration has been observed in several obligate symbioses of myrmecophytes that typically provide hollow nest cavities and nutrition to the occupying ant colony and receive protection against herbivory and against overgrowth by surrounding vegetation in return. Castration is caused by ant species that prune shoots, thereby stimulating new growth, resulting in a higher number and activity of extrafloral nectaries (Yu and Pierce 1998). Pruning also leads to less contact with neighboring plants occupied with competitively superior ant species that appear not to exploit the ant–plant mutualism to their own benefit (Stanton *et al.* 1999).

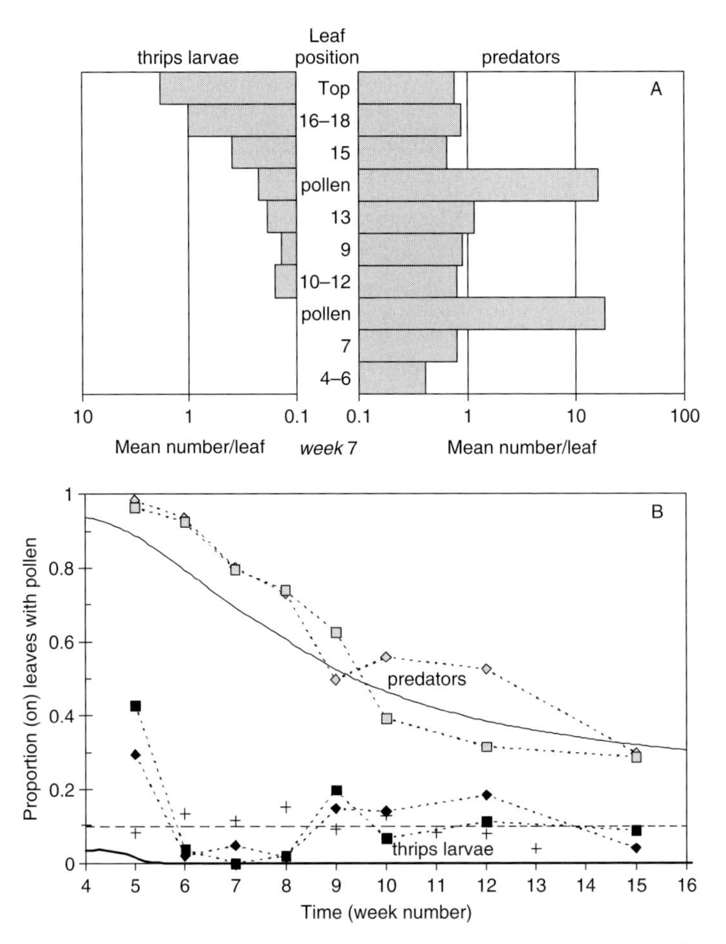

Figure 4.2 Fraction of herbivore and predator population on leaves with pollen in experiments carried out by Van Rijn *et al.* (2002). In panel (A) a snapshot (at week 7) is shown of the vertical distribution of predatory mites and western flower thrips larvae on leaves of a cucumber plant that had pollen on leaves 8 and 14. In panel (B), temporal changes in proportion of thrips and predator populations present on leaves with pollen. Symbols (squares and diamonds) connected by dotted lines are the experimental results from two replicate experiments. Solid lines represent the results of computer simulations, based on the assumption that the prey can balance food gains (leaves, pollen) against risks of being eaten on leaves with pollen versus leaves without pollen and that the predator can assess relative gains (prey vs. pollen plus prey) in these two areas of the plant (black symbols and heavy line for herbivores; gray symbols and thin line for the predators). Crosses indicate the actual proportion of leaves that were supplied with pollen (i.e., 10% of total leaf surface). Note that the predatory mites tend to concentrate on leaves provided with pollen and that the observed distributions over leaves with and without pollen are in agreement with the computer simulations for the predatory mites, but not for the thrips. The thrips distribution seems to fit better to that expected under random movement (i.e., 10% of the thrips population on leaves with pollen).

Variation in the degree of plant protection by different species of ants has also been observed (Gaume and McKey 1999). *Leonardoxa africana* is an understory tree of the coastal rainforests of Cameroon which provides extrafloral nectar and nest sites (swollen, hollow internodes) for one of two plant-specific species of ant. The formicine ant *Petalomyrmex phylax* inhabits three out of each four occupied trees, whereas the myrmicine *Cataulacus mckeyi* is found in the remaining trees. Young leaves patrolled by *Petalomyrmex* suffered significantly less damage than those from which ants were excluded (2% versus 24%). In contrast, young leaves patrolled by *Cataulacus* suffered much greater herbivory (31%), not significantly different from that on ant-excluded leaves (46%). Since the latter ant species excludes the former from trees, this should render the food-for-protection mutualism less efficient in the long run, unless the currently inferior ant becomes superior due to frequency-dependent selection.

There is some evidence for plants acting against ants that cheat. They may reduce space for ant nesting sites (domatia) (Izzo and Vasconcelos 2002) or bar ants from visiting flowers where they would reduce pollination or even castrate the plant (see previous section).

Discussion

Four conclusions emerge from this review. First, the original function of some plant-provided foods may well have been in the regulation of plant physiological processes (exudates, nectars) or reproduction (pollen), but this is not an argument to reject a role of evolution in modulating the availability of these foods to organisms at other trophic levels. Induction of higher-quality nectars and the existence of food bodies are strong arguments against a purely physiological function. Second, food-for-protection mutualisms are widespread and established at a low (but not negligible) direct cost to the plant, making trade-off relations with other vital processes hard to detect. Third, there are theoretical – as yet untested – arguments why many plants invest in foods as a mode of indirect defense and others do not, and why these investments are kept at a low level. However, we do not yet fully understand how plant–predator mutualisms evolve and persist in the face of organisms that exploit the mutualism (tolerant herbivores, robbers, cheaters). Fourth, such exploitative organisms are usually found, when sought for (Bronstein 1994a, b, 1998, 2001).

Various theoretical models show that mutualistic interactions in homogeneously mixed populations of the interacting organisms cannot persist indefinitely in the face of exploitation (e.g., Doebeli and Knowlton 1998). The empirical evidence for the existence of mutualisms therefore poses a challenge to ecological theory. Two possible solutions have been proposed. One rests on the assumption of spatial population structure and stochasticity of colonization processes. If there is a chance for mutualists to escape in space and gain full profit of the mutualism, albeit temporarily, then mutualists may win (Pels *et al.* 2002), or coexist with exploiters under certain restrictive conditions with respect to dispersal ability (Yu *et al.* 2001; Wilson *et al.* 2003). The other possible solution rests on the assumption of asymmetries in the competition among mutualists and exploiters (Bronstein 2001). If mutualists can somehow act against exploiters or gain more benefit from rewards than exploiters, such *punishing* or *rewarding asymmetries* in competition for rewards may tilt the cost–benefit balance in favor of the mutualists, or create ecological coexistence, especially when competition among mutualists and among exploiters is weak (Morris *et al.* 2003). The evolutionary dynamics of this asymmetric competition process is analyzed by Ferrière *et al.* (2002). They show that exploiters, in effect, establish a background against which better mutualists can display any competitive superiority. This can lead to the divergence and coexistence of mutualist and cheater phenotypes. Suicide of the mutualists due to exploiters may still occur at an evolutionary timescale, but now in two forms. If asymmetries in competition for commodities (i.e., food or protection in the context of this chapter) are too weak, selection will favor lower and lower provision of commodities and the mutualists will ultimately go extinct. If the asymmetries are too strong, selection will favor more and more provision of commodities until the costs incurred are so large that the association may become non-viable and the interacting populations go extinct (Ferrière *et al.* 2002). Neither of the evolutionary suicide scenarios is very likely to occur in food-for-protection mutualisms, because asymmetries in competition for food or protection are the rule rather than the exception and because higher investment in food provisioning obeys the law of diminishing returns and may cause the net benefit curve to become humped. Runaway selection for increased investments may therefore come to halt.

These theoretical results should trigger empirical research into the nature of asymmetries in competition among mutualists and exploiters. A particularly clear case of such asymmetry is observed in the case where pollen is consumed by otherwise folivorous thrips and otherwise predaceous mites (van Rijn *et al.* 2002) (Figs. 4.2 and 4.3). When predator and herbivore congregate at leaf sites with pollen, it is the predator that will win the competition for pollen, thereby monopolizing this resource. Thrips can either avoid visiting predator-occupied pollen sites or they will congregate and be eaten by the predators at those sites. The plant can thus gain control over the distribution of herbivores and predators by providing pollen locally, i.e., at specific sites on the plant. This condition is either always fulfilled since pollen occurs only in flowers (or rains down only

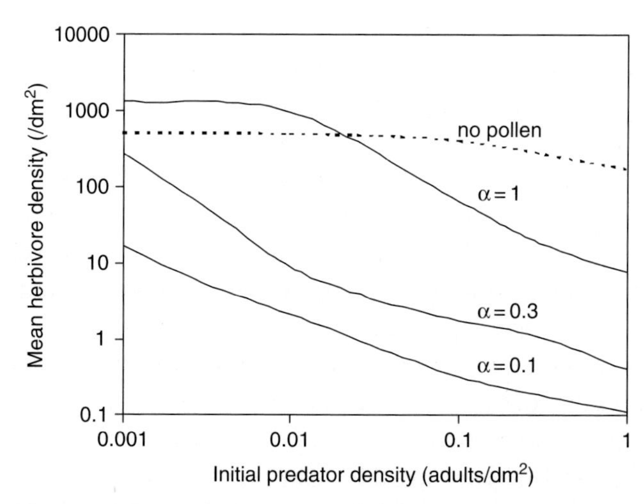

Figure 4.3 Computer simulations showing how mean herbivore (thrips) density during the first 100 days after predator release is affected by absence of pollen (dotted line) and presence of pollen (solid lines), when the same amount of pollen is spread over different proportions of the plant surface ($\alpha = 1, 0.3, 0.1$) (Van Rijn *et al.* 2002). The more concentrated the pollen, the stronger is the impact of the predator on the herbivore's density.

to leaves directly below), or it is achieved by the plant via asynchronous within-plant flowering. Progress in understanding the mutualistic nature of plant–predator interactions will thus depend on unraveling how plants "manipulate" asymmetries in the competition between mutualists and exploiters for plant-provided rewards (e.g., pollen). Similar asymmetries may exist within species or even local populations of individuals that act as mutualists or exploiters, depending on genotype or context. Identifying this level of variability will probably be the most important task for the future.

Ever since Thomas Belt (1874) expressed his ideas on food-for-protection mutualisms, theory and/or experiments have provided a great deal of support, challenge and skepticism, but in retrospect much insight has been gained into the dynamic nature of mutualistic interactions. In this sense, this branch of research stands as a model for analyzing other – recently discovered – indirect plant defenses, such as herbivore-induced plant alarms (Dicke and Sabelis 1988), and other mutualisms in which plants donate food and receive a reward other than protection. To illustrate the latter by a very recent discovery (Hamilton and Frank 2001), consider the impact of grazing or clipping on a grass species common in pastures, *Poa pratensis*. This appears to trigger considerable root exudation of carbon, as measured from a ^{13}C pulse-chase experiment. Such a response to biomass loss seems highly counter-intuitive: why do these grasses not retain the nutrients needed for regrowth? The answer may well be hidden in indirect responses in the soil biosystem: the carbon exuded is quickly assimilated into a prospering microbial population in the rhizosphere and this in turn feeds back on soil inorganic pools, plant nitrogen uptake, leaf nitrogen content, and photosynthesis of the grasses. This shows that plants are not passive participants in decomposition processes, they can stimulate microbial growth, thereby facilitating uptake of nutrients (e.g., nitrogen) that would otherwise limit their own growth (Hamilton and Frank 2001). However appealing this interpretation may be, there are numerous questions to be answered before it can be accepted: Why do grasses exude carbon only after herbivory? Why do nitrogen fixers monopolize the carbon exuded by the plant? Who are the cheaters in this game and to what extent do they jeopardize the persistence of the plant–microbe mutualism? Clearly, many of these questions bear strong similarity to those discussed in this chapter, and their answers should come from understanding the web of – aboveground and belowground – species interactions.

Acknowledgments

We thank Judith Bronstein, Felix Wäckers, and Jan Bruin for many useful and thought-provoking comments.

References

Adler, L. S. 2000. The ecological significance of toxic nectar. *Oikos* **91**: 409–420.

Agrawal, A. A. 1999. Leaf damage and associated cues induced aggressive ant recruitment in a Neotropical ant-plant. *Ecology* **80**: 1713–1723.

Agrawal, A. A. and B. J. Dubin-Thaler. 1999. Induced responses to herbivory in the Neotropical ant–plant association between *Azteca* ants and *Cecropia* trees: response of ants to potential inducing cues. *Behavioral Ecology and Sociobiology* **45**: 47–54.

Agrawal, A. A. and R. Karban. 1999. Why induced defenses may be favored over constitutive strategies in plants. In R. Tollrian and C. D. Harvell (eds.) *The Ecology and Evolution of Inducible Defenses*. Princeton, NJ: Princeton University Press, pp. 45–61.

Agrawal, A. A. and M. T. Rutter. 1998. Dynamic anti-herbivore defense in ant-plants: the role of induced responses. *Oikos* **83**: 227–236.

Baggen, L. R., G. M. Gurr, and A. Meats. 1999. Flowers in tri-trophic systems: mechanisms allowing selective exploitation by insect natural enemies for conservation biological control. *Entomologia Experimentalis et Applicata* **91**: 155–161.

Bakker, F. M. and M. E. Klein. 1992. Transtrophic interactions in cassava. *Experimental and Applied Acarology* **14**: 299–311.

Beattie, A. J. 1985. *The Evolutionary Ecology of Ant–Plant Interactions*. Cambridge, UK: Cambridge University Press.

Becerra, J. X. I. and D. L. Venable. 1989. Extrafloral nectaries: a defense against ant–Homoptera mutualisms. *Oikos* **55**: 276–280.

Beckman, N. and L. E. Hurd. 2003. Pollen feeding and fitness in praying mantids: the vegetarian side of a tritrophic predator. *Environmental Entomology* **32**: 881–885.

Belt, T. 1874. *The Naturalist in Nicaragua*. London: J. M. Dent & Sons.

Breton, L. M. and J. F. Addicott. 1992. Density-dependent mutualism in an aphid–ant interaction. *Ecology* **73**: 2175–2180.

Bronstein, J. L. 1994a. Our current understanding of mutualism. *Quarterly Review of Biology* **69**: 31–51.

1994b. Conditional outcomes in mutualistic interactions. *Trends in Ecology and Evolution* **9**: 214–217.

1998. The contribution of ant–plant protection studies to our understanding of mutualism. *Biotropica* **30**: 150–161.

2001. The exploitation of mutualisms. *Ecology Letters* **4**: 277–287.

Davidson, D. W., R. R. Snelling, and J. T. Longino. 1989. Competition among ants for myrmecophytes and the significance of plant trichomes. *Biotropica* **21**: 64–73.

De la Fuente, M. A. S. and R. J. Marquis. 1999. The role of ant-tended extrafloral nectaries in the protection and benefit of a Neotropical rainforest tree. *Oecologia* **118**: 192–202.

DeVries, P. J. and I. Baker. 1989. Butterfly exploitation of an ant–plant mutualism: adding insult to herbivory. *Journal of the New York Entomological Society* **97**: 332–340.

Del-Claro, K. and P. S. Oliveira. 1993. Ant–Homoptera interaction: do alternative sugar sources distract tending ants? *Oikos* **68**: 202–206.

Di Giusto, B., M. C. Anstett, E. Dounias, and D. B. McKey. 2001. Variation in the effectiveness of biotic defence: the case of an opportunistic ant–plant protection mutualism. *Oecologia* **129**: 367–375.

Dicke, M. and M. W. Sabelis. 1988. How plants obtain predatory mites as bodyguards. *Netherlands Journal of Zoology* **38**: 148–165.

1992. Costs and benefits of chemical information conveyance: proximate and ultimate factors. In B. Roitberg and M. Isman (eds.) *Insect Chemical Ecology: An Evolutionary Approach*. London: Chapman and Hall, pp. 122–155.

Doebeli, M. and N. Knowlton. 1998. The evolution of interspecific mutualisms. *Proceedings of the National Academy of Sciences of the USA* **95**: 8676–8680.

Dyer, L. A., C. D. Dodson, J. Beihoffer, and D. K. Letourneau. 2001. Trade-offs in antiherbivore defenses in *Piper cenocladum*: ant mutualists versus plant secondary metabolites. *Journal of Chemical Ecology* **27**: 581–592.

Eck, G., B. Fiala, K. E. Linsenmair, R. Bin Hashim, and P. Proksch. 2001. Trade-off between chemical and biotic antiherbivore defense in the South East Asian plant genus *Macaranga*. *Journal of Chemical Ecology* **27**: 1979–1996.

Engel, V., M. K. Fischer, F. L. Wäckers, and W. Völkl. 2001. Interactions between extrafloral nectaries, aphids and ants: are there competition effects between plant and homopteran sugar sources? *Oecologia* **129**: 577–584.

Eubanks, M. D., K. A. Nesci, M. K. Petersen, Z. Liu, and H. B. Sanchez. 1997. The exploitation of an ant-defended host plant by a shelter-building herbivore. *Oecologia* **109**: 454–460.

Evans, E. W. and J. G. Swallow. 1993. Numerical responses of natural enemies to artificial honeydew in Utah alfalfa. *Environmental Entomology* **22**: 1392–1401.

Faegri, K. and L. van der Pijl. 1979. *The Principles of Pollination Ecology*. Toronto, Ontario: Pergamon Press.

Fagan, W. F., E. Siemann, C. Mitter, *et al.* 2002. Nitrogen in insects: implications for trophic complexity and species diversification. *American Naturalist* **160**: 784–802.

Feinsinger, P. and L. A. Swarm. 1978. How common are ant-repellent nectars? *Biotropica* **10**: 238–239.

Ferrière, R., J. L. Bronstein, S. Rinaldi, R. Law, and M. Gauduchon. 2002. Cheating and the evolutionary stability of mutualisms. *Proceedings of the Royal Society of London Series B* **269**: 773–780.

Fiala, B. 1990. Extrafloral nectaries vs. ant–Homoptera mutualisms: a comment on Becerra and Venable. *Oikos* **59**: 281–282.

Fiala, B., H. Grunsky, U. Maschwitz, and K. E. Linsenmair. 1994. Diversity of ant–plant interactions: protective efficacy in *Macaranga* species with different degrees of ant association. *Oecologia* **97**: 186–192.

Fischer, R. C., A. Richter, W. Wanek, and V. Mayer. 2002. Plants feed ants: food bodies of myrmecophytic *Piper* and their significance for the interaction with *Pheidole bicornis* ants. *Oecologia* **133**: 186–192.

Galen, C. 1999. Flowers and enemies: predation by nectar-thieving ants in relation to variation in floral form of an alpine wildflower, *Polemonium viscosum*. *Oikos* **85**: 426–434.

Gaume, L. and D. McKey. 1999. An ant–plant mutualism and its host-specific parasite: activity rhythms, young leaf patrolling, and effects on herbivores of two specialist plant-ants inhabiting the same myrmecophyte. *Oikos* **84**: 130–144.

Gaume, L., D. McKey, and S. Terrin. 1998. Ant–plant–homopteran mutualism: how the third partner affects the interaction between a plant-specialist ant and its myrmecophyte host. *Proceedings of the Royal Society of London Series B* **265**: 569–575.

Ghazoul, J. 2001. Can floral repellents pre-empt potential ant-plant conflicts? *Ecology Letters* **4**: 295–299.

Hagen, K. S. 1986. Ecosystem analysis: plant cultivars (HPR), entomophagous species and food supplements. In D. J. Boethel and R. D. Eikenbarry (eds.) *Interactions of Plant Resistance and Parasitoids and Predators of Insects*. Chichester, UK: Ellis Horwood, pp. 151–197.

Hamilton, E. W. and D. A. Frank. 2001. Can plants stimulate soil microbes and their own nutrient supply? Evidence from a grazing tolerant grass. *Ecology* **82**: 2397–2402.

Hatada, A., T. Itioka, R. Yamaoka, and T. Itino. 2002. Carbon and nitrogen contents of food bodies in three myrmecophytic species of *Macaranga*: implications for antiherbivore defense mechanisms. *Journal of Plant Research* **115**: 179–184.

Heil, M., T. Delsinne, A. Hilpert, *et al.* 2002. Reduced chemical defence in ant-plants? A critical re-evaluation of a widely accepted hypothesis. *Oikos* **99**: 457–468.

Heil, M., B. Fiala, B. Baumann, and K. E. Linsenmair. 2000b. Temporal, spatial and biotic variations in extrafloral nectar secretion by *Macaranga tanarius*. *Functional Ecology* **14**: 749–757.

Heil, M., B. Fiala, T. Boller, and K. E. Linsenmair. 1999. Reduced chitinase activities in ant-plants of the genus *Macaranga*. *Naturwissenschaften* **86**: 146–149.

Heil, M., B. Fiala, W. Kaiser, and K. E. Linsenmair. 1998. Chemical contents of *Macaranga* food bodies: adaptations to their role in ant attraction and nutrition. *Functional Ecology* **12**: 117–122.

Heil, M., B. Fiala, U. Maschwitz, and K. E. Linsenmair. 2001b. On benefits of indirect defence: short- and long-term studies of antiherbivore protection via mutualistic ants. *Oecologia* **126**: 395–403.

Heil, M., B. Fiala, G. Zotz, P. Menke, and U. Maschwitz. 1997. Food body production in *Macaranga triloba* (Euphorbiaceae): a plant investment in antiherbivore defence via symbiotic ant partners. *Journal of Ecology* **85**: 847–861.

Heil, M., A. Hilpert, B. Fiala, *et al.* 2002. Nutrient allocation of *Macaranga triloba* ant plants to growth, photosynthesis and indirect defence. *Functional Ecology* **16**: 475–483.

Heil, M., A. Hilpert, B. Fiala, and K. E. Linsenmair. 2001a. Nutrient availability and indirect (biotic) defence in a Malaysian ant-plant. *Oecologia* **126**: 404–408.

Heil, M., T. Koch, A. Hilpert, *et al.* 2001c. Extrafloral nectar production of the ant-associated plant, *Macaranga tanarius*, is an induced, indirect, defensive response elicited by jasmonic acid. *Proceedings of the National Academy of Sciences of the USA* **98:** 1083–1088.

Heil, M., C. Staehelin, and D. McKey. 2000a. Low chitinase activity in *Acacia* myrmecophytes: a potential trade-off between biotic and chemical defenses? *Naturwissenschaften* **87**: 555–558.

Itino, T. and T. Itioka. 2001. Interspecific variation and ontogenetic change in antiherbivore defense in myrmecophytic *Macaranga* species. *Ecological Research* **16**: 765–774.

Itino, T., T. Itioka, A. Hatada, and A. A. Hamid. 2001. Effects of food rewards offered by ant-plant *Macaranga* on the colony size of ants. *Ecological Research* **16**: 775–786.

Ito, F. and S. Higashi. 1991. An indirect mutualism between oaks and wood ants via aphids. *Journal of Animal Ecology* **60**: 463–470.

Izzo, T. J. and H. L. Vasconcelos. 2002. Cheating the cheater: domatia loss minimizes the effects of ant castration in an Amazonian ant-plant. *Oecologia* **133**: 200–205.

Jansen, V. A. A. and M. W. Sabelis. 1995. Outbreaks of colony-forming pests in tri-trophic systems: consequences for pest control and the evolution of pesticide resistance. *Oikos* **74**: 172–176.

Janzen, D. H. 1966. Coevolution of mutualism between ants and acacias in Central America. *Evolution* **20**: 249–275.

Jolivet, P. 1996. *Ants and Plants: An Example of Coevolution*. Leiden, the Netherlands: Backhuys.

Koptur, S. 1992. Extrafloral nectary-mediated interactions between insects and plants. In E. Bernays (ed.) *Insect–Plant Interactions*, vol. 4. Boca Raton, FL: CRC Press, pp. 81–129.

Labeyrie, E., L. Pascal, J. Delabie, *et al.* 2001. Protection of *Passiflora glandulosa* (Passifloraceae) against herbivory: impact of ants exploiting extrafloral nectaries. *Sociobiology* **38**: 317–321.

Lanza, J. 1991. Response of fire ants (Formicidae, *Solenopsis invicta* and *S. geminata*) to artificial nectars with amino-acids. *Ecological Entomology* **16**: 203–210.

Larsen, K. J., L. M. Staehle, and E. J. Dotseth. 2001. Tending ants (Hymenoptera: Formicidae) regulate *Dalbulus quinquenotatus* (Homoptera: Cicadellidae) population dynamics. *Environmental Entomology* **30**: 757–762.

Letourneau, D. K. 1990. Code ant–plant mutualism broken by parasite. *Science* **248**: 215–217.

Linsenmair, K. E., M. Heil, W. M. Kaiser, *et al.* 2001. Adaptations to biotic and abiotic stress: *Macaranga*-ant plants optimize investment in biotic defence. *Journal of Experimental Botany* **52**: 2057–2065.

Mondor, E. B. and J. F. Addicott. 2003. Conspicuous extra-floral nectaries are inducible in *Vicia faba*. *Ecology Letters* **6**: 495–497.

Morales, M. A. 2000. Mechanisms and density dependence of benefit in an ant–membracid mutualism. *Ecology* **81**: 482–489.

Morris, W. F., J. L. Bronstein, and W. G. Wilson. 2003. Three-way coexistence in obligate mutualist-exploiter interactions: the potential role of competition. *American Naturalist* **161**: 860–875.

Ness, J. H. 2003a. *Catalpa bignonioides* alters extrafloral nectar production after herbivory and attracts ant bodyguards. *Oecologia* **134**: 210–218.

2003b. Contrasting exotic *Solenopsis invicta* and native *Forelius pruinosus* ants as mutualists with *Catalpa bignonioides*, a native plant. *Ecological Entomology* **28**: 247–251.

Nomikou, M., A. Janssen, R. Schraag, and M. W. Sabelis. 2002. Phytoseiid predators suppress populations of *Bemisia tabaci* on cucumber plants with alternative food. *Experimental and Applied Acarology* **27**: 57–68.

Novak, H. 1994. The influence of ant attendance on larval parasitism in hawthorn psyllids (Homoptera, Psyllidae). *Oecologia* **99**: 72–78.

Offenberg, J. 2001. Balancing between mutualism and exploitation: the symbiotic interaction between *Lasius* ants and aphids. *Behavioral Ecology and Sociobiology* **49**: 304–310.

Oliveira, P. S. 1997. The ecological function of extrafloral nectaries: herbivore deterrence by visiting ants and reproductive output in *Caryocar brasiliense* (Caryocaraceae). *Functional Ecology* **11**: 323–330.

Oliveira, P. S., V. Rico-Gray, C. Diaz-Castelazo, and C. Castillo-Guevara. 1999. Interaction between ants, extrafloral nectaries and insect herbivores in Neotropical coastal sand dunes: herbivore deterrence by visiting ants increases fruit set in *Opuntia stricta* (Cactaceae). *Functional Ecology* **13**: 623–631.

Pallini, A., A. Janssen, and M. W. Sabelis. 1999. Spider mites avoid plants with predators. *Experimental and Applied Acarology* **23**: 803–815.

Pels, B., A. M. de Roos, and M. W. Sabelis. 2002. Evolutionary dynamics of prey exploitation in a metapopulation of predators. *American Naturalist* **159**: 172–189.

Pemberton, R. W. and J. H. Lee. 1996. The influence of extrafloral nectaries on parasitism of an insect herbivore. *American Journal of Botany* **83**: 1187–1194.

Price, P. W., C. E. Bouton, P. Gross, *et al.* 1980. Interactions among three trophic levels: influence of plants on interactions between insect herbivores and natural enemies. *Annual Review of Ecology and Systematics* **11**: 41–65.

Rico-Gray, V. and G. Castro. 1996. Effect of an ant–aphid interaction on the reproductive fitness of *Paullinia fuscecens* (Sapindaceae). *Southwestern Naturalist* **41**: 434–440.

Risch, S. J. and F. R. Rickson 1981. Mutualism in which ants must be present before plants produce food bodies. *Nature* **291**: 149–150.

Rogers, C. E. 1985. Extrafloral nectar: entomological implications. *Bulletin of the Entomological Society of America* **31**: 15–20.

Rosenzweig, M. L. 2002. The distraction hypothesis depends on relatively cheap extrafloral nectaries. *Evolutionary Ecology Research* **4**: 307–311.

Roulston, T. H., J. H. Cane, and S. L. Buckmann. 2000. What governs protein content of pollen: pollinator preferences, pollen-pistol interactions, or phylogeny? *Ecological Monographs* **70**: 617–643.

Rudgers, J. A., J. G. Hodgen, and J. W. White. 2003. Behavioral mechanisms underlie an ant–plant mutualism. *Oecologia* **135**: 51–59.

Ruhren, S. and S. N. Handel. 1999. Jumping spiders (Salticidae) enhance the seed production of a plant with extrafloral nectaries. *Oecologia* **119**: 227–230.

Sabelis, M. W. and M. C. M. de Jong. 1988. Should all plants recruit bodyguards? Conditions for a polymorphic ESS of synomone production in plants. *Oikos* **53**: 247–252.

Sabelis, M. W. and P. C. J. van Rijn. 1997. Predation by insects and mites. In T. Lewis (ed.) *Thrips as Crop Pests*. Wallingford, UK: CAB-International, pp. 259–354.

Sabelis, M. W., A. Janssen, J. Bruin, *et al.* 1999c. Interactions between arthropod predators and plants: a conspiracy against herbivorous arthropods? In J. Bruin, L. P. S. van der Geest, and M. W. Sabelis (eds.) *Ecology and Evolution of the Acari*. Dordrecht, the Netherlands: Kluwer Academic Publishers, pp. 207–230.

Sabelis, M. W., A. Janssen, A. Pallini, *et al.* 1999b. Behavioural responses of predatory and herbivorous arthropods to induced plant volatiles: from evolutionary ecology to agricultural applications. In A. Agrawal, S. Tuzun, and E. Bent (eds.) *Induced Plant Defenses against Pathogens and Herbivores*. St. Paul, MN: American Phytopathological Society, pp. 269–298.

Sabelis, M. W., M. van Baalen, F. M. Bakker, *et al.* 1999a. Evolution of direct and indirect plant defence against herbivorous arthropods. In H. Olff, V. K. Brown, and R. H. Drent (eds.) *Herbivores: Between Plants and Predators*. Oxford, UK: Blackwell Science, pp. 109–166.

Sabelis, M. W., M. van Baalen, J. Bruin, *et al.* 1999d. The evolution of overexploitation and mutualism in plant–herbivore–predator interactions and its impact on population dynamics. In B. A. Hawkins and H. V. Cornell (eds.) *Theoretical Approaches to Biological Control*. Cambridge, UK: Cambridge University Press, pp. 259–282.

Sabelis, M. W., M. van Baalen, B. Pels, M. Egas, and A. Janssen. 2002. Evolution of exploitation and defence in plant–herbivore–predator interactions. In U. Dieckmann, J. A. J. Metz, M. W. Sabelis, and K. Sigmund (eds.) *The Adaptive Dynamics of Infectious Diseases: In Pursuit of Virulence Management*. Cambridge, UK: Cambridge University Press, pp. 297–321.

Sakata, H. 1994. How an ant decides to prey on or to attend aphids. *Researches in Population Ecology* **36**: 45–51.

Sakata, H. and Y. Hashimoto. 2000. Should aphids attract or repel ants? Effect of rival aphids and extrafloral nectaries on ant–aphid interactions. *Population Ecology* **42**: 171–178.

Schmitz, O. J., P. A. Hamback, and A. P. Beckerman. 2000. Trophic cascades in terrestrial systems: a review of the effects of carnivore removals on plants. *American Naturalist* **155**: 141–153.

Sipura, M. 2002. Contrasting effects of ants on the herbivory and growth of two willow species. *Ecology* **83**: 2680–2690.

Smith L. L., J. Lanza, and G. C. Smith. 1990. Amino-acid concentrations in extrafloral nectar of *Impatiens sultani* increase after simulated herbivory. *Ecology* **71**: 107–115.

Stanton, M. L., T. M. Palmer, T. P. Young, A. Evans, and M. L. Turner. 1999. Sterilization and canopy modification of a swollen thorn acacia tree by a plant-ant. *Nature* **401**: 578–581.

Suttle, K. B. 2003. Pollinators as mediators of top-down effects on plants. *Ecology Letters* **6**: 688–694.

Tilman, D. 1978. Cherries, ants and tent caterpillars: timing of nectar production in relation to susceptibility of caterpillars to ant predation. *Ecology* **59**: 686–692.

Van Rijn, P. C. J. and L. K. Tanigoshi. 1999a. Pollen as food for the predatory mites *Iphiseius degenerans* and *Neoseiulus cucumens* (Acari: Phytoseiidae): dietary range and life history. *Experimental and Applied Acarology* **23**: 785–802.

1999b. The contribution of extrafloral nectar to survival and reproduction of the predatory mite *Iphiseius degenerans* on *Ricinus communis*. *Experimental and Applied Acarology* **23**: 281–296.

Van Rijn, P. C. J., Y. M. van Houten, and M. W. Sabelis. 2002. How plants benefit from providing food to predators when it is also edible to herbivores. *Ecology* **83**: 2664–2679.

Wäckers, F. L. and T. M. Bezemer. 2003. Root herbivory induces an above-ground indirect defence. *Ecology Letters* **6**: 1–4.

Wäckers, F. L. and R. Wunderlin. 1999. Induction of cotton extrafloral nectar production in response to herbivory does not require a herbivore-specific elicitor. *Entomologia Experimentalis et Applicata* **91**: 149–154.

Wäckers, F. L., D. Zuber, R. Wunderlin, and F. Keller. 2001. The effect of herbivory on temporal and spatial dynamics of foliar nectar production in cotton and castor. *Annals of Botany* **87**: 365–370.

Wagner, D. 1997. The influence of ant nests on *Acacia* seed production, herbivory and soil nutrients. *Journal of Ecology* **85**: 83–93.

2000. Pollen viability reduction as a potential cost of ant association for *Acacia constricta* (Fabaceae). *American Journal of Botany* **87**: 711–715.

Wagner, D. and A. Kay. 2002. Do extrafloral nectaries distract ants from visiting flowers? An experimental test of an overlooked hypothesis. *Evolutionary Ecology Research* **4**: 293–305.

Werner, E. E. and S. D. Peacor. 2003. A review of trait-mediated indirect interactions in ecological communities. *Ecology* **84**: 1083–1100.

Willmer, P. G. and G. N. Stone. 1997. How aggressive ant-guards assist seed set in *Acacia* flowers. *Nature* **388**: 165–167.

Wilson, W. G., Morris W. F., and J. L. Bronstein. 2003. Coexistence of mutualists and exploiters on spatial landscapes. *Ecological Monographs* **73**: 397–413.

Yu, D. W. and N. E. Pierce. 1998. A castration parasite of an ant–plant mutualism. *Proceedings of the Royal Society of London Series B* **265:** 375–382.

Yu, D. W., H. B. Wilson, and N. E. Pierce. 2001. An empirical model of species coexistence in a spatially structured environment. *Ecology* **82**: 1761–1771.

Part II ARTHROPODS FEEDING ON PLANT-PROVIDED FOOD

5

Food needs of adult parasitoids: behavioral adaptations and consequences

D. M. OLSON, K. TAKASU, AND W. J. LEWIS

The importance of adult food for parasitoids

The importance of adult food for parasitoids has been recognized for decades. Numerous laboratory studies show that suitable food sources can substantially increase longevity and fecundity of adult hymenopteran and dipteran parasitoids (reviews: Heimpel *et al.* 1997; Lewis *et al.* 1998). It is now appreciated that the consumption of non-host food can influence many other aspects of parasitoid biology such as egg viability, diapause in progeny, foraging decisions, searching efficiency, the onset and rate of egg resorption, primary sex ratio of progeny, flight initiation, and timing of flight. As a consequence non-host food can affect parasitoid and host dynamics, competitive interactions and niche partitioning among parasitoid species, productivity in laboratory cultures, and the probability of parasitoid establishment in classical biological control (Jervis 1998). In those parasitoid species that resorb eggs, starved individuals generally have higher rates of egg resorption than well-fed individuals, but this is not always the case (Olson *et al.* 2000). Egg resorption may increase life expectancy at the cost of reduced or delayed reproduction, which in turn may increase the risk of predation or other mortality factors. Conversely, there is presumably a metabolic cost to increased egg maturation, which decreases life expectancy (e.g., Roitberg 1989). For species that do not resorb eggs (Olson *et al.* 2000), investing in reproduction rather than increased life expectancy may be a viable reproductive strategy, especially in host-rich habitats. In host-poor habitats, however, non-host food will be important to extend life expectancy. For females that resorb eggs and that do not host-feed, obtaining non-host food is crucial to future egg maturation

Plant-Provided Food for Carnivorous Insects, ed. F. L. Wäckers, P. C. J. van Rijn, and J. Bruin.
Published by Cambridge University Press.

and ultimately reproduction. Therefore, expansion of our knowledge of the effects of adult feeding on the ecology of parasitoid species in their specific habitats is fundamentally important for effective pest management.

Food sources utilized

Parasitoids can be separated into four broad categories in terms of adult feeding requirements:

(1) Pro-ovigenic species where adult feeding is needed for maintenance but not for egg production (e.g., Jervis and Kidd 1996). Very few examples exist of truly pro-ovigenic species (Jervis *et al.* 2001).
(2) Synovigenic species that do not host-feed but feed on non-host food for maintenance and egg production (e.g., *Microplitis croceipes*) (Takasu and Lewis 1993).
(3) Synovigenic species in which females host-feed for egg production and both males and females non-host-feed for maintenance (e.g., *Ooencyrtus nezara*) (Takasu and Hirose 1991).
(4) Synovigenic species in which females host-feed for both maintenance and egg production (e.g., *Bracon hebetor*) (Jervis *et al.* 1994; K. Takasu, unpublished data).

For those species that host-feed, the choice whether to use the host for oviposition or egg production, or to invest in current or future reproduction, adds additional complexity to food- and host-searching behaviors. Host materials mainly provide parasitoids with protein, vitamin, and salt resources for reproduction, whereas plant nectars and honeydew provide energy resources mainly from the sugars present, although amino acids are also present (Harborne 1993).

Individual plant species can provide several sources of adult parasitoid food including floral nectar, extrafloral nectar, or homopteran honeydew. Although circumstances indicate pollen feeding by some parasitoids, pollen-feeders seem rare (review: Jervis 1998). Nectar is obtained either from the surfaces of exposed (extra-)floral nectaries or from nectaries that are covered or hidden within a tubular corolla. Honeydew is obtained from plant surfaces or directly from the sap-feeding species (Jervis 1998).

Although it is known that parasitoids utilize sugar sources in the field, a detailed understanding of parasitoid food use has been hindered in part by a lack of the appropriate methodologies that provide proof of actual feeding (Olson *et al.* 2000). However, biochemical techniques utilized for detection of sucrose, glucose, and fructose feeding by mosquitoes in the field (Van Handel 1985a, 1985b) can also be applied to parasitoid species (Olson *et al.* 2000). These tests could be used as analytic tools to determine whether field-collected parasitoids had recently fed on sucrose, glucose, or fructose and to distinguish between individuals that are starving and individuals that are relatively well fed. However, insects in general can vary considerably with respect to the spectrum of nectar and honeydew sugars they can utilize (Wäckers 2001, Chapter 2). Therefore, use of other methodologies to detect a broader range of sugars in field-collected parasitoid species would be useful in determining the most suitable sugar sources for a particular parasitoid species. Using high-performance liquid chromatography (HPLC), Wäckers and Steppuhn (2003) were able to demonstrate that parasitoids collected adjacent to a flowering field border had higher sugar levels than individuals collected in control fields. Moreover, between 55% and 80% of the collected parasitoids contained honeydew-specific sugars, indicating the prevalent use of this sugar source.

Parasitoid adaptations to food foraging

Many species of parasitoids display a multitude of behaviors and respond to a number of stimuli based on their physiological state (e.g., Lewis and Takasu 1990; Browne 1993; Jervis and Kidd 1995). In the conceptual model developed by Lewis *et al.* (1990), the physiological state of individuals acts as a filter in the expression of different behaviors with the range of behavioral expression determined by the response potential of the individual to various stimuli. This response potential is, in turn, determined by a combination of the genotype and degree of phenotypic plasticity of individuals (Lewis *et al.* 1990). An individual can modify behavioral responses through learning, which can improve location of host and food resources that vary in their quantity, quality, accessibility, and detectability (Lewis *et al.* 1998).

Parasitoids must balance their needs for both food and hosts by effectively responding to stimuli associated with each of these resources. Our knowledge of how parasitoids forage for adult food is quite limited as compared to our extensive knowledge of how they seek hosts. Still, several studies have shown innate and directed search by several species in response to food-related signals (Wäckers and Swaans 1993; Patt *et al.* 1997; Stapel *et al.* 1997). These responses can be similar to the responses to host-related signals (e.g., Lewis *et al.* 1990), but may also be specific to the task of food foraging (Wäckers *et al.* 2002; Olson *et al.* 2003b). Learning plays a significant role, as parasitoids are able to use different visual and olfactory cues in accordance with their physiological state and previous experience (Lewis and Takasu 1990; Wäckers and Lewis 1994; Takasu and Lewis 1996; Sato and Takasu 2000; Tertuliano *et al.* 2004). Learning can be very useful as the quantity and quality of food resources often varies across plants or within the plant. This variation may be caused by factors such as the presence of other nectar-feeding species, the spatial and temporal secretion of nectar, and the nutritional value, repellency, or toxicity of different nectars (Jervis *et al.* 1993). Tertuliano *et al.* (2004) found that females that had learned to associate a particular odor with food rewards will continue to elicit food-searching behaviors after several unrewarding experiences with the odor when they are very hungry, whereas females that were less hungry ceased to respond to the learned odor after only two unrewarding experiences. Food-searching responses of the less hungry females were recovered after a single exposure to the odor with a food reward (Tertuliano *et al.* 2004). Thus, parasitoid adults are predicted to respond to food stimuli that are more strongly associated with their current needs and in accordance with prior experience.

Potential food sources, such as plant nectars, honeydews, or hosts, provide various cues to insects. Plants often advertise the presence of nectars to potential pollinators by scents, colors, honey guides, and shapes (Harborne 1993). Herbivores often produce pheromones or other odors, and plants infested by herbivores emit specific volatiles. To locate their resources (food and hosts) parasitoids use chemical, visual, and tactile stimuli, or their combination (Otten *et al.* 2001; Vet *et al.* 2002), but chemical cues usually dominate (Wäckers and Lewis 1994). However, plants vary in their ability to attract different species. For some parasitoid species, extrafloral nectaries are more easily perceived than sucrose droplets or honeydew (Stapel *et al.* 1997), and some species show innate preferences for particular (floral) colors and odors (Wäckers 1994; Wäckers and Lewis 1994). Once the nectar has been detected, parasitoids could increase future food-searching efficiency through their demonstrated ability to learn various cues associated with the more profitable sites (e.g., Lewis and Takasu 1990; Wäckers and Lewis 1994). The degree that plants advertise the presence of suitable food or the strength of the associated cue(s) would likely influence the food-searching efficiency of parasitoids.

The quantity, quality, and accessibility of available sugars may also affect the parasitoid's ability to utilize them, and sugar concentration, composition, and degree of access can vary with the plant species (see, e.g., Wäckers, Chapter 2). Although there are distinct differences between insects in their gustatory response to particular sugars (Wäckers 1999) as well as their ability to utilize them (Ferreira *et al.* 1998), few studies examined the suitability as parasitoid food sources of individual carbohydrates, other than sucrose and the hexose components glucose and fructose. Although most parasitoids studied are able to utilize sucrose, glucose, and fructose, which are the most prevalent sugars in nectars and honeydew (reviews: Wäckers 2001; Beach *et al.* 2003), even a relatively low concentration of an unsuitable sugar can strongly diminish the suitability of the total sugar source (Wäckers 2001).

Although it is not clearly known whether parasitoids show specificity in the food sources they exploit (Jervis and Kidd 1996), several taxa have specialized mouthparts, referred to as a concealed nectar extraction apparatus (CNEA), for reaching floral nectaries (Quicke 1997; Jervis 1998). The CNEAs of parasitoids vary in length and are primarily utilized to extract nectar contained in long or deep tubular flower corollas that are not accessible to larger parasitoids or those lacking a CNEA. Those species lacking a CNEA appear capable of everting their labiomaxillary complex far enough to exploit nectar contained in very short, narrow, tubular flower corollas or for host-feeding from the ovipositor puncture wound (Jervis 1998). Therefore, the physical structure of the nectar-producing parts of the plant, the morphology of parasitoid mouthparts, and parasitoid size will influence the accessibility of non-host food sources for parasitoid species.

For host-feeders, hemolymph is usually obtained through a hole made by the ovipositor, but ichneumonids often puncture their hosts with their mandibles (Jervis and Kidd 1996). Host-feeders that attack concealed hosts such as gall-makers construct feeding tubes, which are formed by ovipositor secretions, and others utilize a CNEA to obtain host materials directly (e.g., Heimpel and Rosenheim 1995; Heimpel *et al.* 1997). Matching of food sources that are suitable in terms of quantity, quality, and accessibility with the behavioral and morphological adaptations of a particular parasitoid species is a recognized challenge faced by many researchers and biological control practitioners.

Balancing food and host needs

Spatial and temporal availability of food and hosts

The patterns of non-host food distribution on plants may be broadly categorized as: (1) host plants continuously bear non-host food for parasitoids; (2) host plants bear honeydew, floral, or extrafloral nectaries with limited availability (in time and/or space); (3) host plants do not provide adult food, but grow in habitats rich in non-host food on neighboring plants; and (4) host plants do not provide adult food and grow in habitats poor in food sources. In the latter situation, commuting by food foragers between food and host location may be costly in terms of increased risks associated with moving and time and energy expended. The cost may be offset by a higher energy gain and reproductive potential once the food source has been acquired. However, it is still unclear to what extent parasitoids commute between host and food sites, although this factor is likely to vary with the species and the relative spatial and temporal distribution of food and hosts (Lewis *et al.* 1998). See Heimpel and Jervis (Chapter 9) for some initial evidence based on gut analyses.

The distribution of food can influence a parasitoid's allocation of time and energy devoted to foraging for food. The absence of food within a host patch decreases retention time and parasitism efficiency of some non-host-feeding synovigenic parasitoids (Takasu and Lewis 1993, 1995; Stapel *et al.* 1997). For those species that require non-host food for egg maturation and maintenance, habitats with both host and food resources most readily available in space and time may provide the best means to maximize their reproductive potential.

The parasitoid's feeding state could also affect relative responses to host and food stimuli. In the non-host-feeding species *M. croceipes*, hunger state strongly affects the individual's foraging choice. Hungry females prefer food-associated odors versus host-associated odor, whereas well-fed females prefer host-associated odors (Lewis and Takasu 1990; Takasu and Lewis 1993). In the host-feeding parasitoid *Pimpla luctuosa*, host-feeding effectively stimulates egg maturation, but does not improve longevity (Iizuka and Takasu 1999). In this species food-deprived females often prefer nectar-associated stimuli rather than host-associated stimuli, whereas food-deprived females that have also been deprived of hosts for a long time nevertheless respond very strongly to host odors (Iizuka and Takasu 1999). *Pimpla luctuosa* may obtain other needed resources by host-feeding, but may also be investing more strongly in reproduction because of the perception of lower life expectancy. For those species that can increase both longevity and egg maturation by host feeding (e.g., *B. hebetor*), females may respond to host stimuli more strongly than food stimuli regardless of hunger state.

Landscape patterns

The landscape could dramatically affect the distribution and reproductive potential of parasitoids through the spatial and temporal availability of appropriate food, hosts, and shelter (see Wilkinson and Landis (Chapter 10) for an in-depth overview). For example, microclimate effects are often harsher in frequently disturbed areas (e.g., arable lands) than later-successional habitats such as hedges or forests. The early growing stage of crops can include conditions of high heat, low relative humidity, and the absence of hosts and food sources (e.g., Dyer and Landis 1996), which precludes many insect species from early colonization. Under these conditions, adult parasitoids would need to obtain food, alternate hosts, and shelter somewhere else. The rate of colonization of parasitoids into a crop area would likely depend on the availability of these alternate resources during the early growing stage of the crop (e.g., Mayse and Price 1978), as well as on the distance a species must travel to colonize.

The density and diversity of plants, and their spatial and temporal distribution can also influence the abundance of parasitoid food as well as the ability of parasitoids to locate hosts and food. Floral and extrafloral nectars and honeydew can vary in their presence because of differences in plant species (e.g., those lacking extrafloral nectaries) and their floral bloom times. For example, buckwheat (*Fagopyrum esculentum*) wilts in warm weather and ceases to secrete floral nectar in the afternoon, which coincides with low insect visitation (Pellet 1976; Bugg and Ellis 1990).

Associated plants can facilitate herbivore–enemy interactions or they can interfere with them through, for example, masking of attractive odors or direct repellent effects of odors from associated plants (e.g., Price *et al.* 1980). The important role that plants play in the herbivore–enemy interaction by providing needed hosts, food, shelter, and resource cues suggests that understanding plant–plant interactions is necessary to foster appropriate foraging environments for parasitoid species.

Consequences for plant–parasitoid interactions

Plants have phenological, physiological, and/or biochemical adaptations that allow them to improve their chance of survival and reproduction relative to their environment. Some traits of individual plants can modify interactions between herbivores and predators or parasitoids by operating directly on the herbivore, the carnivore, or both. Plants may interact with their higher trophic levels through chemicals (e.g., toxins, digestibility-reducers, and nutrient balance) or physical attributes (e.g., pubescence and tissue toughness) (Price *et al.* 1980). Many plants may also become functional allies with beneficial species (or members of the third trophic level) by providing food in the form of pollen and floral and extrafloral nectar. Plant provisioning of these and other food sources may increase the plant's short-term and long-term benefits through positive effects of beneficial insects on seed protection and dispersal. Interestingly, Wäckers *et al.* (2001) found that herbivory can induce secretion of additional nectar at and near the site of damage, thereby providing both hosts and adult food to parasitoids. The fact that nectar production was inducible in two unrelated species (cotton and castor bean) suggests that this active plant response may be a more widespread phenomenon. Although we are gaining increasing knowledge of the role that individual plants play in their own defense through traits affecting the third trophic level (e.g., Turlings and Wäckers 2004), the effects on the plant population level are less understood (Price *et al.* 1980).

It has been well established that many agricultural and non-agricultural plant species also respond to herbivory and other activities by emitting volatiles that are attractive to members of the third trophic level (review: Dicke and Van Loon 2000), thereby serving as a potential indirect plant defense. In two studies

(Thaler 1999; Olson *et al.* 2003a), the benefits to the plant and the parasitoid from volatile emissions were clear in terms of increased rates of parasitism of herbivore species, but more studies are needed to make any generalizations. However, two studies have shown that excess nitrogen in cotton plants interacts with herbivore damage and decreases attraction of parasitoids (A. M. Cortesero, unpublished data), while increasing attraction of some herbivore and carnivore species (Olson *et al.* 1999). In addition, Loughrin *et al.* (1995) showed that a naturalized cotton variety damaged by herbivores had an almost sevenfold production of volatiles compared to several newly developed commercial varieties. Such lowered signal production could reduce plant recruitment of parasitoids, thereby decreasing their life expectancy and reproductive potential and increasing mortality through predation or other factors.

Conclusion

Foraging efficacy of parasitoids is dependent on a variety of interacting factors, and among them food resources are very important in shaping these interactions. Non-host food, such as plant nectars and honeydew, and host materials are vital for survival and reproduction of parasitoids. Parasitoids have morphological adaptations geared for their specific food and host resources, and forage for food as well as hosts using multiple sensory and learning mechanisms to increase their foraging efficiency. Plants can affect parasitoid performance by providing cues indicating the presence of suitable food and hosts, although associated plants, nutrient, and water availability, and choice of cultivar may interfere with or diminish these signals. Herbivores as well as other species also exploit plant nectars and cues and have adaptations that help them gain resources and avoid predation (e.g. Dicke and Van Loon 2000; Romeis *et al.*, Chapter 7). Other foraging species may also affect a parasitoids ability to gain food resources. Thus, foraging environments are composed of multiple interactions that can determine how much of a parasitoid's potential is expressed. A fuller understanding of these interactions is crucial to managing effective biological control strategies as is discussed more fully in other chapters.

References

Beach, J. P., L. Williams, D. L. Hendrix, and L. D. Price. 2003. Different food sources affect the gustatory response of *Anaphes iole*, an egg parasitoid of *Lygus* spp. *Journal of Chemical Ecology* **29**: 1203–1222.

Browne, L. B. 1993. Physiologically induced changes in resource-oriented behavior. *Annual Review of Entomology* **38**: 1–25.

Bugg, R. L. and R. T. Ellis. 1990. Insects associated with cover crops in Massachusetts. *Biological Agriculture and Horticulture* **7**: 47–68.

Dicke, M. and J. J. A. van Loon. 2000. Multitrophic effects of herbivore-induced plant volatiles in an evolutionary context. *Entomologia Experimentalis et Applicata* **97**: 237–249.

Dyer, L. E. and D. A. Landis. 1996. Effects of habitat, temperature, and sugar availability on longevity of *Eriborus terebrans* (Hymenoptera: Ichneumonidae). *Environmental Entomology* **25**: 1192–1201.

Ferreira, C., B. B. Torres, and W. R. Terra. 1998. Substrate specification of midgut beta-glycosidases from insects of different orders. *Comparative Biochemistry and Physiology B* **119**: 219–225.

Harborne, J. B. 1993. *Introduction to Ecological and Biochemistry*, 4th edn. London: Academic Press.

Heimpel, G. E. and J. A. Rosenheim. 1995. Dynamic host feeding in the parasitoid *Aphytis melinus*: the balance between current and future reproduction. *Journal of Animal Ecology* **64**: 153–167.

Heimpel, G. E., J. A. Rosenheim, and D. Kattari. 1997. Adult feeding and lifetime reproductive success in the parasitoid *Aphytis melinus*. *Entomologia Experimentalis et Applicata* **83**: 305–315.

Iizuka, T. and K. Takasu. 1999. Balancing between host- and food-foraging by the host-feeding pupal parasitoid *Pimpla luctuosa* Smish (Hymenoptera: Ichneumonidae). *Entomological Science* **2**: 67–73.

Jervis, M. A. 1998. Functional and evolutionary aspects of mouthpart structure in parasitoid wasps. *Biological Journal of the Linnean Society* **63**: 461–493.

Jervis, M. A. and N. A. C. Kidd. 1995. Incorporating physiological realism into models of parasitoid feeding behaviour. *Trends in Ecology and Evolution* **10**: 434–436.

1996. Phytophagy. In M. A. Jervis and N. A. C. Kidd (eds.) *Insect Natural Enemies: Practical Approaches to their Study and Evaluation*. London: Chapman and Hall, pp. 375–394.

Jervis, M. A., G. E. Heimpel, P. N. Ferns, J. A. Harvey, and N. A. C. Kidd. 2001. Life-history strategies in parasitoid wasps: a comparative analysis of 'ovigeny'. *Journal of Animal Ecology* **70**(3): 442–458.

Jervis, M. A., N. A. C. Kidd, and H. E. Almey. 1994. Post-reproductive life in the parasitoid *Bracon hebetor* (Say) (Hym., Braconidae). *Journal of Applied Entomology* **117**: 72–77.

Jervis, M. A., N. A. C. Kidd, M. G. Fitton, T. Huddleston, and H. A. Dawah. 1993. Flower visiting by hymenopteran parasitoids. *Journal of Natural History* **27**: 67–105.

Lewis, W. J. and K. Takasu. 1990. Use of learned odours by a parasitic wasp in accordance with host and food needs. *Nature* **348**: 635–636.

Lewis, W. J., J. O. Stapel, A. M. Cortesero, and K. Takasu. 1998. Understanding how parasitoids balance food and host needs: importance to biological control. *Biological Control* **11**: 175–183.

Lewis, W. J., L. E. M. Vet, J. H. Tumlinson, J. C. van Lenteren, and D. R. Papaj. 1990. Variations in parasitoid foraging behavior: essential element of a sound biological control theory. *Environmental Entomology* **19**: 1183–1193.

Loughrin, J. H., A. Manukian, R. R. Heath, and J. H. Tumlinson. 1995. Volatiles emitted by different cotton varieties damaged by feeding beet armyworm larvae. *Journal of Chemical Ecology* **21**: 1217–1227.

Mayse, M. A. and P. W. Price. 1978. Seasonal development of soybean arthropod communities in east central Illinois. *Agro-ecosystems (Amsterdam)* **4**: 387–405.

Olson, D. M., H. Fadamiro, J. G. Lundgren, and G. E. Heimpel, 2000. Effects of sugar feeding on carbohydrate and lipid metabolism in a parasitoid wasp. *Physiological Entomology* **25**: 17–26.

Olson, D. M., Hodges, T. and Lewis, W. J. 2003a. Foraging efficacy of a larval parasitoid in a cotton patch: influence of chemical cues and learning. *Journal of Insect Behavior* **16**: 613–624.

Olson, D. M., S. C. Phatak, and W. J. Lewis. 1999. Influence of nitrogen levels on cotton plant/insect interactions in a conservation tillage system. Proc. 22nd Annual Southern Conservation Tillage Conference for Sustainable Agriculture, Tifton, GA, pp. 119–130.

Olson, D. M., G. C. Rains, T. Meiners, *et al.* 2003b. Parasitic wasps learn and report diverse chemicals with unique conditionable behaviors. *Chemical Senses* **28**: 545–549.

Otten, H., F. L. Wäckers, and S. Dorn. 2001. Efficacy of vibrational sounding in the parasitoid *Pimpla turionellae* (Hymenoptera: Ichneumonidae) is affected by female size. *Animal Behaviour* **61**: 671–677.

Patt, J. M., G. C. Hamilton, and J. H. Lashomb. 1997. Foraging success of parasitoid wasps on flowers: interplay of insect morphology, floral architecture and searching behavior. *Entomologia Experimentalis et Applicata* **83**: 21–30.

Pellet, F. C. 1976. *American Honey Plants*. Hamilton, IL: Dadant and Sons.

Price, P. W., C. E. Bouton, P. Gross, *et al.* 1980. Interactions among three trophic levels: influence of plants on interactions between insect herbivores and natural enemies. *Annual Review of Ecology and Systematics* **11**: 41–65.

Quicke, D. L. J. 1997. *Parasitic Wasps*. London: Chapman and Hall.

Roitberg, B. D. 1989. The cost of reproduction in rosehip flies, *Rhagoletis basiola*: eggs are time. *Evolutionary Ecology* **3**: 156–165.

Sato, M. and K. Takasu. 2000. Food odor learning by both sexes of the pupal parasitoid *Pimpla alboannulatus* Uchida (Hymenoptera: Ichneumonidae). *Journal of Insect Behavior* **13**: 263–272.

Stapel, J. O., A. M. Cortesero, C. M. De Moraes, J. H. Tumlinson, and W. J. Lewis. 1997. Extrafloral nectar, honeydew, and sucrose effects on searching behavior and efficiency of *Microplitis croceipes* (Hymenoptera: Braconidae) in cotton. *Environmental Entomology* **26**: 617–623.

Takasu, K. and Y. Hirose. 1991. Host searching behavior in the parasitoid *Ooencyrtus nezarae* (Hymenoptera: Encrytidae) as influenced by non-host food deprivation. *Applied Entomology and Zoology* **26**: 415–417.

Takasu, K. and W. J. Lewis. 1993. Host- and food-foraging of the parasitoid *Microplitis croceipes:* learning and physiological state effects. *Biological Control* **3**: 70–74.

1995. Importance of adult food source to host searching of the larval parasitoid *Microplitis croceipes*. *Biological Control* **5**: 25–30.

1996. The role of learning in adult food location by the larval parasitoid, *Microplitis croceipes* (Hymenoptera: Braconidae). *Journal of Insect Behavior* **9**: 265–281.

Tertuliano, M., D. M. Olson, G. C. Rains, and W. J. Lewis. 2004. Influence of handling and conditioning protocol on learning and memory of *Microplitis croceipes*. *Entomologia Experimentalis et Applicata* **110**: 165–172.

Thaler, J. S. 1999. Jasmonate-inducible plant defences cause increased parasitism of herbivores. *Nature* **399**: 686–688.

Turlings, T. C. J. and F. L. Wäckers. 2004. Recruitment of predators and parasitoids by herbivore-injured plants. In R. T. Cardé and J. Millar (eds.) *Advances in Chemical Ecology of Insects*. Cambridge, UK: Cambridge University Press, pp. 21–75.

Van Handel, E. 1985a. Rapid determination of glycogen and sugars in mosquitoes. *Journal of the American Mosquito Control Association* **1**: 299–301.

1985b. Rapid determination of total lipids in mosquitoes. *Journal of the American Mosquito Control Association* **1**: 302–304.

Vet, L. E. M., L. Hemerik, M. E. Visser, and F. L. Wäckers. 2002. Flexibility in host search and patch use strategies of insect parasitoids. In E. E. Lewis, J. F. Cambell, and M. V. K. Sukhdeo (eds.) *The Behavioural Ecology of Parasites*. Wallingford, UK: CAB International, pp. 39–64.

Wäckers, F. L. 1994. The effect of food deprivation on the innate visual and olfactory preferences in the parasitoid *Cotesia rubecula*. *Journal of Insect Physiology* **40**: 641–649.

1999. Gustatory response by the hymenopteran parasitoid *Cotesia glomerata* to a range of nectar and honeydew sugars. *Journal of Chemical Ecology* **25**: 2863–2877.

2001. A comparison of nectar- and honeydew sugars with respect to their utilization by the hymenopteran parasitoid *Cotesia glomerata*. *Journal of Insect Physiology* **47**: 1077–1084.

Wäckers, F. L. and W. J. Lewis. 1994. Olfactory and visual learning and their combined influence on host site location by the parasitoid *Microplitis croceipes* (Cresson). *Biological Control* **4**: 105–112.

Wäckers, F. L. and A. Steppuhn. 2003. Characterizing nutritional state and food source use of parasitoids collected in fields with high and low nectar availability. *IOBC/WPRS Bulletin* **26**: 203–208.

Wäckers, F. L. and C. P. M. Swaans. 1993. Finding floral nectar and honeydew in *Cotesia rubecula*: random or directed? *Proceedings of the Section Experimental and Applied Entomology of the Netherlands Entomological Society* **4**: 67–72.

Wäckers, F. L., C. Bonifay, and W. J. Lewis. 2002. Conditioning of appetitive behavior in the hymenopteran parasitoid *Microplitis croceipes*. *Entomologia Experimentalis et Applicata* **103**: 135–138.

Wäckers, F. L., D. Zuber, R. Wunderlin, and F. Keller. 2001. The effect of herbivory on temporal and spatial dynamics of foliar nectar production in cotton and castor. *Annals of Botany* **87**: 365–370.

6

Effects of plant feeding on the performance of omnivorous "predators"

MICKY D. EUBANKS AND JOHN D. STYRSKY

Introduction

Thousands of arthropod species consume both prey and plant food (Whitman *et al.* 1994; Alomar and Wiedenmann 1996; Coll and Guershon 2002). Omnivorous arthropods include species that are usually thought of as being exclusively herbivorous (e.g., grasshoppers and thrips), exclusively predaceous (e.g., ants and spiders), and species whose catholic tastes are better known (e.g., true bugs and carabid beetles). These omnivorous taxa all share the ability to utilize both prey and plant food, but differ widely in the timing and extent to which they utilize these resources, and in their morphologies, physiologies, and behaviors. Some omnivorous arthropods have been referred to as life-history omnivores (Wäckers and Van Rijn, Chapter 1). Life-history omnivores utilize prey and plant food at different stages or periods of their life (Coll and Guershon 2002; Wäckers and Van Rijn, Chapter 1). Other omnivores, however, include both prey and plant food in their diet throughout their life, although the relative importance of the two may vary with age. These species could be referred to as lifelong omnivores. Big-eyed bugs (Heteroptera: Geocoridae) are excellent examples of insects that consume prey and plant food in all life stages. We focus on the experimental studies of "predaceous" omnivores and attempt to draw conclusions regarding the effects of plant feeding on omnivorous predators and their interactions with other species.

Potential effects of plant feeding on predator–prey interactions

From an ecological perspective, the most important question regarding plant feeding by omnivorous predators is: what effect does plant feeding have

Plant-Provided Food for Carnivorous Insects, ed. F. L. Wäckers, P. C. J. van Rijn, and J. Bruin.
Published by Cambridge University Press.

on the suppression of prey by omnivores? Classical community ecology predicts that omnivores may be more likely than strict predators to suppress populations of their prey because animals that feed at multiple trophic levels are unlikely to starve or emigrate when prey are scarce (Crawley 1975; Pimm and Lawton 1977, 1978; Walde 1994; Eubanks and Denno 1999). As a result, omnivores may continue to capture and consume prey at low prey densities and may drive them to local extinction. In contrast, typical predators either starve or emigrate when prey are scarce, allowing prey to escape predation at low densities and populations to rebound. Thus, feeding at more than one trophic level may deny prey density-related refugia from predation. Consequently, plant feeding by omnivorous predators should promote top–down control and may increase the likelihood of trophic cascades (Fig. 6.1A) (Dayton 1984; Holt 1984; Polis *et al.* 1989; Polis 1991; Holt and Lawton 1994; Polis and Strong 1996; Holt and Polis 1997; Eubanks and Denno 1999, 2000).

The potential impact of omnivores on prey, however, not only depends on their ability to persist during periods of prey scarcity, but also on the extent that feeding on plants and alternative prey decreases the consumption of focal prey (i.e., the functional response) (Abrams 1987; Cottrell and Yeargan 1998; Eubanks and Denno 2000). For example, feeding on high-quality plant food may result in relatively persistent and large omnivore populations, but plants might also provide a preferred food source and omnivores may consume fewer prey when high-quality plant food is available (Fig. 6.1B). Thus, the persistence afforded omnivores by feeding on multiple trophic levels may not necessarily translate into enhanced suppression of prey as predicted by theory.

Our goal was to use empirical data from the literature to determine if plant feeding by omnivores ultimately increases, decreases, or does not affect their impact on prey populations. Since very few studies have actually addressed this specific question, we primarily detail how plant feeding affects aspects of an omnivore's biology and ecology that are likely to promote increased or decreased prey suppression. We ask a series of simple questions: Does plant feeding increase development rate and survival of immature omnivorous predators? Does plant feeding increase the longevity and fecundity of adults? Does plant feeding really allow omnivores to persist when prey are scarce? Given these potential benefits, does plant feeding affect the dispersal and distribution of omnivorous predators? Similarly, does plant feeding really affect the population size of omnivores? And finally, does plant feeding affect the per capita consumption of prey by omnivores and their ability to suppress prey populations?

In the first part of this review, we summarize data from 26 studies that empirically quantified the effect of plant feeding on five life-history traits of

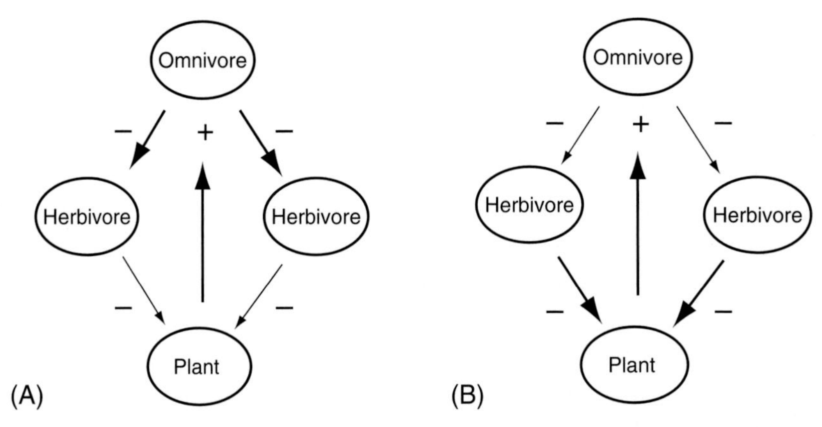

Figure 6.1 Two possible ways that plant feeding could mediate omnivore–herbivore interactions and influence prey suppression. (A) Plant feeding by omnivores results in large, persistent omnivore populations. Larger omnivore populations could translate into increased predation pressure on herbivore populations. Under this scenario, the positive, direct effect of plants on omnivores drives the system. Plants, therefore, would have an indirect, negative effect on the herbivores that consume them. The result would be smaller herbivore populations due to plant feeding by omnivores. (B) In this scenario, plant feeding by omnivores results in large and persistent omnivore populations. Plant feeding, however, reduces the number of prey consumed by individual omnivores to the extent that herbivore populations are larger. Plants, therefore, indirectly benefit the herbivores that are consuming the plants and this indirect effect drives the system. (Arrowheads indicate the recipient of the effect, line thickness indicates strength of effect, and plus and minus signs indicate positive and negative effects, respectively.)

omnivorous predators: nymphal development time and survival to adulthood, and adult female mass, longevity, and fecundity (Table 6.1). We consider only those studies that specifically tested for effects of plant food as a supplement to strict diets of prey alone, thus allowing an evaluation of the influence of phytophagy in predaceous omnivores. To broaden the review, we considered plant food to encompass not only leaves but also developing seeds, pollen, extra-floral nectar, and homopteran honeydew. Eighteen of the 26 studies focussed on species in the heteropteran families Pentatomidae, Lygaeidae, Miridae, and Anthocoridae, one study focussed on a lacewing species (Neuroptera: Chrysopidae), one study focussed on a coccinellid beetle (Coleoptera), and six studies focussed on several mite species (Acari: Phytoseiidae). Our purpose in this review is to evaluate the benefit of plant feeding in omnivorous predators; therefore, we consider only those taxa in which phytophagy is considered a feeding habit secondary to predation. Although other groups of arthropod predators are also significantly phytophagous (e.g., Carabidae), we found no

Table 6.1 *Influence of plant food as a supplement to prey-only diets on selected life-history traits of zoophytophagous insect and mite "predators"; values represent the percentage change (positive or negative) in a particular trait with the addition of plant food to a prey-only diet*

			Nymph		Adult female			
Species	Prey food	Supplemental plant food	Development rate[a]	Survival[b]	Mass	Longevity	Fecundity[c]	Reference
INSECTS								
Pentatomidae								
Podisus maculiventris	*Leptinotarsa* larvae	Potato leaf	10.0[*]	27.5	−6.6[d]	–	–	Ruberson *et al.* 1986
Podisus maculiventris	Mealworms	Green bean leaf	NE[e]	–	–	–	–	Crum *et al.* 1998
		Green bean pod	NE[e]	–	–	–	–	
		Tomato leaf	6.5[*f]	–	−9.8[*]	–	39.0[*]	
		Celery leaf	4.9[f]	–	−5.7	–	37.0[*]	
		Basil leaf	7.4[*f]	–	−8.2[*]	–	40.0[*]	
Podisus maculiventris	*Manduca* larvae	Green bean pod	1.0[e]	–	–	–	–	Weiser and Stamp 1998
Tynacantha marginata	Mealworms	*Eucalyptus* leaf	−4.6[g]	16.0[h]	17.8	6.4	177.7[*h]	Moreira *et al.* 1997
Lygaeidae								
Geocoris bullatus	Pea aphids	Sunflower seed	–	20.0	–	–	PFR	Tamaki and Weeks 1972
		Seed + green bean	–	80.0	–	–	PFR	
Geocoris pallens	Pea aphids	Sunflower seed	–	8.0	–	–	0.0	
		Seed + green bean	–	82.8	–	–	PFR	
Geocoris punctipes	Tubermoth eggs	Green bean pod	−5.0[d]	−6.7[d]	4.4[d]	–	111.2[*i]	Dunbar and Bacon 1972

Table 6.1 *(cont.)*

Species	Prey food	Supplemental plant food	Nymph		Adult female			Reference
			Development rate[a]	Survival[b]	Mass	Longevity	Fecundity[c]	
	Pea aphids	Green bean pod	PFR	PFR	PFR	-	PFR[j]	
Geocoris punctipes	*Heliothis* eggs	Green bean pod	15.3	2.5	-8.4^{j}	-	$-6.5^{k,l}$	Cohen and DeBolt 1983
	Lygus eggs	Green bean pod	2.5	4.0	0.7^{n}	-	$22.0^{k,m}$	
Geocoris punctipes	*Spodoptera* eggs	*Amaranthus* leaf	0.9	23.4*	13.0*	-	$-8.5^{k,n}$	Naranjo and Stimac 1985
		Chenopodium leaf	0.5	33.4*	8.8	-	$-1.4^{k,n}$	
		Ambrosia leaf	1.4	26.7*	15.2*	-	$-2.8^{k,n}$	
		Bidens leaf	3.2	6.7	10.9	-	$-4.2^{k,n}$	
		Heterotheca leaf	2.3	10.0	8.7	-	$-12.7^{k,n}$	
		Solidago leaf	1.4	16.7	−2.2	-	$1.4^{k,n}$	
		Cassia leaf	1.8	10.0	19.6*	-	$11.3^{k,n}$	
		Crotalaria leaf	−1.4	16.7*	4.3	-	$-21.1^{k,n}$	
		Desmodium leaf	1.4	6.7	0.0	-	$-5.6^{k,n}$	
		Glycine leaf	0.9	26.7*	2.2	-	$-9.8^{k,n}$	
		Richardia leaf	0.9	6.7	4.3	-	$-12.7^{k,n}$	
		Green bean pod	2.3	0.0	8.7	-	$-12.7^{k,o}$	
Geocoris punctipes	Pea aphids	Lima bean pod	PFR	PFR	-	PFR	-	Unpublished data from experiments described in Eubanks and Denno 1999[d]
		Lima bean leaf	PFR	PFR	-	PFR	-	
		Pod + leaf	PFR	PFR	-	PFR	-	

	Corn earworm eggs	Lima bean pod	15.4	8.0	-	35.0	-	
		Lima bean leaf	3.4	−10.0	-	1.8	-	
		Pod + leaf	29.1	15.0	-	27.0	-	
	Aphids + eggs	Lima bean pod	8.6	25.0	-	124.0	-	
		Lima bean leaf	−0.5	10.0	-	5.9	-	
		Pod + leaf	12.1	30.0	-	182.0	-	
Miridae								
Deraecoris signatus	Cotton aphids	Cotton tips and squares[p]	4.6*	10.0	-	13.3	68.5*	Chinajariyawong and Harris 1987
Dicyphus hesperus	*Ephestia* eggs	Tomato leaf	-	91.0*	-	-	-	Gillespie and McGregor 2000
Anthocoridae								
Orius albidipennis	*Ephestia* eggs	Flower pollen	-	-	-	−0.8	38.7*[q]	Cocuzza *et al.* 1997
Orius laevigatus	*Ephestia* eggs	Flower pollen	-	-	-	3.0	2.3[r]	
Orius albidipennis	*Ephestia* eggs	Flower pollen	1.2	10.0	-	-	-	Vacante *et al.* 1997
Orius laevigatus	*Ephestia* eggs	Flower pollen	−0.8	5.0	-	-	-	
Orius insidiosus	Greenbug	Green bean pod	15.4*[s]	−1.0[j]	-	1.2	8.3	Bush *et al.* 1993
	Cotton aphid	Green bean pod	15.2*[t]	14.0[j]	-	2.2	12.1	
	Heliothis eggs	Green bean pod	10.5*[u]	−1.0[j]	-	61.1*	43.1*	
Orius insidiosus	Thrips	Pollen	7.0	28.2*	-	−3.5	89.7	Kiman and Yeargan 1985
		Green bean pod	5.1	8.9	-	9.8	3.4	
	Mites	Pollen	0.7	−3.0	-	−1.9	8.1	
	Mites	Green bean pod	0.7	−21.9*	-	10.3	−57.2	

Table 6.1 (*cont.*)

			Nymph		Adult female			
Species	Prey food	Supplemental plant food	Development rate[a]	Survival[b]	Mass	Longevity	Fecundity[c]	Reference
	Heliothis eggs	Pollen	−0.7	−0.2	-	17.8	3.2	
		Green bean pod	−4.5	−7.0	-	6.7	−0.8	
Orius laevigatus	Thrips	Pine pollen	-	-	-	−13.0	−18.0[*t]	J. Hulshof, pers. comm.
Orius minutus	Green peach aphid	*Corylus* pollen	1.6[t]	26.7	2.4	12.0	−39.4	Schmidt-Tiedemann and Sell 1997
		Capsicum pollen	3.2[*t]	40.0	12.5[*]	147.0[*]	85.7	
		Cucumis pollen	7.5[*t]	48.3	11.6[*]	65.0	−0.6	
Orius tristicolor	Thrips	Pollen	10.6[*]	0.0	-	-	-	Salas-Aguilar and Ehler 1977
		Green bean pod	−8.2[*]	−7.5[d]	-	-	-	
Chrysopidae								
Chrysoperla plorabunda	Cotton aphid	Extrafloral nectar	-	0.0[w]	-	-	-	Limburg and Rosenheim 2001
Coccinellidae								
Coccinella transversalis	*Helicoverpa* larvae	Honeydew	-	-	-	-	400.0[*v]	Evans 2000
MITES								
Phytoseiidae								
Amblyseius hibisci	Spider mites	Honeydew	-	30.2[e]	-	-	51.6[e]	McMurtry and Scriven 1964
Amblyseius swirskii	*Coccus* crawlers	Honeydew	-	NS	-	-	PFR	Ragusi and Swirski 1977

	Saissetia crawlers	Honeydew	–	4.2	–	–	NR	
Amblyseius swirskii	Spider mites	*Ricinus* pollen	–	–	–	0.0	−10.1	Momen and El-Saway 1993
		Ricinus pollen	–	–	–	0.0	-24.9^{*}	
Euseius fustis	Spider mites	Maize pollen	–	−4.7	–	148.1^{*}	458.3^{*}	Bruce-Oliver *et al.* 1996
		Castor bean pollen	–	15.0	–	88.9	301.7^{*}	
Euseius hibisci	Pacific mite	Aphid honeydew	6.9^{w}	6.7	–	–	38.2^{*x}	Zhimo and McMurtry 1990
		Whitefly honeydew	5.7^{w}	0.0	–	–	51.9^{*x}	
	Citrus red mite	Aphid honeydew	3.2^{w}	−10.0	–	–	39.1^{x}	
		Whitefly honeydew	2.8^{w}	−3.3	–	–	51.6^{x}	
Euseius stipulatus	Pacific mite	Aphid honeydew	1.9^{w}	−3.3	–	–	16.9^{x}	
		Whitefly honeydew	2.6^{w}	3.3	–	–	13.5^{x}	
	Citrus red mite	Aphid honeydew	1.8^{w}	3.3	–	–	42.3^{x}	
		Whitefly honeydew	2.4^{w}	6.7	–	–	50.0^{*x}	
Euseius tularensis	Pacific mite	Aphid honeydew	2.7^{w}	15.2	–	–	44.8^{*x}	
		Whitefly honeydew	2.6^{w}	16.0	–	–	40.0^{*x}	
	Citrus red mite	Aphid honeydew	1.0^{w}	13.3	–	–	59.0^{*x}	
		Whitefly honeydew	4.4^{w}	16.6	–	–	68.9^{*x}	
Typhlodromalus peregrinus	Broad mites	*Malephora* pollen	9.4^{h}	–	–	−8.9	-31.6^{*}	Fouly *et al.* 1995
		Quercus pollen	-0.8^{h}	–	–	−30.7	-38.4^{*}	
		Typha pollen	4.2^{h}	–	–	−27.5	-41.1^{*}	

*Statistically significant ($P \leq 0.05$) effects of supplemental plant food as reported in the cited references.
NE, No effect; indicates no statistically significant difference between treatments (authors did not report relevant data).
NS, No survival; indicates no nymphs survived in either treatment.

NR, No reproduction; indicates no females reproduced in either treatment.
PFR, Plant food required; indicates that nymphs and/or adults did not survive and/or reproduce on diets of prey alone.
[a] Measured as days from eclosion to adult stage unless noted otherwise.
[b] Measured as percentage survival from eclosion to adult stage unless noted otherwise; percentage change in nymphal survival calculated as the difference between percent survival in the two diet treatments.
[c] Measured as total number of eggs per female unless noted otherwise.
[d] Difference in treatments not tested statistically.
[e] Measured as rate of development (per day) from the third to the fifth instar.
[f] Measured as rate of development (per day) from third instar to adult stage.
[g] Data for males and females averaged.
[h] Measured as oviposition period, in days; net reproductive rate also increased significantly, by 135.6%.
[i] Oviposition period, in days, also increased, by 8.4%; egg viability also increased, by 91.4%.
[j] Males and females combined.
[k] Measured as number of eggs per female per day.
[l] Egg viability also decreased, by 10.9%.
[m] Egg viability also increased, by 15.9%.
[n] Egg viability decreased by between 0.6% and 20.3%.
[o] Egg viability increased by 1.7%.
[p] Data averaged from separate studies of cotton tips and cotton squares as plant supplements.
[q] Oviposition period, in days, also increased, by 14.2%.
[r] Oviposition period, in days, also increased, by 10.4%.
[s] Females only.
[t] Measured as total number of eggs per female per day for 3 days.
[u] Measured through first 5 days of life; no mortality occurred in either diet treatment.
[v] Measured as total number of eggs per female per day for 7 days.
[w] Measured as hours from laying to adult stage; difference in survival non-significant.
[x] Measured as number of eggs laid per female per day for 10 days.

studies of such groups that quantified the influence of phytophagy on the five life-history traits of interest.

In the second part of this review, we summarize data from the very few studies that examine the effect of phytophagy on some important ecological traits of omnivorous "predators" including predator persistence, dispersal, and population size. Much of this review draws from our own work with the big-eyed bug, *Geocoris punctipes*. We strongly believe that there is a need for more work on the ecology of omnivorous predators and we end our chapter by highlighting several unanswered questions.

We note here the limitations to our review. First, few studies have specifically tested hypotheses regarding the consequences of phytophagy in omnivorous predators, and not all groups of omnivorous predators have received equal attention. Second, the available data are further limited by lack of independence: most of the studies that we use in our review are of closely related species and many of the studies include some of the same species. For example, 50 individual experiments tested effects of plant feeding on nymphal development rate of omnivorous insects, but these 50 experiments focussed on just nine species, primarily in the heteropteran families Lygaeidae and Anthocoridae. Because of the paucity of data, we treated all experiments independently and did not attempt a meta-analysis. Third, to evaluate our hypotheses, we report the number of experiments, or cases, in which plant feeding increased, decreased, or did not affect the life-history traits of interest. We caution, though, that not all changes in the value of a trait represent statistically significant changes. In several instances, we could not determine statistical significance because authors did not report the relevant information. Statistically significant changes in the value of life-history traits, when known, are indicated by asterisks in Table 6.1. Although there are several limitations to the data, our review highlights questions that have been well addressed in the literature and, more importantly, highlights questions that have not been well addressed.

Effects of plant feeding on development and survival of immature omnivores

Supplementing prey diets with plant food generally increased the development rate as well as the survival of immatures in both insects and mites. Immature insect predators reared on diets supplemented with plant food developed faster than those reared on prey alone in 36 out of 50 cases (72%) (Fig. 6.2A). In these cases, however, development rate increased from 0.7% to 29.1% with a mean increase of only 6.4% (±1.1%) (Fig. 6.2B). Similarly, in 34 out of 50 cases (68%), survival of immature insect predators reared on prey diets supplemented

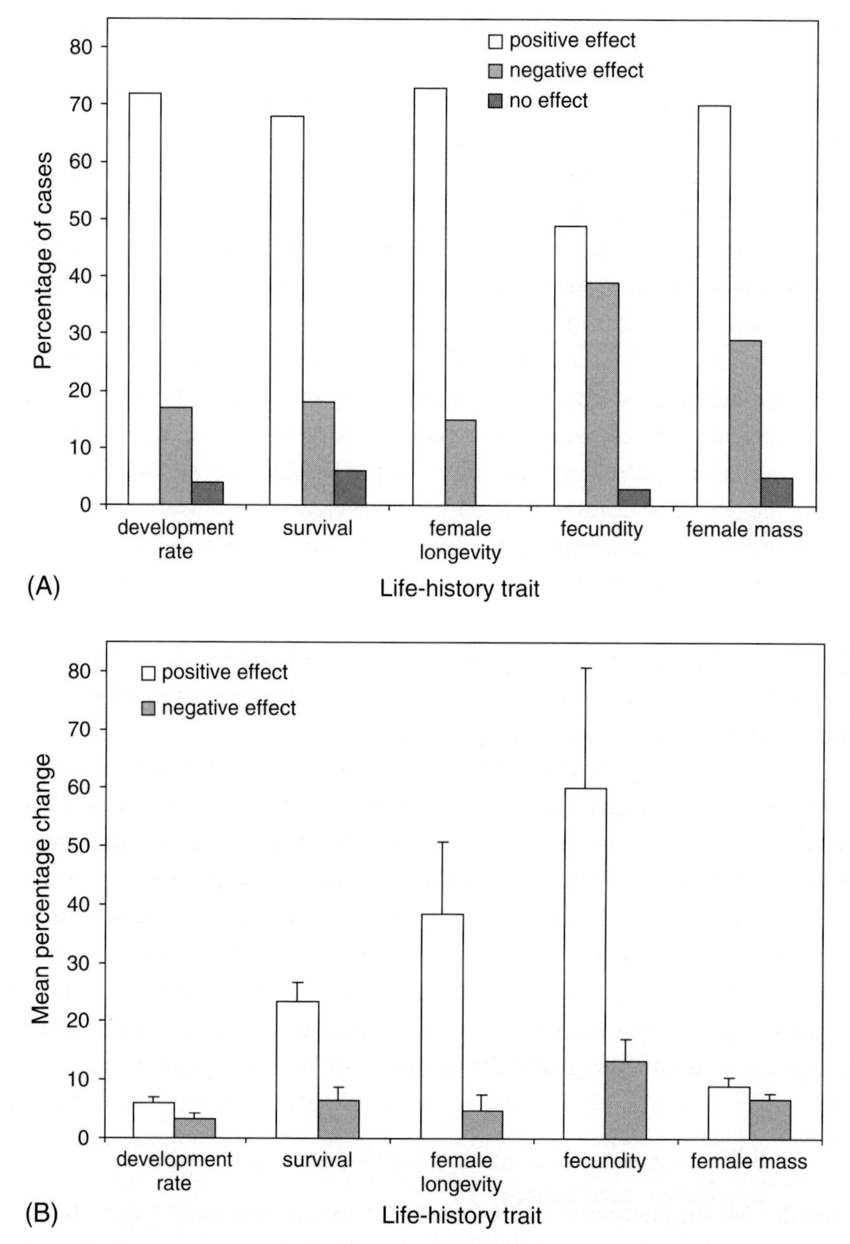

Figure 6.2 Summary of studies that investigated the influence of plant food as a supplement to prey-only diets on selected life-history traits of zoophytophagous insect predators. (A) The percentage of cases (individual experiments) in which supplemental plant food had a positive effect, negative effect, or no effect on the traits of interest. (B) The mean (± 1 standard error) percentage change in trait values (positive and negative) caused by the addition of plant food to prey-only diets. Both panels based on data presented in Table 6.1.

with plant food was greater than that for those reared on prey alone (Fig. 6.2A). Supplemental plant food had a considerably greater positive effect on immature survival than on development rate: survival increased from 2.5% to 91%, with an average increase of 23.4% (±3.3%) (Fig. 6.2B). In a few cases (4 out of 50; 8%), nymphal predators could not survive without the addition of plant food to their diets (Table 6.1). In these four cases, pea aphids were provided as prey, insects relatively low in nutritional quality (low nitrogen content) (Dunbar and Bacon 1972; Eubanks and Denno 1999).

The stronger positive impact on immature insect survival than on development rate probably reflects the benefit of supplemental plant food to immature insect predators as a source of water rather than as a source of nutrition. Indeed, the percentage increase in immature insect survival was roughly equivalent across three categories of supplemental plant food: leaves, pods/seeds, and pollen (Table 6.2). In contrast, the percentage increase in insect development rate was greatest when prey diets were supplemented with pods or seeds (Table 6.2), a category of plant food that is higher in quality (in terms of nitrogen content) than leaves. The nitrogen content of pollen, however, is also greater than that of leaves, but the average percentage increase in development rate in insects that received pollen as a diet supplement was only slightly greater than that of insects that received leaves (Table 6.2).

The impact of plant food supplements of differing quality on immature insect predator development rate and survival has been studied specifically in big-eyed bugs. When fed the same prey diet of corn earworm eggs (*Helicoverpa zea*), big-eyed bug nymphs developed significantly faster (15.4%) when supplemented with lima bean pods, a high-quality (nitrogen-rich) plant food, than when provided lima bean leaves, a lower-quality (nitrogen-poor) plant food (3.4%) (see Table 6.1; M. D. Eubanks and R. F. Denno, unpublished data). Similarly, when fed the same prey diet of corn earworm eggs and pea aphids (*Acyrthosiphon pisum*), big-eyed bug nymph survival increased by 25% when nymphs were additionally provided lima bean pods, but by just 10% when they were additionally provided lima bean leaves (see Table 6.1; Eubanks and Denno, unpublished data). In a related study, however, the benefit to overall big-eyed bug survival (nymph + adult) provided by supplemental plant food was a function of prey quality (Eubanks and Denno 1999). Supplementing prey-only diets with either lima bean pods or leaves had no significant effect on the overall survival of big-eyed bugs when they were fed high-quality prey (corn earworm eggs). In contrast, when big-eyed bugs were fed low-quality prey (pea aphids), supplemental lima bean pods significantly increased overall survival whereas supplemental lima bean leaves did not. Eubanks and Denno (1999) concluded that high-nitrogen plant food (lima bean pods) could ameliorate to some extent the costs imposed by feeding

Table 6.2 *Summary of the influence of three categories of supplemental plant food on selected life-history traits of zoophytophagous insect "predators"; values are drawn from the data presented in Table 6.1*

			Positive effect		Negative effect	
Life-history trait	Supplemental plant food	Number of experiments	Percent of cases	Change (%)[a]	Percent of cases	Change (%)[a]
Immature development rate	Leaf	20	75.0	3.1 ± 0.7	15.0	-2.2 ± 1.2
	Pod/seed	20	65.0	10.2 ± 2.3	15.0	-5.9 ± 1.2
	Pollen	9	77.8	4.5 ± 1.5	22.2	-0.75 ± 0.05
Immature survival	Leaf	17	88.2	21.9 ± 5.4	5.9	-10.0
	Pod/seed	22	54.5	24.9 ± 8.0	27.3	-7.5 ± 3.1
	Pollen	9	66.7	26.4 ± 6.8	22.2	-1.6 ± 1.4
Adult female longevity	Leaf	4	75.0	4.7 ± 1.5	-	-
	Pod/seed	12	83.3	45.9 ± 19.2	-	-
	Pollen	9	55.6	49.0 ± 26.8	44.4	-4.8 ± 2.8
Adult female fecundity	Leaf	15	40.0	51.1 ± 26.2	60.0	-8.8 ± 2.1
	Pod/seed	15	40.0	33.4 ± 16.6	26.6	-19.3 ± 12.9
	Pollen	9	66.7	38.0 ± 16.7	33.3	-19.3 ± 11.2
Adult female mass	Leaf	16	62.5	10.5 ± 1.9	31.3	-6.5 ± 1.3
	Pod/seed	5	60.0	4.6 ± 2.3	20.0	-8.4
	Pollen	3	100.0	8.8 ± 3.2	-	-

[a] Mean $\pm$ 1 SE.

on low-nitrogen prey. Prey quality and supplemental plant food quality interacted to affect development rate and survival in immature insect predators in other studies as well (e.g., see Kiman and Yeargan 1985; Bush *et al.* 1993 in Table 6.1) and is probably not uncommon.

In a few studies, supplementing prey-only diets with plant food detrimentally affected immature insect predator development rate and survival. Supplemental plant food decreased development rate in eight out of 50 cases (17%) (Fig. 6.2A), but only by a mean of 3.2 ± 1.0% (range: 0.5% to 8.2%) (Fig. 6.2B). Supplemental plant food decreased survival in 9 out of 50 cases (18%) (Fig. 6.2A), but only by a mean of 6.5 ± 2.2% (range: 0.2% to 21.9%) (Fig. 6.2B). Immature insect survival was most adversely affected in predatory anthocorids: seven out of nine cases in which supplemental plant food decreased nymphal survival were experiments with *Orius* species, six of which were with *O. insidiosus* (Table 6.1).

As did immature insect predators, immature predatory mites developed faster and survived longer on prey diets that were supplemented with plant food than on prey alone. Development rate increased in 14 out of 15 cases (93%) and survival increased in 11 out of 16 cases (65%) (Fig. 6.3A). Despite the very consistent positive influence, supplemental plant food increased mite development rate on average by only 3.7 ± 6.1% (range: 1.0% to 9.4%) (Fig. 6.3B). The benefit of supplemental plant food to nymphal mite survival was on average only slightly greater: 11.9 ± 2.4% (range: 3.3% to 30.2%) (Fig. 6.3B).

As in the experiments with insects, the quality of the supplemental plant food that was provided to nymphal predatory mites seemed to determine the magnitude of the effect of plant food on both development rate and survival. In all cases, pollen, extrafloral nectar, or homopteran honeydew was used to supplement prey-only diets (Table 6.1). Though pollen was provided as a supplement in fewer studies, the mean percentage increase in development rate was more than two times higher than when extrafloral nectar or honeydew were provided as a supplement (Table 6.3). Similarly, in the case that supplemental pollen had a positive effect on nymphal mite survival (out of two cases available), the percentage increase in survival was 30% greater than that provided by supplemental extrafloral nectar or honeydew (Table 6.3). Pollen had a greater positive effect on mite development rate than on survival, and a greater positive effect overall, probably because pollen has a higher nitrogen content than either extrafloral nectar or honeydew.

In just a few studies, supplementing prey-only diets with plant food negatively affected predatory mite development rate and survival. Supplemental plant food decreased mite development rate in only 1 out of 15 cases (6.7%) by 0.8% (Figs. 6.3A and B). Supplemental plant food decreased nymphal mite survival in 4 out of 17 cases (24%) (Fig. 6.3A) but by only a mean of 5.3 ± 1.6% (Fig. 6.3B).

Table 6.3 *Summary of the influence of two categories of supplemental plant food on selected life-history traits of zoophytophagous "predatory" mites; values are drawn from the data presented in Table 6.1*

			Positive effect		Negative effect	
Life-history trait	Supplemental plant food	Number of experiments	Percent of cases	Change (%)[a]	Percent of cases	Change (%)[a]
Immature development rate	Pollen	3	66.7	6.8 ± 2.6	33.3	−0.8
	Nectar/honeydew	12	100.0	3.2 ± 0.5	-	-
Immature survival	Pollen	2	50.0	15.0	50.0	−4.7
	Nectar/honeydew	15	66.7	11.6 ± 2.7	20.0	−5.5 ± 2.2
Adult female longevity	Pollen	7	28.6	118.5 ± 29.6	42.9	22.4 ± 6.8
	Nectar/honeydew	-	-	-	-	-
Adult female fecundity	Pollen	7	28.6	380.0 ± 78.3	71.4	29.2 ± 5.5
	Nectar/honeydew	15	86.7	43.7 ± 4.2	-	-

[a] Mean ± 1 SE.

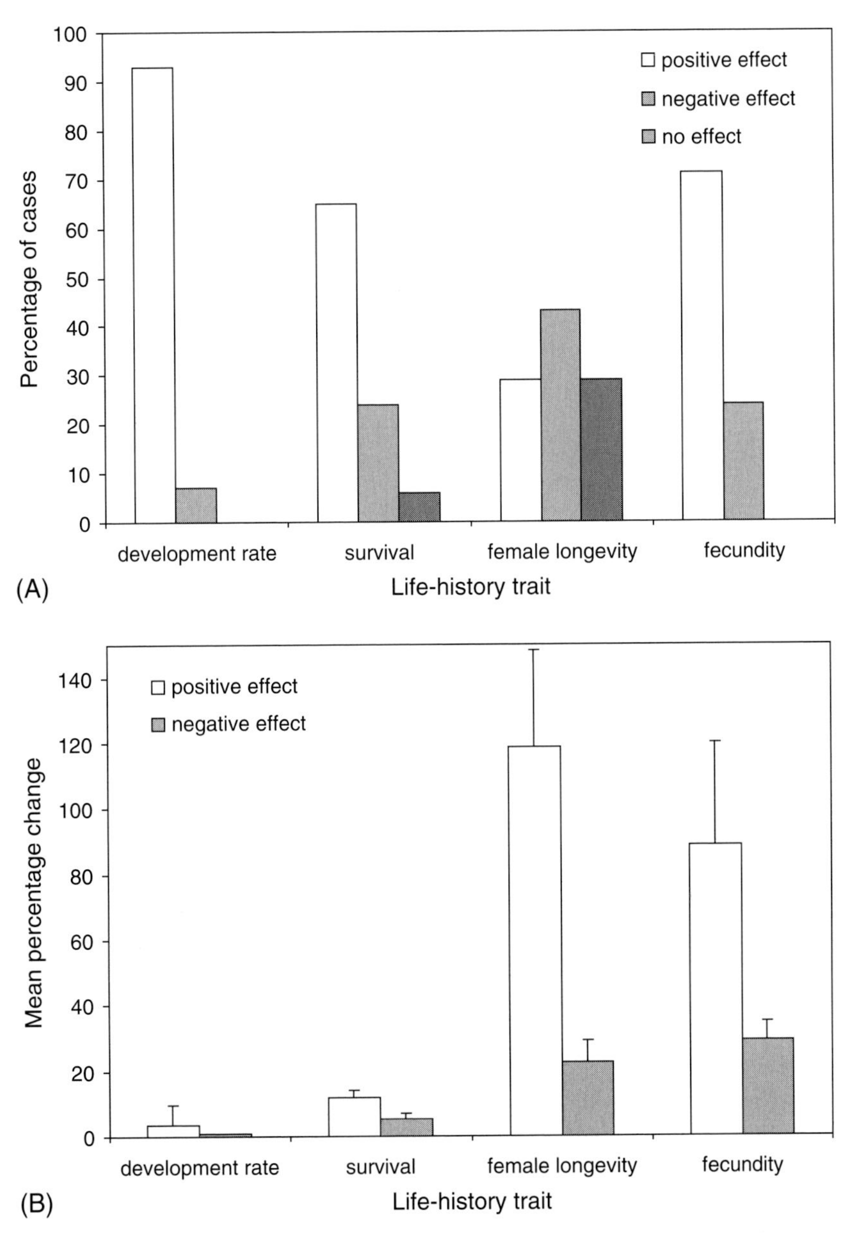

Figure 6.3 Summary of studies that investigated the influence of plant food as a supplement to prey-only diets on selected life-history traits of zoophytophagous predatory mites. (A) The percentage of cases (individual experiments) in which supplemental plant food had a positive effect, negative effect, or no effect on the traits of interest. (B) The mean ($\pm$ 1 standard error) percentage change in trait values (positive and negative) caused by the addition of plant food to prey-only diets. Both panels based on data presented in Table 6.1.

Effects of plant feeding on adult longevity and fecundity

In addition to benefiting immature development and survival, plant feeding also positively influenced life-history traits in adult predatory insects and mites. In insects, supplementing prey-only diets with plant food increased adult female longevity in 19 out of 26 cases (73%) (Fig. 6.2A), but the magnitude of the increase in longevity was highly variable. The consumption of supplemental plant food increased longevity from 1.2% to 182% with a mean of 38.5% (± 20.5%) (Fig. 6.2B). In 4 out of 26 cases (15%) the addition of plant food to prey-only diets actually decreased female longevity (Fig. 6.2A), but on average only by 4.8 ± 2.8% (Fig. 6.2B).

Whereas female longevity increased in most cases with the addition of plant food to prey-only diets, in only 20 out of 41 cases (49%) did supplemental plant food increase adult female fecundity in insect predators (Fig. 6.2A). For some insect predators, however, plant food is required for reproduction (4 out of 41 cases, 10%). In two studies of plant-supplement effects on fecundity in three predaceous lygaeids, adult female *Geocoris bullatus*, *G. pallens*, and *G. punctipes* were unable to produce eggs when maintained on a diet of pea aphids alone (Dunbar and Bacon 1972; Tamaki and Weeks 1972). If these four cases are included in the total number of cases in which supplemental plant food benefited female fecundity, then the total percentage of cases that reported such an effect increases to 59%. Despite this slim majority, the mean percentage increase in fecundity in these cases is very high: 60.1 ± 20.5% (Fig. 6.2B). In fact, the mean percentage increase in fecundity in these cases is four times as great as the mean percentage decrease (13.4 ± 3.8%) in cases in which the addition of plant food to prey-only diets negatively affected fecundity (16 out of 41 cases; 39%) (Fig. 6.2B). We note here, however, that 10 of these 16 cases are from one study of *G. punctipes* in which residual pesticide on the plant food supplement may have caused the decrease in fecundity (Naranjo and Stimac 1985) (Table 6.1).

Finally, several studies also measured the effects of plant food supplements to prey-only diets on adult female mass as a correlate of female fecundity. Whereas a consistent positive effect of supplemental plant food on fecundity was lacking, a majority of cases (16 out of 23; 70%) demonstrated a positive influence of supplemental plant food on female mass (Fig. 6.2A). Plant food supplements increased female mass on average by 9.1 ± 1.0%) (Fig. 6.2B). In contrast, the consumption of supplemental plant food decreased female mass in 6 out of 23 cases (26%) (Fig. 6.2A) by a mean of 6.8 ± 1.1% (Fig. 6.2B), mostly in studies of predaceous pentatomids (Table 6.1).

The quality of supplemental plant food, as defined by nitrogen content, was not consistently correlated with the magnitude of its positive effect on adult female longevity, fecundity, and mass in insect predators. Whereas all three categories of supplemental plant food (leaves, pods/seeds, pollen) equally benefited immature insect survival, pods/seeds and pollen provided a far greater benefit to adult insect longevity than leaves (Table 6.2). In contrast, leaves as a diet supplement provided a greater benefit to female fecundity than either pods/seeds or pollen (Table 6.2). This result is surprising considering that adult females require nitrogen for egg production and that leaves typically have less nitrogen than pods/seeds or pollen. Although the magnitude of the percentage increase in fecundity was greater for adults provided with leaf supplements, leaf supplements actually benefited fecundity in a minority of cases (6 out of 15; 40%). Diet supplements consisting of pods/seeds or pollen, on the other hand, benefited fecundity in a majority of cases (Table 6.2). Each of the categories of supplemental plant food increased adult female mass in a majority of cases but leaves and pollen provided the greatest benefit (Table 6.2).

The effect of supplemental plant food on longevity differed between adult predatory mites and adult insect predators. Whereas supplemental plant food increased insect longevity in a great majority of cases (Fig. 6.2A), plant food supplements increased mite longevity in only two of seven cases (29%) (Fig. 6.3A). In these two cases, however, supplemental plant food increased female longevity by a mean of 118.5 $\pm$ 29.6% (Fig. 6.3B), a far greater increase in longevity than in insect predators (compare Figs. 6.2B and 6.3B). Supplemental plant food decreased mite longevity in three of seven cases, but on average only by 22.4 $\pm$ 6.8% (Fig. 6.3B). In the other two cases, mite longevity was unaffected by the addition of plant food to a prey-only diet.

The effect of supplemental plant food on fecundity also differed between adult predatory mites and insect predators. Supplemental plant food decreased female fecundity in insects in almost as many cases as it increased fecundity, but plant food increased female fecundity far more often in mites. Supplemental plant food increased mite fecundity in 15 of 21 cases (71%) (Fig. 6.3A), and in one experiment, adult female mites could not produce eggs without a honeydew supplement (Ragusa and Swirski 1977) (Table 6.1). In these 15 cases, the benefit of supplemental plant food to mite fecundity was substantial: female mites given access to plant food increased fecundity by 88.5 $\pm$ 31.7% (Fig. 6.3B). Supplemental plant food decreased mite fecundity in only 5 of 21 cases (24%) (Fig. 6.3A), on average by only 29.2 $\pm$ 5.5% (Fig. 6.3B).

We could find only three studies that tested for the effect of phytophagy on adult female longevity in predatory mites (Table 6.1). Pollen was used as the plant food supplement in each of these studies so we cannot discuss how plant food quality influences mite longevity (Table 6.3). We did find, however, several studies that tested for the effects of supplemental pollen and extrafloral nectar/honeydew on mite fecundity. As might be expected based on nitrogen content, mites that were provided with pollen as a diet supplement showed a dramatically greater increase in fecundity (380%) than did mites that were provided with nectar/honeydew (Table 6.3). Surprisingly, however, supplemental pollen benefited mite fecundity in only two of seven cases (29%) (Bruce-Oliver *et al.* 1996) (Table 6.3, see also Table 6.1). Pollen as a plant food supplement had no effect on the fecundity of *Amblyseius swirskii* in one study (Momen and El-Saway 1993) and a negative effect on the fecundity of *Typhlodromalus peregrinus* in another study (Fouly *et al.* 1995).

Effects of plant feeding on population persistence without prey

This is one of the most important questions regarding the ecology of omnivorous predators. As discussed earlier, classical ecological theory predicts that omnivores may be more likely to suppress prey populations than strict predators because feeding at multiple trophic levels should allow omnivores to persist when prey are scarce (Crawley 1975; Pimm and Lawton 1977, 1978; Walde 1994; Eubanks and Denno 1999). Given the potential ecological consequences of omnivore persistence at low prey densities, it is surprising that few studies have addressed this topic.

Some studies indirectly support the idea that plant feeding allows omnivores to survive periods without prey. Notably, many minute pirate bugs (e.g., *O. insidiosus*) can complete development from egg hatching to adulthood on a pollen-only diet, although development rate and fecundity are frequently higher if prey are also available (Fauvel 1974; Kiman and Yeargan 1985; McCaffrey and Horsburgh 1986). Based on these results, most minute pirate bugs presumably can survive extensive periods without prey. Similarly, Cottrell and Yeargan (1998) reported that the ladybird beetle *Coleomegilla maculata* can complete development on only pollen and therefore should be able to survive extended prey-free periods.

In many phytoseiid mites, nymphs can fully develop when fed only pollen and adult females can produce eggs when provided with pollen (Peña 1992; Abou-Setta *et al.* 1997; Van Rijn and Tanigoshi 1999a; Nomikou *et al.* 2001). Indeed, in a review of the roles of phytoseiid mites in biological control, McMurtry and Croft (1997) claim that most species in this family reproduce on pollen, several at a rate equivalent to that on a diet of prey alone. This ability could allow persistence when prey are scarce. Predatory mites also consume extrafloral nectar and honeydew, both of which provide about the same benefit in adults as pollen. A diet of nectar alone, however, typically does not permit nymphal development (see Van Rijn and Tanigoshi 1999b).

Studies of the predaceous stink bug, *Podisus maculiventris*, show that adults survive almost 60 days without prey when provided with plant material (leaves of several crop species or green beans) (Valicente and O'Neil 1995; Wiedenmann *et al.* 1996). The effects of plant food on stink bug survival in the absence of prey, however, are analogous to providing the animals with water, so it appears that plant food primarily allows this insect to avoid desiccation (Valicente and O'Neil 1995). In contrast, other studies of this species suggest that plant food does not always result in extended survival in the absence of prey (Crum *et al.* 1998; Weiser and Stamp 1998). Without detailed experiments that directly ask this question, it is difficult to say whether plant feeding really allows this predator to survive extended prey-free periods.

In a study of big-eyed bugs (*G. punctipes*), Eubanks and Denno (1999) explicitly tested the hypothesis that plant feeding allows this omnivore to survive extended periods without prey. They conducted two experiments (one with nymphs, one with adults) to assess the effects of: (1) the length of the prey-free period (zero, 4 or 10 days without prey), (2) the plant part provided during the prey-free period (pods, leaves, or a water control), and (3) the prey species consumed prior to prey removal (moth eggs or pea aphids) on the ability of big-eyed bugs to survive prey-free periods.

Both big-eyed bug nymphs and adults were significantly more likely to survive prey-free periods when plant food was available versus a water control. Both previous prey and plant quality affected survival during prey-free periods. Big-eyed bug nymphs previously fed moth eggs were significantly more likely to survive 4-day periods without prey than big-eyed bugs previously fed aphids. They were also significantly more likely to survive prey-free periods when provided lima bean pods than when provided lima bean leaves or water, especially those previously fed aphids. The interactive effect of plant and prey quality was even greater during the 10-day prey-free period. Nymphs previously fed moth eggs only survived the 10-day period when they were provided plant food. Nymphs previously fed pea aphids only survived 10-day periods without prey when they were provided lima bean pods. There was also an interactive effect of prey and plant quality on the ability of big-eyed bug adults to survive periods without prey. First, almost all adult bugs survived a 4-day period without prey. It appears that they have enough stored energy to deal with these relatively short periods of prey scarcity. Adults previously fed moth eggs also faired well when faced with 10-day periods without prey. Approximately 95% of these survived regardless of the presence of plant food. Adults previously fed pea aphids, however, were more dependent on plant resources to sustain themselves: 90% of the adults provided with pods survived the 10-day period, but only 55% of those provided with lima bean leaves or water survived 10 days without prey.

Effects of plant feeding on dispersal and distribution

Many studies suggest that variation in plant food affects the dispersal and distribution of omnivores. Hoverflies are life-history omnivores that consume nectar and pollen as adults. Adult hoverflies are strongly attracted to flowering plants as well as artificial honeydew and are more likely to remain in areas with large numbers of flowering plants (Cowgill *et al.* 1993; Evans and Swallow 1993; Hickman and Wratten 1994, 1996; Hickman *et al.* 1995, 2001; Sutherland *et al.* 1999). The distribution of omnivorous mites in apple orchards and other habitats is strongly affected by the distribution of pollen (Kennett *et al.* 1979; Addison *et al.* 2000). Minute pirate bugs are also attracted to pollen and artificial nectar (Kiman and Yeargan 1985; Read and Lampman 1989; Evans and Swallow 1993) and are frequently more abundant on or around flowering plants (Coll 1996; Eubanks and Denno 1999). This is not surprising given that these omnivorous mites and bugs can complete their development on a diet of pollen (Kiman and Yeargan 1985; Van Rijn and Tanigoshi 1999a). Likewise, the ladybird beetle *C. maculata* is also attracted to pollen and is more abundant in areas with flowering plants (Coll and Bottrell 1991, 1992; Cottrell and Yeargan 1998; Harmon *et al.* 2000). Few of these studies, however, have actually documented dispersal associated with variation in plant quality. Most of these studies demonstrate a correlation between the abundance of flowering plants and omnivores, but as we have already seen, the survival, development, fecundity, and longevity of omnivorous predators are all affected by pollen. In the only study to investigate the effects of variation in plant quality on the dispersal of an omnivorous predator, Eubanks and Denno (1999) found that marked big-eyed bugs were significantly more likely to remain on lima bean plants with pods than on plants without pods (100% recaptured on plants with pods after 4 hours versus 36% recaptured on plants without pods after 4 hours). Although more detailed mark-and-recapture studies would be useful, it is probably safe to conclude that variation in plant food, especially flowers, has a strong effect on the dispersal of most omnivorous predators.

Effects of plant feeding on population size

Although plant feeding affects the survival, development, fecundity, longevity, and dispersal of omnivorous predators, variation in the effects of plant food on the biology of omnivores (discussed in each section) and variation in factors such as intraguild predation (Rosenheim *et al.* 1993, 1995) and competition for limited prey (Polis and McCormick 1986) probably limits any effect of plant feeding on omnivore abundance. Nevertheless, some studies do suggest that variation in plant food is more tightly linked to variation in the size of omnivore populations than variation in prey abundance. Populations of omnivorous mites are typically larger on plants containing wind-blown pollen (McMurtry and Scriven 1966b; Kennett *et al.* 1979; Addison *et al.* 2000). Big-eyed bug populations were consistently larger in lima bean plots containing plants with pods than in plots containing pod-free plants (Eubanks and Denno 1999). Furthermore, variation in the numbers of pods per plant was the most important predictor of big-eyed bug densities, whereas neither variation in the numbers of flowers per plant nor the density of herbivores (potential prey) was a significant predictor of big-eyed bug densities. Densities of big-eyed bug adults decreased rapidly after pod loss, suggesting that big-eyed bug adults emigrated from lima bean plots as soon as pod loss occurred. In contrast, the relatively slow reduction in nymphal densities suggests that nymphal survival slowly decreased in pod-free plots or that new nymphs were not recruited into the population because female big-eyed bugs were less abundant in pod-free plots.

The densities of omnivorous insects other than big-eyed bugs (minute pirate bugs, damsel bugs, and ladybird beetles) were also lower in pod-free plots than in control plots. However, because plants without pods flowered again late in the season, Eubanks and Denno (1999) were able to separate the effects of pods and flowers on the abundance of omnivores. Populations of these omnivores decreased following the loss of flowers and not after the loss of pods or changes in herbivore densities. As previously discussed, many omnivorous insects, especially minute pirate bugs and ladybird beetles, are pollen-feeders and are strongly attracted to flowers (Kiman and Yeargan 1985; Read and Lampman 1989; Coll and Bottrell 1991, 1992; Coll 1996). As with big-eyed bugs, variation in plant quality, flower density in this case, played an important role in the population size of minute pirate bugs and ladybird beetles.

Cottrell and Yeargan (1998) manipulated pollen abundance in corn by removing tassels from corn plants in large field plots and monitored the population size of *C. maculata* beetles in plots with relatively high pollen production and plots with no pollen production over two years. Ladybird beetle eggs and larvae were significantly more abundant in pollen plots during the second year of the study, but not in the first year. In another study of *C. maculata*, Harmon *et al.* (2000) found that populations of these beetles were larger in alfalfa fields with flowering dandelions and they attributed this to pollen attraction. Although more work on this question is needed, it appears that preferred plant food frequently results in larger populations of omnivorous predators.

Effects of plant feeding on per capita prey consumption

Several studies have attempted to quantify the effect of supplemental plant food on the per capita consumption of prey by omnivorous predators. A few studies have found that prey consumption increases when omnivores are provided with plant food. For example, moth egg consumption by the mirid *Dicyphus hesperus* increased when these omnivorous bugs were also allowed to feed on tomato leaves (Gillespie and McGregor 2000). Most studies of the effects of plant food on prey consumption by omnivores, however, have found that the presence of plant food decreases prey consumption. Several functional response experiments have found that the presence of pollen significantly reduces prey consumption by omnivorous mites (McMurtry and Scriven 1966a, 1966b; Wei and Walde 1997). Likewise, Cottrell and Yeargan (1998) found more intense moth egg predation by *C. maculata* in pollen-free corn plots than in plots with pollen. Cottrell and Yeargan concluded that pollen acted as a preferred, alternative food and significantly reduced per capita predation rates by *C. maculata*.

Eubanks and Denno (2000) conducted a similar study of the effects of plant feeding on the consumption of prey by big-eyed bugs. They quantified the effects of variation in plant quality (the presence or absence of bean pods) on the functional response of big-eyed bugs to aphids and moth eggs in the laboratory and the consumption of prey by big-eyed bugs on caged bean plants in the field. They found that the presence of pods significantly reduced the consumption of both prey species by big-eyed bugs in laboratory experiments. Pods, therefore, had an indirect, positive effect on both prey species. Likewise, big-eyed bugs significantly reduced the size of pea aphid populations on caged, podless lima bean plants in the field, but the suppression of aphid populations was significantly reduced when caged plants had pods. Because they frequently saw big-eyed bugs feeding on lima bean pods, Eubanks and Denno (2000) concluded that big-eyed bugs spent considerable time feeding on pods and became satiated as a result of pod feeding. Pods, therefore, serve as preferred alternative prey for big-eyed bugs in this system (Murdoch 1969; Holt and Lawton 1994).

Effects of plant feeding by omnivores on suppression of prey populations

Understanding the relative strengths of plant food on the numerical and functional responses of omnivores seems key to understanding the effects of plant feeding on prey population suppression by omnivores. Cottrell and Yeargan (1998) demonstrated that the positive effects of plant food on omnivore survival and distribution do not always translate into enhanced prey suppression. In this study, pollen-feeding was very beneficial to *C. maculata* ladybird beetles, but dramatically decreased prey consumption. The increase in the population size of omnivorous beetles due to pollen-feeding was not great enough to overcome the reduction in per capita prey consumption due to pollen-feeding and this resulted in reduced prey suppression.

Other studies, however, have demonstrated that the positive effects of plant feeding on omnivore performance and distribution can be more important then the negative effects of plant feeding on per capita prey consumption. For example, pollen-feeding by the omnivorous mite *Amblyseius hibisci* reduced the number of pest mites consumed by this species, but the positive effect of pollen on the numerical response of this omnivore was such that prey suppression was greatest on plants that contained pollen (McMurtry and Scriven 1966a, 1966b). Likewise, big-eyed bugs were more abundant and suppressed aphid and moth populations to a greater extent in bean plots with pods than in bean plots without pods, even though the presence of pods significantly reduced the per capita effects of big-eyed bugs on prey. Herbivore populations were suppressed such that plants should experience significantly less damage. Pod-feeding by big-eyed bugs, therefore, is likely to induce a trophic cascade whereby plant feeding by the omnivore ultimately benefits the plant.

These results are consistent with the verbal model presented by Polis and Strong (1996). Omnivory dispersed the direct effects of consumption throughout the food web rather than focussing them at a particular trophic level (e.g., pod-feeding reduced the consumption of prey by big-eyed bugs). However, because the ability to feed at multiple trophic levels resulted in relatively large, persistent, and often less variable omnivore populations, omnivory ultimately produced intense omnivore–prey interactions and promoted prey suppression (Eubanks and Denno 1999, 2000) (Fig. 6.1A).

Future work

It seems clear that plant feeding by "predaceous" arthropods is widespread and that plant feeding profoundly affects the biology and ecology of these animals. It also seems clear that variation in plant and prey quality is important. High-nitrogen plant food such as pollen, seeds, and pods seems to have the largest impact on omnivore performance, especially when high-quality plant food supplements poor-quality prey. Although studies of additional species may eventually produce exceptions, we believe that this is a general trend. What is less clear is the impact of plant feeding on prey suppression by omnivorous predators. It is difficult to make predictions about the role of plant food in the suppression of prey by omnivores because so few studies have adequately addressed this question (but see Van Rijn and Sabelis, Chapter 8). We feel that this is the most important unanswered question regarding omnivorous predators. Consequently, future research should focus on quantifying the effect of plant feeding on omnivore–prey interactions. We feel that additional studies of the effect of plant feeding on omnivore biology (survival, development, etc.) have relatively little to offer at this point. In contrast, studies that focus on the ecology of omnivore–plant–prey interactions have much to offer. In particular, we feel that understanding the diet preferences of omnivorous predators and the effects of plant food on the numerical response of omnivores is key to understanding the interaction between omnivores and prey. If understanding the functional and numerical response of omnivorous predators can allow us to predict the outcome of omnivore–prey–plant interactions, then relatively simple experiments may provide a powerful means of disentangling this complex trophic interaction.

Acknowledgments

We thank Moshe Coll, Helen Hull-Sanders, Jan Hulshof, Ian Kaplan, Paul van Rijn, and Michael S. Singer for constructive criticisms of early drafts. M. D. E. acknowledges funding from the National Science Foundation (DEB-9423543 and DEB-9903601), the US Department of Agriculture Southern Region Integrated Pest Management, and the Department of Entomology and Plant Pathology at Auburn University.

References

Abou-Setta, M. M., A. H. Fouly, and C. C. Childers. 1997. Biology of *Proprioseiopsis rotendus* (Acari: Phytoseiidae) reared on *Tetranychus urticae* (Acari: Tetranychidae) or pollen. *Florida Entomologist* **80**: 27–34.

Abrams, P. A. 1987. The functional responses of adaptive consumers of two resources. *Theoretical Population Biology* **32**: 262–288.

Addison, J. A., J. M. Hardman, and S. J. Walde. 2000. Pollen availability for predaceous mites on apple: spatial and temporal heterogeneity. *Experimental and Applied Acarology* **24**: 1–18.

Alomar, O. and R. N. Wiedenmann (eds.). 1996. *Zoophytophagous Heteroptera: Implications for Life History and Integrated Pest Management*. Lanham, MD: Entomological Society of America.

Bush, L., T. J. Kring, and J. R. Ruberson. 1993. Suitability of greenbugs, cotton aphids, and *Heliothis virescens* eggs for development and reproduction of *Orius insidiosus*. *Entomologia Experimentalis et Applicata* **67**: 217–222.

Bruce-Oliver, S. J., M. A. Hoy, and J. S. Yaninek. 1996. Effect of some food sources associated with cassava in Africa on the development, fecundity and longevity of *Eusius fustis* (Pritchard and Baker) (Acari: Phytoseiidae). *Experimental and Applied Acarology* **20**: 73–85.

Chinajariyawong, A. and V. E. Harris. 1987. Inability of *Deraeocoris signatus* (Distant) (Hemiptera: Miridae) to survive and reproduce on cotton without prey. *Journal of the Australian Entomological Society* **26**: 37–40.

Cocuzza, G. E., P. De Clercq, M. Van de Veire, *et al.* 1997. Reproduction of *Orius laevigatus* and *Orius albidipennis* on pollen and *Ephestia kuehniella* eggs. *Entomologia Experimentalis et Applicata* **82**: 101–104.

Cohen, A. C. and J. W. Debolt. 1983. Rearing *Geocoris punctipes* on insect eggs. *Southwestern Entomologist* **8**: 61–64.

Coll, M. 1996. Feeding and ovipositing on plants by an omnivorous insect predator. *Oecologia* **105**: 214–220.

Coll, M. and D. G. Bottrell. 1991. Microhabitat and resource selection of the European corn borer (Lepidoptera: Pyralidae) and its natural enemies in Maryland field corn. *Environmental Entomology* **20**: 526–533.

1992. Mortality of European corn borer larvae by enemies in different corn microhabitats. *Biological Control* **2**: 95–103.

Coll, M. and M. Guershon. 2002. Omnivory in terrestrial arthropods: mixing plant and prey diets. *Annual Review of Entomology* **47**: 267–297.

Cottrell, T. E. and K. V. Yeargan. 1998. Effect of pollen on *Coleomegilla maculata* (Coleoptera: Coccinellidae) population density, predation, and cannibalism in sweet corn. *Environmental Entomology* **27**: 1402–1410.

Cowgill, S. E., S. D. Wratten, and N. W. Sotherton. 1993. The effect of weeds on the numbers of hoverfly (Diptera: Syrphidae) adults and the distribution and composition of their eggs in winter wheat. *Annals of Applied Biology* **123**: 499–515.

Crawley, M. J. 1975. The numerical response of insect predators to changes in prey density. *Journal of Animal Ecology* **44**: 877–892.

Crum, D. A., L. A. Weiser, and N. E. Stamp. 1998. Effects of prey scarcity and plant material as a dietary supplement on an insect predator. *Oikos* **81**: 549–557.

Dayton, P. 1984. Properties structuring some marine communities: are they general? In D. Strong, D. Simberloff, L. Abele, and A. Thistle (eds.) *Ecological Communities: Conceptual Issues and the Evidence*. Princeton, NJ: Princeton University Press, pp. 181–197.

Dunbar, D. M. and O. G. Bacon. 1972. Feeding, development, and reproduction of *Geocoris punctipes* (Heteroptera: Lygaeidae) on eight diets. *Annals of the Entomological Society of America* **65**: 892–895.

Eubanks, M. D. and R. F. Denno. 1999. The ecological consequences of variation in plants and prey for an omnivorous insect. *Ecology* **80**: 1253–1266.

2000. Host plants mediate omnivore–herbivore interactions and influence prey suppression. *Ecology* **81**: 936–947.

Evans, E. W. 2000. Egg production in response to combined alternative foods by the predator *Coccinella transversalis*. *Entomologia Experimentalis et Applicata* **94**: 141–147.

Evans, E. W. and J. G. Swallow. 1993. Numerical responses of natural enemies to artificial honeydew in Utah alfalfa. *Environmental Entomology* **22**: 1392–1401.

Fauvel, G. 1974. Sur l'alimentation pollinique d'un anthocoride prédateur, *Orius (Heterorius) vicinus* Rib (Hémiptère). *Annales de Zoologie-Ecologie Animale* **6**: 245–258.

Fouly, A. H., M. M. Abou-Setta, and C. C. Childers. 1995. Effects of diet on the biology and life tables of *Typhlodromalus peregrinus* (Acari: Phytoseiidae). *Environmental Entomology* **24**: 870–874.

Gillespie, D. R. and R. R. McGregor. 2000. The functions of plant feeding in the omnivorous predator *Dicyphus hesperus*: water places limits on predation. *Ecological Entomology* **25**: 380–386.

Harmon, J. P., A. R. Ives, J. E. Losey, A. C. Olson, and K. S. Rauwald. 2000. *Coleomegilla maculata* (Coleoptera: Coccinellidae) predation on pea aphids promoted by proximity to dandelions. *Oecologia* **125**: 543–548.

Hickman, J. M. and S. D. Wratten. 1994. Use of *Phacelia tanacetifolia* (Hydrophyllaceae) as a pollen resource to enhance hoverfly (Diptera: Syrphidae) populations in sweetcorn fields. *Bulletin OILB/SROP* **17**: 156–167.

1996. Use of *Phacelia tanacetifolia* strips to enhance biological control of aphids by hoverfly larvae in cereal fields. *Journal of Economic Entomology* **89**: 832–840.

Hickman, J. M., G. L. Lovei, and S. D. Wratten. 1995. Phenology and ecology of hoverflies (Diptera: Syrphidae) in New Zealand. *Environmental Entomology* **24**: 595–600.

Hickman, J. M., S. D. Wratten, P. C. Jepson, and C. M. Frampton. 2001. Effect of hunger on yellow water trap catches of hoverfly (Diptera: Syrphidae) adults. *Agricultural and Forest Entomology* **3**: 35–40.

Holt, R. D. 1984. The ecological consequences of shared natural enemies. *Annual Review of Ecology and Systematics* **25**: 495–520.

Holt, R. D. and J. H. Lawton. 1994. Apparent competition and enemy-free space in insect host–parasitoid communities. *American Naturalist* **142**: 623–645.

Holt, R. D. and G. A. Polis. 1997. A theoretical framework for intraguild predation. *American Naturalist* **149**: 745–764.

Kennett, C. E, D. L. Flaherty, and R. W. Hoffmann. 1979. Effect of wind-borne pollens in the population dynamics of *Amblyseius hibisci* (Acari: Phytoseiidae). *Entomophaga* **24**: 83–98.

Kiman, Z. B. and K. V. Yeargan. 1985. Development and reproduction of the predator *Orius insidiosus* (Hemiptera: Anthocoridae) reared on diets of selected plant material and arthropod prey. *Annals of the Entomological Society of America* **78**: 464–467.

Limburg, D. D. and J. A. Rosenheim. 2001. Extrafloral nectar consumption and its influence on survival and development on an omnivorous predator, larval *Chrysoperla plorabunda* (Neuroptera: Chrysopidae). *Environmental Entomology* **30**: 595–604.

McCaffrey, J. P. and R. L. Horsburgh. 1986. Biology of *Orius insidiosus* (Heteroptera: Anthocoridae): a predator in Virginia apple orchards. *Environmental Entomology* **15**: 984–988.

McMurtry, J. A. and B. A. Croft. 1997. Life-styles of phytoseiid mites and their roles in biological control. *Annual Review of Entomology* **42**: 291–321.

McMurtry, J. A. and G. T. Scriven. 1964. Studies on the feeding, reproduction, and development of *Amblyseius hibisci* (Acarina: Phytoseiidae) on various food substances. *Annals of the Entomological Society of America* **57**: 649–655.

1966a. The influence of pollen and prey density on the numbers of prey consumed by *Amblyseius hibisci* (Acarina: Phytoseiidae). *Annals of the Entomological Society of America* **59**: 149–157.

1966b. Studies on predator–prey interactions between *Amblyseius hibisci* and *Oligonychus punicae* (Arcarina: Phytoseiidae, Tetranychidae) under greenhouse conditions. *Annals of the Entomological Society of America* **59**: 793–800.

Momen, F. M. and S. A. El-Saway. 1993. Biology and feeding behaviour of the predatory mite, *Amblyseius swirskii* (Acari, Phytoseiidae). *Acarologia* **34**: 199–204.

Moreira, L. A., J. C. Zanuncio, M. C. Picanço, and R. N. C. Guedes. 1997. Effect of *Eucalyptus* feeding on the development, survival and reproduction of *Tynacantha marginata* (Heteroptera: Pentatomidae). *Revista de Biologia Tropical* **45**: 253–257.

Murdoch, W. W. 1969. Switching in general predators: experiments on predator and stability of prey populations. *Ecological Monographs* **39**: 335–354.

Naranjo, S. E. and J. L. Stimac. 1985. Development, survival, and reproduction of *Geocoris punctipes* (Hemiptera: Lygaeidae): effects of plant feeding on soybean and associated weeds. *Environmental Entomology* **14**: 523–530.

Nomikou, M., A. Janssen, R. Schraag, and M. W. Sabelis. 2001. Phytoseiid predators as potential biological control agents for *Bemisia tabaci*. *Experimental and Applied Acarology* **25**: 271–291.

Peña, J. E. 1992. Predator–prey interactions between *Typhlodromalus peregrinus* and *Polyphagotarsonemus latus*: effects of alternative prey and other food resources. *Florida Entomologist* **75**: 241–248.

Pimm, S. L. and J. H. Lawton. 1977. The number of trophic levels in ecological communities. *Nature* **268**: 329–331.

1978. On feeding on more than one trophic level. *Nature* **275**: 542–544.

Polis, G. A. 1991. Complex trophic interactions in deserts: an empirical critique of food web theory. *American Naturalist* **138**: 123–155.

Polis, G. A. and S. J. McCormick. 1986. Scorpions, spiders and solpugids: predation and competition among distantly related taxa. *Oecologia* **71**: 111–116.

Polis, G. A. and D. R. Strong. 1996. Food web complexity and community dynamics. *American Naturalist* **147**: 813–846.

Polis, G. A., C. A. Myers, and R. D. Holt. 1989. The ecology and evolution of intraguild predation: potential competitors that eat each other. *Annual Review of Ecology and Systematics* **20**: 297–330.

Ragusa, S. and E. Swirski. 1977. Feeding habits, post-embryonic and adult survival, mating virility, and fecundity of the predaceous mite *Amblyseius swirskii* (Acarina: Phytoseiidae) on some coccids and mealybugs. *Entomophaga* **22**: 383–392.

Read, C. D. and R. L. Lampman. 1989. Olfactory responses of *Orius insidiosus* (Hemiptera: Anthocoridae) to volatiles of corn silks. *Journal of Chemical Ecology* **15**: 1109–1115.

Rosenheim, J. A., L. R. Wilhoit, and C. A. Armer. 1993. Influence of intraguild predation among generalist insect predators on the suppression of an herbivore population. *Oecologia* **96**: 439–449.

Rosenheim, J. A., H. K. Kaya, L. E. Ehler, J. J. Marois, and B. A. Jaffee. 1995. Intraguild predation among biological-control agents: theory and evidence. *Biological Control* **5**: 303–335.

Ruberson, J. R., M. J. Tauber, and C. A. Tauber. 1986. Plant feeding by *Podisus maculiventris* (Heteroptera: Pentatomidae): effect on survival, development, and preoviposition period. *Environmental Entomology* **15**: 894–897.

Salas-Aguilar, J. and L. E. Ehler. 1977. Feeding habits of *Orius tristicolor*. *Annals of the Entomological Society of America* **70**: 60–62.

Schmidt-Tiedemann, A. and P. Sell. 1997. The suitability of different kinds of pollen for feeding the predaceous anthocorid *Orius minutus* (L.) (Heteroptera: Anthocoridae). *Mededelingen van de Faculteit Landbouwwetenschappen, Universiteit van Gent* **62**(2b): 473–482.

Sutherland, J. P., M. S. Sullivan, and G. M Poppy. 1999. The influence of floral character on the foraging behavior of the hoverfly, *Episyrphus balteatus*. *Entomologia Experimentalis et Applicata* **93**: 157–164.

Tamaki, G. and R. E. Weeks. 1972. Biology and ecology of two predators, *Geocoris pallens* Stal and *G. bullatus* (Say). *US Department of Agriculture Technical Bulletin* **1446**.

Vacante, V., G. E. Cocuzza, P. De Clercq, M. Van de Meire, and L. Tirry. 1997. Development and survival of *Orius albidipennis* and *O. laevigatus* (Het.: Anthocoridae) on various diets. *Entomophaga* **42**: 493–498.

Valicente, F. H. and R. J. O'Neil. 1995. Effects of host plants and feeding regimes on selected life history characteristics of *Podisus maculiventris* (Say) (Heteroptera: Pentatomidae). *Biological Control* **5**: 449–461.

Van Rijn, P. C. J. and L. K. Tanigoshi. 1999a. Pollen as food for the predatory mites *Iphiseius degenerans* and *Neosiulus cucumeris* (Acari: Phytoseiidae): dietary range and life history. *Experimental and Applied Acarology* **23**: 785–802.

1999b. The contribution of extrafloral nectar to survival and reproduction of the predatory mites *Iphiseius degenerans* and *Ricinus communis*. *Experimental and Applied Acarology* **23**: 281–296.

Walde, S. J. 1994. Immigration and the dynamics of a predator–prey interaction in biological control. *Journal of Animal Ecology* **63**: 337–346.

Wei, Q. and S. J. Walde. 1997. The functional response of *Typhlodromus pyri* to its prey, *Panonychus ulmi*: the effect of pollen. *Experimental and Applied Acarology* **21**: 677–684.

Weiser, L. A. and N. A. Stamp. 1998. Combined effects of allelochemicals, prey availability, and supplemental plant material on growth of a generalist insect predator. *Entomologia Experimentalis et Applicata* **87**: 181–189.

Whitman, D. W., M. S. Blum, and F. Slansky Jr. 1994. Carnivory in phytophagous insects. In T. N. Ananthakrishnan (ed.) *Functional Dynamics of Phytophagous Insects*. Lebanon, NH: Science Publishers, pp. 161–205.

Wiedenmann, R. N., J. C. Legaspi, and R. J. O'Neil. 1996. Impact of prey density and facultative plant feeding on the life history of the predator *Podisus maculiventris* (Heteroptera: Pentatomidae). In O. Alomar and R. N. Wiedenmann (eds.) *Zoophytophagous Heteroptera: Implications for Life History and Integrated Pest Management*. Lanham, MD: Entomological Society of America, pp. 57–93.

Zhimo, Z. and J. A. McMurtry. 1990. Development and reproduction of three *Euseius* (Acari: Phytoseiidae) species in the presence and absence of supplementary foods. *Experimental and Applied Acarology* **8**: 233–242.

7

Nectar- and pollen-feeding by adult herbivorous insects

JÖRG ROMEIS, ERICH STÄDLER, AND FELIX L. WÄCKERS

Introduction

Among herbivorous insects with a complete metamorphosis the larval and adult stages usually differ significantly in their biology, food requirements, and ecology (Schoonhoven *et al.* 1998). Often it is the larval stage that is strictly herbivorous, causing damage to a plant, whereas frequently the adult has a different diet, disperses, selects suitable environments (host plants), and reproduces. Studies on herbivore nutritional ecology generally focus on plant feeding by the damaging larval stages. However, the nutritional ecology and foraging behavior of adult stages can also be crucial to our understanding of plant–herbivore interactions. Both as pollinators and as parasites, adult herbivores can impose a strong selective force in the evolution of plant-provided food supplements (Brody 1997). Here we describe the use of plant-provided foods by adult herbivores to provide insight into this often neglected aspect of plant–herbivore interactions.

Adult insects carry over energy reserves and nutrients acquired during larval development. The level of these reserves can vary markedly among species and may be complemented with nutrients obtained through adult feeding (Boggs 1981, 1997a,b; Tsitsipis 1989; May 1992). Some species primarily depend on larval reserves throughout their adult life and require little or no additional feeding (Barbehenn *et al.* 1999). Such non-feeding adults are relatively common among Lepidoptera (Miller 1996) but have also been reported among Diptera (Drew and Yuval 2000). Females of some species receive nutrients during matings (Wheeler 1996). This transfer of so-called nuptial gifts from males to females has been reported for Lepidoptera (Boggs 1990, 1995), Diptera (Aluja *et al.* 1993), and Coleoptera (Fox 1993; Takakura 1999).

Plant-Provided Food for Carnivorous Insects, ed. F. L. Wäckers, P. C. J. van Rijn, and J. Bruin.
Published by Cambridge University Press.

The majority of species, however, have to feed as well to meet their nutritional requirements.

Nutrients obtained through adult feeding are principally used as energy source and to sustain physiological processes including egg maturation. Many herbivorous species primarily rely on carbohydrates as a source of energy that can be mobilized quickly, but other species use lipids or proteins (Beenakkers *et al.* 1984; Barbehenn *et al.* 1999). In most species, both carbohydrates and amino acids or proteins are needed for egg maturation (Tsitsipis 1989; Boggs 1997b). As gametogenesis in females requires more energy and nutrients than in males, females often have a higher demand for nutrients (Wheeler 1996).

Adult herbivores frequently cover their nutritional requirements through feeding on plant-provided food supplements such as floral nectar and pollen. If the adult herbivore acts as a pollinator, this interaction is often mutualistic (Faegri and Van der Pijl 1979; Kevan and Baker 1999). Many species, however, collect nectar and/or pollen without contributing to pollination. In such cases, the cost to the plant may range from a mere loss of the food item to the destruction of reproductive organs. Non-pollinators that remove the foods without physically damaging the plant are referred to as nectar/pollen thieves. This strategy is found among Lepidoptera (Wiklund *et al.* 1979; Jennersten 1984) and Diptera (Larson *et al.* 2001). Insects that bypass the flower opening and damage floral tissue in the process are referred to as nectar/pollen robbers, a strategy described among others in several Coleoptera (Tahhan and Van Emden 1989; Clement 1992) (terminology *sensu* Inouye 1980). Floral herbivory can indirectly reduce the reproductive success of the attacked plants if the damaged flower or the entire plant becomes less attractive to pollinators (Krupnick and Weis 1999; Krupnick *et al.* 1999). A particular category of costs may arise if the presence of adult food increases a plant's chance of being exposed to herbivory (see also Sabelis *et al.*, Chapter 4). For example, Lukefahr *et al.* (1965) reported that several species of Lepidoptera deposited more eggs on cotton varieties possessing extrafloral nectaries as compared to nectariless varieties. Similarly, the presence of *Pseudococcus longispinus* (Sternorrhynchae: Pseudococcidae) honeydew on avocado crops resulted in enhanced oviposition by adults of the moth *Cryptoblabes gnidiella* (Lepidoptera: Pyralidae) visiting this food source (Swirski *et al.* 1980), resulting in increased larval damage to the fruits.

Biological control workers have long been aware that food supplements such as nectar and/or pollen may be used to sustain the effectiveness of parasitoids and predators. This notion has stimulated the supplementing of low-diversity agroecosystems with natural (flowering herbs, extrafloral nectary plants) or artificial (sprays) food sources (Gurr *et al.* 1998; Landis *et al.* 2000; Wilkinson and Landis, Chapter 10). The fact that herbivores may utilize these food sources as well obviously needs to be addressed when designing these programs. Understanding the food selection process and the food needs of adult herbivores can help identify selective food sources, which can enhance the effectiveness of biocontrol agents without benefiting the pest (Rogers 1985; Gurr *et al.* 1998; Wäckers 1999; Romeis and Wäckers 2000; Landis *et al.* 2000).

In this chapter we focus on the role of pollen and floral or extrafloral nectar as food sources for adult herbivores, even though insects may obtain a number of other rewards from plants (Faegri and Van der Pijl 1979; Kevan and Baker 1999; Wäckers 2004, Chapter 2). We review the use of plant-derived foods by adult Lepidoptera, Diptera, and Coleoptera, as these insect orders include important pest species. Several other orders feature herbivores that also use nectar and pollen as a food source, such as Heteroptera (Schuster *et al.* 1976) and Thysanoptera (Kirk 1997).

Lepidoptera

Lepidoptera as pests and pollinators

Most of the estimated total of 119 000 lepidopteran species are herbivorous during their larval stage (Schoonhoven *et al.* 1998). The adult stages of many species visit flowers to feed on nectar and/or pollen (Norris 1936; Gilbert and Singer 1975). As such, Lepidoptera are one of the most important groups of pollinators. Many species have a strong mutualistic relationship with plants, reflected in specific adaptions in both Lepidoptera and flowers. Butterfly flowers are generally brightly colored, often possess floral guides, may or may not have a strong smell, have long tubular corollas, and often provide a landing platform (e.g., Asteraceae) (Faegri and Van der Pijl 1979; Kevan and Baker 1999; Dobson 1994; Proctor *et al.* 1996). Flowers visited by nocturnal moths are frequently pale and strongly scented. These flowers may not have landing platforms as many moths feed while hovering (Faegri and Van der Pijl 1979; Kevan and Baker 1999; Proctor *et al.* 1996). Flowers visited by butterflies and non-hovering (settling) moths tend to produce nectar that is sucrose dominant and rich in amino acids, whereas nectar from hawkmoth-visited flowers contains a comparably low amount of amino acids (Watt *et al.* 1974; Baker and Baker 1975, 1982, 1983).

Lepidoptera often exhibit distinct flower preferences (Wiklund 1977; Jennersten 1984; Boggs 1987; Erhardt and Thomas 1991; Porter *et al.* 1992; Tooker *et al.* 2002). They may select flowers on the basis of visual and/or olfactory cues (Scherer and Kolb 1987; Dobson 1994; Andersson 2003), nectar accessibility (Boggs 1987; Corbet 2000), nectar composition (Baker and Baker 1975; Masters 1991), or nectar concentration (Boggs 1988; May 1985). Innate flower preferences may be modified through associative learning (Kelber and Pfaff 1997), allowing adults to focus on the most profitable nectar sources. Both operant and classical conditioning have been reported (Lewis 1993; Cunningham *et al.* 1998; Andersson 2003). Learning is particularly important because nectar foraging can comprise a considerable proportion of an adult's time (Shreeve 1992).

Sources of nutrition

Floral nectar is the principal carbohydrate source for Lepidoptera (Norris 1936; Gilbert and Singer 1975; Boggs 1987), although certain species also obtain carbohydrates through extrafloral nectar (Lukefahr and Rhyne 1960; Beach *et al.* 1985), honeydew (Waldbauer *et al.* 1980; Johnson and Stafford 1985; Miller 1989), and plant sap and fruits (Norris 1936), and some are known to feed on non-plant foods including dung, bird droppings, or even blood (Norris 1936; Boggs 1987). Several species have been reported to leave their oviposition sites to search for nectar sources (Topper 1987; Raulston *et al.* 1998; Esquivel and Lingren 2002).

Most Lepidoptera feature an elongated proboscis that facilitates the exploitation of nectar at the base of long-tubed flowers (Gilbert and Singer 1975). Proboscis length is often correlated to corolla tube length of the flower species visited (Gilbert and Singer 1975). The physical dimensions of the proboscis may limit uptake of viscous nectar and thus determine which nectar concentrations yield the highest energy reward (Kingsolver and Daniel 1979; Boggs 1988). Apart from viscosity, also the amount of sugar in a solution may affect the dynamics of fluid ingestion, as was shown in a recent study of nectar-feeding by the hovering hawkmoth, *Macroglossum stellatarum* (Sphingidae) (Josens and Farina 2001).

Pollen-feeding has only been described for a few lepidopteran species (Boggs 1987). The best-known examples involve butterflies of the genera *Heliconius* and *Laparus* (Nymphalidae) (Gilbert 1972). They collect pollen with their proboscis and mix it with an exuded fluid (saliva). This mixture is then agitated for several hours by uncoiling and recoiling of the proboscis. As pollen grains germinate, they release amino acids and other nutritive compounds into the saliva that is subsequently ingested by the butterflies (Gilbert 1972; Boggs 1987). Studies on *Heliconius charitonia* have demonstrated that essential amino acids are transferred directly from pollen to the butterfly eggs (O'Brian *et al.* 2003). Other unrelated butterfly species seem to use pollen in a similar way (DeVries 1979).

The proboscis of pollen-feeding butterflies appears to exhibit morphological adaptations (Gilbert 1972; Krenn and Penz 1998). A comparison of the probosces of pollen-feeding and non-pollen-feeding *Heliconius* species revealed that in pollen-feeding species the proximal region, where the pollen loads are accumulated, possesses longer and more bristle-shaped sensilla (Krenn and Penz 1998).

Behavioral observations of pollen-feeding butterflies did not reveal any unique flower-handling behavior (Penz and Krenn 2000). A special case of pollen-feeding Lepidoptera are micropterygid moths of the genus *Sabatinca*. These ancient insects have chewing mouthparts and feed on the lipid-rich pollenkitt of *Zygogynum* spp. (Winteraceae), on which they also act as pollinators (Thien *et al.* 1985).

Even though the majority of flower-visiting Lepidoptera do not feed directly on pollen, they may nevertheless benefit from pollen-derived nutrients including amino acids and proteins when they feed on pollen-contaminated nectar (Baker and Baker 1975; Erhardt and Baker 1990).

Gustatory response

Once a food source has been located, the insect uses contact-chemoreceptor (gustatory) sensilla to assess its composition and to decide on its acceptance. Both carbohydrates and amino acids are known to play a role at this stage.

Carbohydrates

Several behavioral studies have shown that Lepidoptera of different families perceive sugars with contact-chemoreceptors on their tarsae (Table 7.1), their proboscis (Fig. 7.1), and/or their antennae (Frings and Frings 1956; Ramaswamy 1987; Blaney and Simmonds 1988, 1990; Erhardt 1991, 1992; Rusterholz and Erhardt 1997; Erhardt and Rusterholz 1998; Romeis and Wäckers 2000).

Electrophysiological investigations revealed that the proboscis of moths bear sensilla with receptor neurones most sensitive to the sugars sucrose, fructose, and glucose (Städler and Seabrook 1975; Blaney and Simmonds 1988). The neurones of tarsal sensilla of four noctuid moths were equally sensitive to stimulation by these three sugars (Blaney and Simmonds 1990). All species tested in behavioral studies showed a response following tarsal stimulation with sucrose and fructose, whereas the response to glucose varied among species; some species responded to other mono-, di-, and trisaccharides (Table 7.1).

Based on an extensive data set, Baker and Baker (1982, 1983) concluded that butterflies tend to prefer nectar with a high sucrose-to-hexose ratio. This conclusion has been supported by choice and no-choice experiments, reporting that butterflies show a stronger gustatory response to sucrose over fructose and to fructose over glucose (Kusano and Adachi 1969; Kusano and Sato 1980; Erhardt 1991, 1992; Rusterholz and Erhardt 1997; Romeis and Wäckers 2000). Interestingly, a similar sugar preference has been reported for saprophagous butterfly species (Ômura and Honda 2003). However, other studies reported that butterflies preferentially visit flowers with nectar dominated by the

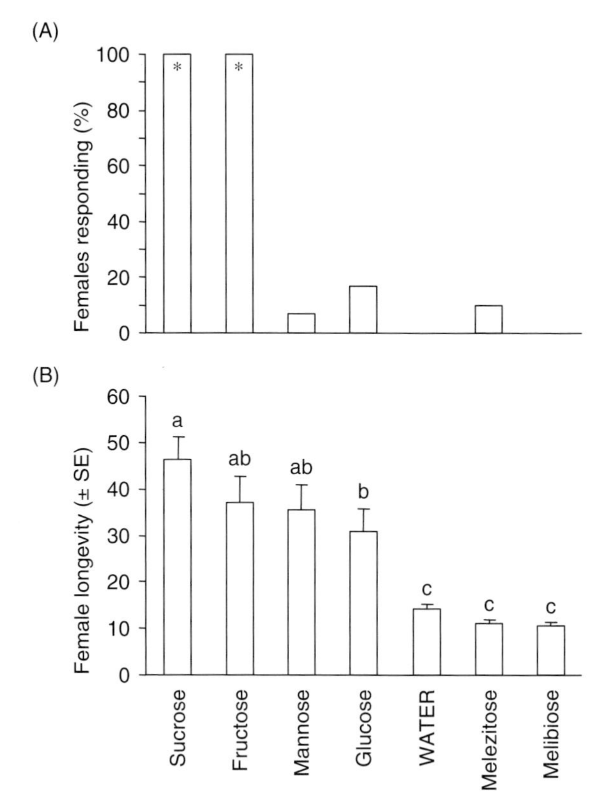

Figure 7.1 Comparison of individual sugars with respect to (A) gustatory response of *Pieris brassicae* females after proboscis stimulation, and (B) effect on butterfly longevity. Different letters above bars indicate significant differences ($P < 0.05$). Asterisks indicate gustatory stimulation (significant difference to the water control, $P < 0.05$). Data after Romeis and Wäckers (2000, 2002).

monosaccharides fructose and glucose (Watt *et al.* 1974; Rusterholz and Erhardt 2000). Sex and physiological state have been found to affect the responsiveness and preference for individual sugars in some species (Kusano and Sato 1980; Erhardt 1992; Rusterholz and Erhardt 2000).

Amino acids

Variable results have been reported on the perception of and preference for amino acids in no-choice and choice experiments. So far, only few studies have shown that butterflies respond to the presence of amino acids. In choice experiments *Jalmenus evagoras* (Lycaenidae) was more likely to feed on a water–amino acid solution than on plain water (Hill and Pierce 1989). Females of *Pieris rapae* (Pieridae) preferred a sugar plus amino acid solution (nectar mimic) over a plain sugar solution, and they ate more of the nectar

Table 7.1 *Sugars which elicit a proboscis extension response in water-satiated adult Lepidoptera after tarsal stimulation*

	Pieris brassicae[a]	*Pieris rapae crucivora*[a]	*Pyrameis atalanta*[b]	*Spodoptera littoralis*[b]	*Heliothis virescens*[a]	*Helicoverpa zea*[c]
Monosaccharides						
Fructose	+	+	+	+	– (+)	– (+)
Glucose	– ((–))	(–)	+	+	– (+)	– (–) ((–))
Galactose	–	(–)	– (–)	–		– (–)
Mannose	–	(–)	(–)	–		
Fucose		(+)	+			+
Rhamnose		(–)		+	(–)	– (–)
Sorbose		(–)		+		– (–)
Xylose		(–)	–(–)	–		
Ribose		(–)			(–)	– (–)
Arabinose		(+)	– (+)	–		
Erythrose				–		
Disaccharides						
Sucrose	+	+	+	+	+	+
Trehalose	–	(–)	– (–) ((+))			
Maltose	–	(–)	+	–?((+))		– (–)
Melibiose	–	(–)		–		

Lactose		(–)	(–)	–		
Cellobiose		(–)	– (–)	–		
Turanose		(–)		+		– (–)
Trisaccharides						
Raffinose	–	(–)	+	+	–	– (–)
Melezitose	–	(–)	+	+		– (–)
Reference	Romeis and Wäckers 2000	Kusano and Sato 1980	Weis 1930	Salama *et al.* 1984	Ramaswamy 1987	Lopez *et al.* 1994

[a] Only females tested.

[b] A mixture of females and males tested.

[c] Only males tested.

+ indicates a response by a minimum of 25% of the insects. If not otherwise stated, sugar concentrations tested were: 0.5 M, (1 M), ((2 M)).

mimic (Alm *et al.* 1990). Similar results were reported for females of *Inachis io* (Nymphalidae) (Erhardt and Rusterholz 1998; Mevi-Schütz and Erhardt 2002), *Araschnia levana* (Nymphalidae) (Mevi-Schütz and Erhardt 2003a), and *Coenonympha pamphilus* (Satyridinae) (Mevi-Schütz *et al.* 2003). In contrast, other species do not seem to display any preference (Hill 1989; Erhardt 1991, 1992; Romeis and Wäckers 2000). Lack of response may be due to the particular amino acids tested (Inouye and Waller 1984; Lanza and Krauss 1984) or to the nutritional status of the butterfly (Mevi-Schütz and Erhardt 2003b,c; Mevi-Schütz *et al.* 2003). The response to amino acids can also depend on the time of year, as only spring-generation females of *Lysandra bellargus* (Lycaenidae) showed a preference for flowers with nectar rich in amino acids (Rusterholz and Erhardt 2000).

Blaney and Simmonds (1990) identified tarsal sensilla with neurones sensitive to amino acids in four species of noctuid moths. The neural and behavioral responses to the tested amino acids varied among species, but differences in sensitivity to alanine, phenylalanine, leucine, and lysine were minor. Gustatory perception of carbohydrates and amino acids may interact: the labial palps of ants have neurones whose sensitivity to glucose is enhanced by the amino acid glycine (Wada *et al.* 2001). Such synergistic effects might also exist in the sensory systems of adult butterflies and may have been overlooked before. Blaney and Simmonds (1988) found that amino acids act synergistically with sucrose as phagostimulants and this seems an indication that the same neurones react to both types of stimulants.

Little is known about the factors eliciting a gustatory response towards pollen in Lepidoptera. Yet some species are able to recognize pollen and even differentiate among pollen from different plant species, as reported for the sunflower moths *Homoeosoma electellum* and *H. nebulellum* (Pyralidae) (Delisle *et al.* 1989; Le Metayer *et al.* 1993). Both moths prefer sunflower plants covered with pollen for oviposition, as pollen is an essential food for neonate larvae. The chemoreceptors for pollen recognition are located on the antennae of *H. electellum* (Delisle *et al.* 1989) and on the ovipositor of *H. nebulellum* (Le Metayer *et al.* 1993).

Nutritional requirements and food utilization

Carbohydrates are thought to be the primary source of energy for adult Lepidoptera (Boggs 1987). The nutritional requirements of adult Lepidoptera may vary among species, or between sexes and often depend on the physiological status (e.g., mating status, age, migration) of the insects. Species whose feeding is restricted to protein-poor nutrient sources, such as nectar, are believed to cover their adult protein requirements mainly by metabolizing their larval reserves (Boggs 1987). In addition, females of some species receive nutrients from the males at mating (Boggs 1990, 1995). Also, pre-ovigenic species are expected to be less dependent on adult food than synovigenic species that mature yolked eggs after emergence (Boggs 1986, 1997a; Karlsson 1994). Sex-specific flower preferences reflect the differences in nutrient requirements between sexes (Wiklund and Åhrberg 1978; Ohsaki 1979; Pivnik and McNeil 1987; Rusterholz and Erhardt 2000). In addition, the time allocated to feeding may differ between sexes: females of *Pieris brassicae* (Pieridae) spend more time feeding than males (Dover 1997). In several other species, males were found to have a higher nectar uptake rate than females resulting in a greater energy intake (May 1985; Boggs 1988). Laboratory studies on different butterfly species revealed that while females often show a distinct preference for a nectar mimic that contains amino acids, males lack such a preference (Alm *et al.* 1990; Erhardt and Rusterholz 1998; Mevi-Schütz and Erhardt 2002, 2003a,b; Mevi-Schütz *et al.* 2003). Studies by Boggs *et al.* (1981) on *Heliconius* species revealed that females in general collected more pollen than males and that older individuals collected more pollen than younger individuals, which probably is associated with respective nitrogen requirements. Also the mating status of female moths can affect their behavioral response to flower signals, mated females being more responsive than unmated individuals (Wiesenborn and Baker 1990; Tingle and Mitchell 1992). In the case of *Pieris napi*, where nutrients are transferred at mating, mated females showed a decreased preference for nectar mimics containing amino acids (Mevi-Schütz and Erhardt 2003b). Differences in nutrient requirements between sexes and among species are therefore affecting the insects' behavioral or even sensory thresholds for food compounds, as discussed above.

It is difficult to impart a clear picture of the effects of specific nectar/pollen compounds on longevity and/or fecundity in adult Lepidoptera. Unmated females in general lay significantly fewer eggs than fertilized females (David and Gardiner 1962; Shorey 1963; Wiklund and Karlsson 1984), thus requiring less energy to maintain basal metabolic functions (Rockstein and Miquel 1973). Furthermore, adult food can have an effect on the mating success of Lepidoptera (Luckefahr and Martin 1964; Jensen *et al.* 1974; Howell 1981; Lederhouse *et al.* 1990). Nevertheless, the available studies allow for some generalizations.

Carbohydrates

The three main nectar sugars – sucrose, fructose, and glucose – generally enhance longevity in Lepidoptera (Lukefahr and Martin 1964; Gunn and Gatehouse 1985; Willers *et al.* 1987; Carroll and Quiring 1992; Leahy and Andow 1994; McEwen and Liber 1995). Some additional nectar and honeydew sugars were found to increase longevity in some species of *Pieris* (Kusano and Adachi

1969; Kusano and Nishide 1978; Kusano and Sato 1980; Romeis and Wäckers 2002) (Fig. 7.1), while other sugars (maltose, melibiose, melezitose, trehalose) had little or even a negative effect on *P. brassicae* longevity (Romeis and Wäckers 2002). Several studies have shown that longevity increases with an increase in the concentration of the sugar solution provided (Kozhantshikov 1938; Shorey 1963; Kusano and Adachi 1969; Hill 1989; Hill and Pierce 1989).

Sugar feeding can also have a stimulatory effect on lifetime fecundity, due to a longer oviposition period (Kozhantshikov 1938; Lukefahr and Martin 1964; Cheng 1972; Gunn and Gatehouse 1985; Willers *et al.* 1987; Carroll and Quiring 1992; Leahy and Andow 1994; McEwen and Liber 1995; Romeis and Wäckers 2002), a higher daily oviposition rate (Cheng 1972; Willers *et al.* 1987; Carroll and Quiring 1992; Leahy and Andow 1994; Romeis and Wäckers 2002), and/or a shorter pre-oviposition period (Cheng 1972; Miller 1987).

It is often observed that the weight or size of lepidopteran eggs decreases with female age (Wiklund and Karlsson 1984; Begon and Parker 1986), but the effect of adult food is unclear. Studies on *Ostrinia nubilalis* (Crambidae) did show that females are able to maintain their egg weight when fed fructose, sucrose, glucose, or honey, while egg weight declines with age in water-fed females (Miller 1988; Leahy and Andow 1994). Also egg fertility has been reported to decrease with female age (David and Gardiner 1962; Howell 1981; Wiklund and Persson 1983; Carroll and Quiring 1992). This is probably due to either sperm depletion or to an age-associated decline in fertilization efficiency or oocyte production (Engelmann 1970). Currently there is no indication that egg fertility is affected in females with access to a carbohydrate source when compared to water-fed females (Howell 1981; Gu and Danthanarayana 1990).

Amino acids

Amino acids in general do not seem to enhance longevity of nectar-feeding adult Lepidoptera (Murphy *et al.* 1983; Moore and Singer 1987; Hill 1989; Hill and Pierce 1989; Romeis and Wäckers 2000; Mevi-Schütz and Erhardt 2003c; see Gunn and Gatehouse 1985 for an exception). Similarly, longevity of *Spodoptera exempta* (Noctuidae) and *O. nubilalis* was not increased when the insects were fed proteinaceous material (Gunn and Gatehouse 1985; Leahy and Andow 1994). Neither does consumption of amino acids or proteinaceous material necessarily benefit fecundity (Murphy *et al.* 1983; Moore and Singer 1987; Gunn and Gatehouse 1985; Hill 1989; Hill and Pierce 1989; Leahy and Andow 1994; Romeis and Wäckers 2002; Mevi-Schütz and Erhardt 2003c). Nevertheless, radiotracer studies in two nymphalid butterflies showed that glucose and amino acids ingested by the adults are used for egg production (Boggs 1997b). O'Brian *et al.* (2000, 2002) reported that essential amino acids present in eggs of the hawkmoth *Amphion floridensis* derive entirely from the female's larval reserves while the non-essential amino acids in eggs are to a large extent synthesized from nectar sugars. The fact that the non-essential amino acids in butterfly eggs are partly synthesized from nectar sugars has also been reported for the distantly related pollen-feeding *Heliconius charitonia*, suggesting that this synthesis might be a widespread phenomenon in nectar-feeding Lepidoptera (O'Brian *et al.* 2003). Contradictory results have been reported with respect to the question whether amino acids affect egg weight in *Euphydryas editha* (Nymphalidae) (Murphy *et al.* 1983; Moore and Singer 1987). Hill (1989) and Hill and Pierce (1989) found in their studies of Lycaenidae that egg weight is not affected by adult food.

Most of the above studies indicate that amino acids play only a minor role in the nutrition of the majority of nectar-feeding Lepidoptera. However, the various mixtures of amino acids that have been studied may not have matched the nutritional requirements of the butterfly species tested, nor were test insects in most studies allowed to respond to these "unbalanced" food sources by compensatory feeding, a mechanism described in locusts and caterpillars (Raubenheimer and Simpson 1999). Therefore, more field and semi-field experiments should be conducted with insects that have access to natural nectar sources. Field-collected *Coenonympha pamphilus* (Satyridae) were unable to sustain a constant egg production and egg weight despite the fact that they had free access to their natural nectar sources (Wickman and Karlsson 1987). More recent studies revealed that female butterflies may select for amino-acid-rich nectar as a response to their nutritional requirements. Spring-generation females of *L. bellargus* prefered to visit flowers with nectar rich in amino acids, perhaps in relation with the nutritional needs for egg production and development (Rusterholz and Erhardt 2000). Studies on *A. levana* and *C. pamphilus* showed that larval food quality can affect the nectar preference of females (Mevi-Schütz and Erhardt 2003a; Mevi-Schütz *et al.* 2003). Females reared on a nutritionally poor diet showed a clear preference for an amino-acid-rich nectar mimic, suggesting that they compensate for poor larval conditions through selective adult feeding. A study on *P. napi* also revealed that butterflies are able to compensate for a lack of nutrients by selecting nectar rich in amino acids (Mevi-Schütz and Erhardt 2003b). Females of this polyandrous species benefit from nuptial gifts transferred during mating, and mating was found to decrease the nectar amino acid preference in this species.

In contrast to nectar-feeding Lepidoptera, the pollen-feeding *Heliconius* species require their pollen meals for survival and continuous oogenesis (Boggs 1987). Nothing is known about the nutritive benefit of pollenkitt feeding in *Sabatinca* spp. (Thien *et al.* 1985).

Diptera

Diptera as pests and pollinators

Schoonhoven *et al.* (1998) estimated that of the total of 119 000 species of Diptera about 30% are herbivorous. The larvae (maggots) attack all plant parts from the roots to flowers and seeds (Skevington and Dang 2002).

Numerous species of each major group of Diptera (Nematocera, Brachycera, Cyclorrapha) have been recorded to visit flowers and to feed on pollen or nectar (Faegri and Van der Pijl 1979; Kevan and Baker 1999; Skevington and Dang 2002). Some dance flies (Empididae) are obligate pollen-feeders (Grimaldi 1999). Larson *et al.* (2001) concluded that Diptera are the second most important order, after the Hymenoptera (Apoidea), of flower-visiting and flower-pollinating insects and their activity as pollinators can be considered basic to angiosperm evolution (Labandeira 1998; Grimaldi 1999). Their associations with flowers seem to be mostly opportunistic. The Diptera are divided into long- and short-tongued species. The length of the mouthparts affects the range of flower species visited (Faegri and Van der Pijl 1979; Kevan and Baker 1999; Proctor *et al.* 1996). Flowers pollinated by Diptera are in general regular and simple, often clustered into dense inflorescenses, with light, dull colors and imperceptible to heavy-sweet scents (Faegri and Van der Pijl 1979). Noticeable exceptions are those flowers that release volatile protein hydrolysates to attract saprophagous Diptera (Kaib 1974; Stensmyr *et al.* 2002).

Diptera often have large eyes with many ommatidia and it is likely that at least all the higher Diptera have color vision (Proctor *et al.* 1996). Studies by Judd and Borden (1991) suggest that visual signals (color) are involved in flower finding by the onion fly *Delia antiqua* (Anthomyiidae). Visual cues may also play a role in the location of feeding sites in fruit flies (Katsoyannos 1989).

Sources of nutrition

The cyclorrhaphous Diptera include Tephritidae (fruit flies), Anthomyiidae (root maggot flies), and Psilidae (rust flies). The labellum is the prominent organ for imbibing fluids in their sucking or sponging mouthparts (McAlpine 1981). Many of these flies feed on exposed fluids such as honeydew (Downes and Dahlem 1987), floral (Hendrichs and Reyes 1987) and extrafloral nectar (Nishida 1958; Kazi 1976), but they may also consume fruit tissue (fruit flies), plant sap, bird feces or plant surface tissue (Bateman 1972; Tsitsipis 1989; Drew and Yuval 2000), and pollen (Deyrup 1988; Wacht *et al.* 2000). *Delia* spp. are known to feed on floral nectar from plants with easily accessible nectaries (e.g., Rosaceae, Umbelliferae) (Kästner 1929; Finch and Coaker 1969; Finch 1971, 1974; Finch *et al.* 1986). Adult *Rivellia* sp. (Platystomatidae), pests of pulse crops, feed on extrafloral nectar from cowpea (Siddappaji and Gowda 1980). Various adult fruit flies collect food material at some distance in habitats adjacent to their reproductive host (Kazi 1976; McQuate *et al.* 2003).

The cabbage root fly *Delia radicum* has been recorded to feed on the flowering heads of grasses (Finch and Coaker 1969; Finch 1971, 1974). But the flies' crop was found to contain no pollen. Apparently, the flies were able to utilize the sugars and amino acids available on the surface of the grass pollen and anthers (Finch and Coaker 1969; Finch 1974). The olive fruit fly *Bactrocera* (*Dacus*) *oleae* (Tephritidae) is thought to ingest smaller pollen grains (less than 150 μm in diameter) either directly or passively when trapped in honeydew (Tsiropoulos 1977). However, based on morphological studies of the mouthparts of four *Bactrocera* species, this observation has recently been questioned by Vijaysegaran *et al.* (1997). Their studies revealed that the flies are not able to ingest particles larger than 0.5 μm. Detailed studies by Haslett (1983) demonstrated that pollen grains are ruptured and release their contents during passage through the digestive system of adult hoverflies (Syrphidae).

Some Diptera are specialized pollen-feeders. Females of *Poecilognathus punctipennis* (Bombyliidae) feed avidly on pollen of two plant species (Commelinaceae) (Deyrup 1988). They rake pollen from the anthers with their front tarsi and transfer the pollen to the tip of the proboscis for ingestion. This typical behavior suggests that this species is specialized to exploit pollen as a food source. Another prominent example is the adult eucalypt nectar fly *Drosophila flavohirta* (Drosophilidae), in which both sexes gather pollen with the proboscis and seem to gain the nutrients by "external digestion". This feeding type is similar to that described for *Heliconius* butterflies (Nicolson 1994). Adult midges (Ceratopogonidae) are reported to pierce and suck the fluid content of pollen grains (Downes 1955), just like thrips do (Kirk 1984).

Gustatory response

Carbohydrates

The gustatory response of Diptera has been studied extensively using blowflies (Dethier 1976), mosquitoes (Schmidt and Friend 1991), and syrphids (Wacht *et al.* 2000) as models. Even though considerable differences exist among species, Diptera appear to respond to a broader range of sugars than insects from other orders (Wäckers 1999). In sharp contrast to the detailed investigations on calliphorid flies, we know relatively little about the gustatory response of herbivorous Diptera to individual nectar compounds. *Anastrepha suspensa* (Tephritidae) consumed more of 4% solutions of sucrose, fructose, maltose, and glucose than of water, suggesting a phagostimulatory effect (Sharp and Chambers 1984). Behavioral studies revealed that sucrose, fructose, and glucose elicit a gustatory response in the Mexican fruit fly *Ceratitis capitata* (Tephritidae)

(Gothilf *et al.* 1971). In accordance with these behavioral observations, the authors found that a sugar-sensitive receptor neurone on the labellum was most sensitive to sucrose, fructose, and glucose and less so to other sugars. Studies by Canato and Zucoloto (1998) revealed that adult flies submitted to sucrose deprivation during the larval phase suffered a profound reduction in discrimination threshold for this sugar. Sucrose-sensitive neurones were found on the tarsi as well as the labellum of the carrot fly *Psila rosae* (Psilidae) (E. Städler, unpublished data) (Fig. 7.2) and *D. radicum* (Städler 1978). In the related *Delia floralis*, Blaney and Simmonds (1994) found sensilla with sucrose-sensitive sensory neurones in both the labellum and tarsal sensilla (Fig. 7.2). Electrophysiological recordings from tarsi of *Rhagoletis pomonella* (Tephritidae) and other fruit flies revealed that the D-sensillum contains a receptor neuron for sucrose (Angioy *et al.* 1978a,b; Bowdan 1984; Städler and Schöni 1991). Duan and Prokopy (1993), studying tarsal acceptance thresholds in *R. pomonella*, showed that sucrose and fructose were the most stimulative of five sugars tested, whereas sensitivity to glucose, melezitose, and maltose was lower. Finch and Coaker (1969) found certain sugars (rhamnose, sorbose) to be a feeding repellent to *D. radicum*.

Amino acids

Protein hydrolysates have been found to cause a phagostimulatory response in fruit flies (Galun *et al.* 1980; Sharp and Chambers 1984; Galun 1989). Several individual amino acids are phagostimulatory for *C. capitata* (Galun *et al.* 1980), *A. suspensa* (Sharp and Chambers 1984), and *B. oleae* (Tsiropoulos 1984). According to unpublished results obtained by Simmonds (personal communication), the amino acids γ-aminobutyric acid (GABA), glycine, and L-alanine stimulate the sucrose receptor neurone of *D. floralis* too, whereas D-alanine was almost inactive.

Based on tarsal acceptance thresholds in *R. pomonella*, Duan and Prokopy (1993) showed that sugars and certain amino acids have a synergistic effect on the feeding response. This is probably due to synergistic actions of sugars and amino acids on sugar receptor neurones as observed and reviewed for blowflies (Dethier 1976). Similarly, the combination of ribonucleotides and sucrose act as synergistic phagostimulants for *C. capitata* (Galun 1989).

Nutritional requirements and food utilization

Most adult Diptera are primarily or entirely dependent on carbohydrate-rich foods as an energy source, especially for flight (Hocking 1953). In contrast to the Lepidoptera, egg production in most Diptera completely depends on protein feeding by adult females (Wheeler 1996). In addition, male calling

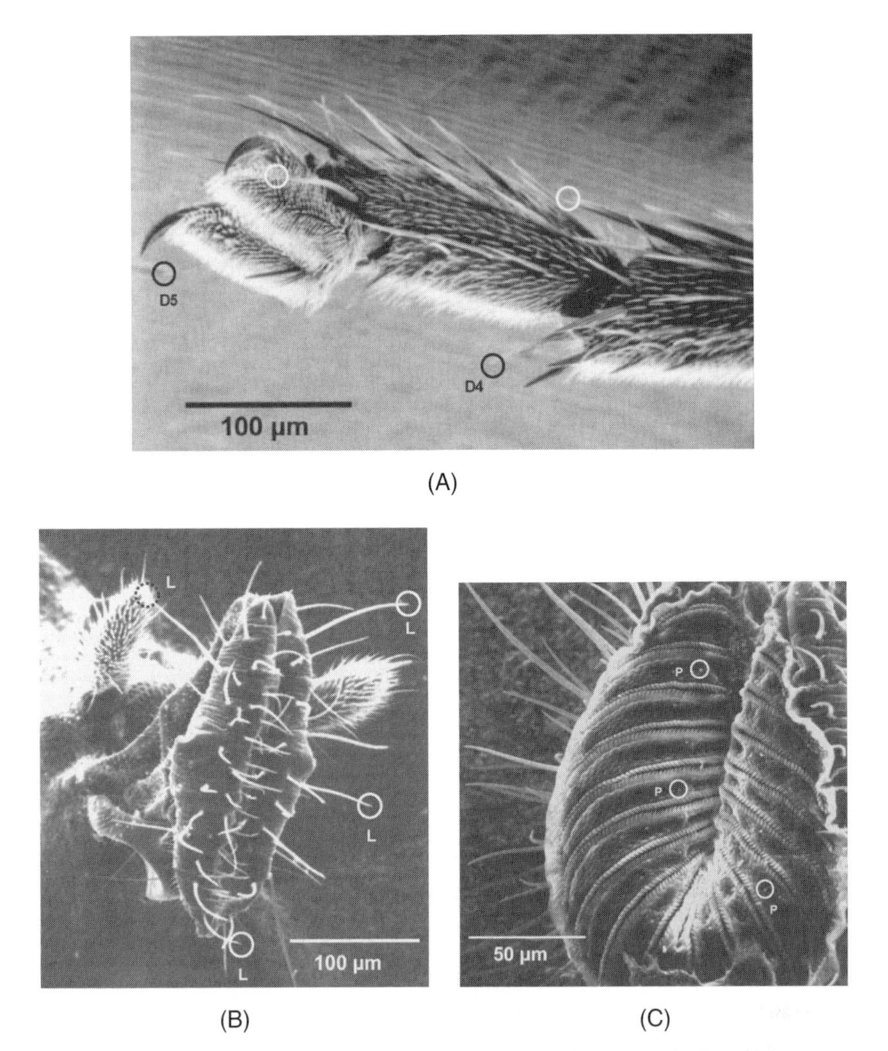

Figure 7.2 (A) Lateral view of the tarsus (two distal segments) of a cabbage root fly female, *Delia radicum*. The tips of D-sensillae containing a sugar-sensitive receptor neurone have been circled. (B) Proboscis of the female carrot fly, *Psila rosae*, in the withdrawn, inactive position, in ventral view. Upon tarsal stimulation with a sugar solution the proboscis will be extended and the tips of contact-chemoreceptor sensilla, the longest first, will touch the surface. Circled are the tips of the longest sensilla containing a sugar-sensitive receptor neurone. (C) Proboscis of *P. rosae* with the inner surface of the labellum open revealing the pseudotrachea (ridged grooves). Additional contact-chemoreceptor sensilla, papillae (P), can be found between the pseudotrachea. In the blowfly, *Phormia regina* (Calliphoridae), these papillae have been investigated by Dethier and Hanson (1965) and were shown to contain chemoreceptor neurones sensitive to different sugars, including sorbitol, sorbose, inositol, fucose, arabinose, glucose, sucrose, and fructose.

activity, pheromone production, pheromone composition, male attractivity, and expression of sexual behaviors were found to be affected by the quality of the adult food in various fruit fly species (see literature cited in Drew and Yuval 2000).

With respect to adult nutritional requirements, the Tephritidae and Anthomyiidae are the best studied of herbivorous Diptera. The nutritional requirements of fruit flies are not only covered by adult diet, but also by nutrients obtained from symbiotic microorganisms or transferred from the larval stages (Tsitsipis 1989; Drew and Yuval 2000). Most information on the nutritional requirements of adult fruit flies has come from research on artificial diets (Tsitsipis 1989). All species of fruit flies require carbohydrates as an energy source and water for survival. Most species need a variety of additional nutrients, including proteinaceous substances and vitamins, for sexual maturation and normal egg production (Bateman 1972; Tsitsipis 1989; Drew and Yuval 2000).

Carbohydrates

Besides the three major nectar sugars (sucrose, fructose, glucose), melibiose, lactose, trehalose, raffinose, and melezitose increased longevity of *D. radicum* compared to water-fed control flies (Finch and Coaker 1969). Some sugars were found to be inert (xylose, galactose), and a few were even toxic (arabinose, ribose, cellobiose). The effect of sucrose on longevity depended on the concentration of the solution: it increased with increasing molarity up to 0.1 M, but decreased at a higher concentration (0.5 M) (Finch and Coaker 1969). Also in *D. antiqua*, sucrose feeding increased adult survival (McDonald and Borden 1996). Tsiropoulos (1980b) reported that sucrose, fructose, glucose, mannose, melibiose, trehalose, maltose, and melezitose increased longevity of female and male *B. oleae* significantly, compared to water-fed flies. Female *C. capitata* utilized the three major nectar sugars as well as galactose, mannose, and maltose (Gothilf *et al.* 1971). Studies on digestive enzyme activity of *R. pomonella* showed that sucrose, melezitose, and trehalose were hydrolyzed by gut extracts, whereas no hydrolysis of lactose, melibiose, and maltose was observed (Ross *et al.* 1977).

Körting (1940) found that female *P. rosae* need sucrose and water for normal oviposition. Studies on the narcissus bulb fly *Lampetia equestris* (Syrphidae) revealed that sucrose, fructose, glucose, and maltose enhance the flies' fecundity (Doucette and Eide 1955). Olive fruit flies and *C. capitata* maintained on a sucrose-only diet were still able to lay eggs continuously, indicating that they could use protein reserves from the larval period (Tsiropoulos 1980a; Cangussu and Zucoloto 1992). This is in contrast to other *Bactrocera* (*Dacus*) species (Hagen 1952, in Tsiropoulos 1980a), *Rhagoletis completa* (Tephritidae) (Tsiropoulos 1978), and *D. radicum* (Finch and Coaker 1969; Finch 1971) that are unable to mature eggs on a carbohydrate-only diet.

Protein/amino acids

Supplemental protein was found to increase longevity in *D. antiqua* (McDonald and Borden 1996). Although Keiser and Schneider (1969) reported that supplemental protein had no effect on the longevity of various *Bactrocera* (*Dacus*) fruit fly species, McQuate *et al.* (2003) found a significant increase in survival of *Bactrocera dorsalis* provided with sucrose and corn pollen when compared to flies kept on sucrose only. In *B. oleae*, access to pollen from certain plant species significantly increased longevity, fecundity, and egg viability and shortened the pre-oviposition period (Tsiropoulos 1977). The importance of (mixtures of) certain amino acids for the survival of female and male *B. oleae* was confirmed in a subsequent study (Tsiropoulos 1983).

Protein is indispensable for egg production in *B. dorsalis* and *Bactrocera cucurbitae* (see literature in Tsitsipis 1989). Although female *B. oleae* are able to mature a few eggs autogenously when fed carbohydrates, both fecundity and maturation rates are increased by the addition of protein hydrolyzate to the diet (Tsiropoulos 1980a). A recent study by McQuate *et al.* (2003) revealed that *B. cucurbitae* is not able to utilize corn pollen to enhance fecundity.

Studies by Finch and Coaker (1969) and Finch (1971) on *D. radicum* revealed that females must feed on carbohydrate-rich food sources, but not on protein, to mature the initial batch of eggs. However, sources of protein or free amino acids were needed to lay the second and subsequent batches. Likewise, access to yeast hydrolyzate significantly increased the number of eggs laid in *P. rosae* (Städler 1971) and *D. antiqua* (McLeod 1964). Adult flies of the genus *Anastrepha* depend on a protein-rich diet for egg production and viability (Jácome *et al.* 1999; Aluja *et al.* 2001; Cresoni-Pereira and Zucoloto 2001).

Studies on digestive enzyme activity indicated that *R. pomonella* is unable to digest unhydrolyzed protein and must therefore feed on free amino acids (Ross *et al.* 1977). Hendrichs *et al.* (1993) confirmed that fecundity increased when carbohydrates and yeast hydrolyzate were available as food, while access to pollen caused no effect. Access to a protein source not only impacted longevity and fecundity in *C. capitata*, but also affected male mating success (Shelly *et al.* 2002).

Coleoptera

Beetles as pests and pollinators

According to Schoonhoven *et al.* (1998), about 35% of the estimated 349 000 species of Coleoptera are herbivorous. Both adults and larvae may attack roots, vegetative, and reproductive plant parts.

Whereas Coleoptera are of marginal importance as pollinators in cool-temperate climates, they are more important in tropical areas (Faegri and Van der Pijl 1979; Proctor *et al.* 1996). Many beetle-pollinated flowers are found among the Annonaceae, Araceae, and Cyclanthaceae. Beetle-pollinated flowers are often relatively robust, open bowl-shaped and with a typically strong, fruity, or aminoid smell (Faegri and Van der Pijl 1979; Kevan and Baker 1999). Although beetle-pollinated flowers lack special structures, there are many examples of strong mutualistic interactions between plants and coleopteran pollinators (Gottsberger 1989; Anstett 1999).

Sources of nutrition

Floral feeding occurs in many Coleoptera, especially among Chrysomelidae and Bruchidae (Samuelson 1994; Kevan and Baker 1999). Chrysomelids are facultative pollen-eaters and important pollinators (Faegri and Van der Pijl 1979; Crowson 1981; Samuelson 1994). Pollen-feeding beetles respond to flower characteristics such as color (Nolte 1959) and volatiles (Dobson 1994; Metcalf *et al.* 1998). Adults of some beetles, including the boll weevil *Anthonomus grandis* (Curculionidae) (Jones 1997) and the pollen beetle *Meligethes aeneus* (Nitidulidae) (Ruther and Thiemann 1997), respond to volatiles from a variety of plant species, enabling them to locate pollen also from non-host sources. Beetles have biting and chewing-type mouthparts and some show morphological characteristics that are believed to be adaptations to floral feeding such as forward projection and elongation of mouthparts, up-tilting of the head, and elongation of the prothorax (Faegri and Van der Pijl 1979; Crowson 1981; Samuelson 1994; Kevan and Baker 1999).

Although many species of Coleoptera reportedly take up pollen grains it is generally unknown to what extent they are able to digest pollen (Samuelson 1994; Roulston and Cane 2000). Only few studies present direct evidence that pollen grains are cracked within the digestive tract of the beetles (Cate and Skinner 1978; Rickson *et al.* 1990; Tran and Huignard 1992; Jones *et al.* 1993). Alternatively, pollen nutrients may be extracted by diffusion (Roulston and Cane 2000), or may be released when pollen grains germinate in the crop (Crowson 1981; Samuelson 1994). One interesting example is an adult Oedemeridae that has a rather large sac-like diverticulum in place of the usual crop (Crowson 1981, p. 170). This gut part was found densely packed with germinating pollen grains, while the pollen grains from posterior parts of the gut appeared as completely empty shells, suggesting that in this species germination is part of digestion. Despite our limited knowledge of the mechanisms of pollen utilization, there is little doubt that beetles obtain nutritional benefits from pollen-feeding.

Adult *A. grandis* boll weevils, regardless of sex or physiological status, feed actively on pollen of *Gossypium* spp. (Malvaceae), their reproductive host, as well as on non-host pollen (Benedict *et al.* 1991; Jones *et al.* 1993; Jones 1997; Hardee *et al.* 1999; Jones and Coppedge 1999). All three major corn rootworm species feed on maize pollen and on non-host pollen, i.e., the western corn rootworm *Diabrotica virgifera virgifera* (Chrysomelidae), the northern corn rootworm *D. barberi* (Ludwig and Hill 1975; Hill and Mayo 1980; Hollister and Mullin 1999; Moeser and Vidal 2001), and the Mexican corn rootworm *D. virgifera zeae* (Jones and Coppedge 2000). However, the use of non-maize pollen is more common in *D. barberi* than in *D. virgifera virgifera* (Ludwig and Hill 1975; Hill and Mayo 1980; Siegfried and Mullin 1990; Metcalf *et al.* 1998). The requirement for pollen is known to affect the beetles' dispersal behavior. Field studies have shown that *D. barberi* is present in maize fields mainly when pollen is available, and that during the remaining periods, beetles leave the field to feed on non-host pollen and return to maize for oviposition (Cinereski and Chiang 1968; Lance *et al.* 1989; Naranjo 1991). *Diabrotica virgifera virgifera*, on the other hand, is less likely to abandon maize in search for pollen as it can survive well by feeding on other maize plant parts (Lance *et al.* 1989; Elliott *et al.* 1990). Nevertheless, recent field studies in Hungary revealed that *D. virgifera virgifera* in maize fields does consume non-maize pollen during maize flowering and that this pollen becomes even more important after flowering has ceased in maize (Moeser and Vidal 2001). Similar to the boll weevil and corn rootworms, pollen beetles of the genus *Meligethes* (Nitidulidae) feed on pollen from many different plant families, but they oviposit almost exclusively on flowers of Brassicaceae (Fritzsche 1957; Free and Williams 1978). Pollen-feeding has also been reported in *Bruchus atrolineatus* (Bruchidae), which leaves the fields with larval host plants to feed on non-host pollen and returns for oviposition (Alzouma and Huignard 1981). Field studies with *Bruchidius dorsalis* (Bruchidae) revealed that males visit non-host flowers for feeding during the non-flowering period of the host plants. Females, on the other hand, were found only on host plants irrespective of the flowering phenology (Takakura 2004). This difference in the feeding behavior between sexes could indicate that females do not require plant nutrients since they receive highly nutritious gifts from males in the form of seminal fluid which was found to greatly enhance fecundity (Takakura 1999).

Flower-visiting beetles are likely to ingest some nectar, but direct nectar-feeding by beetles seems rare (Samuelson 1994). Whereas the pollen-eating habit developed very early during evolution (Gottsberger 1977), nectar-feeding in Coleoptera is probably a recent development (Faegri and Van der Pijl 1979; Kevan and Baker 1999). Most reports on nectar feeding in beetles concern bruchids. In the field

Bruchus dentipes and *Bruchus rufimanus* were reported to feed on extrafloral nectaries of faba beans (*Vicia faba*; Fabaceae) (Scholz and Dörner 1976; Tahhan and Van Emden 1989). In the laboratory, *B. dentipes* has been observed to rob nectar from the base of the corolla of faba bean flowers (Tahhan and Van Emden 1989). Both sexes of the pea weevil *Bruchus pisorum* collected nectar via small slits at the base of the corolla tubes on their host *Pisum sativum* (Clement 1992). The intensity of nectar robbing was equal among the sexes. Koptur and Lawton (1988) reported *Apion* weevils (Curculionidae) feeding on stipular extrafloral nectaries of *Vicia sativa.*

Gustatory response

Carbohydrates

A study on the gustatory response of the sweetclover weevil *Sitona cylindricollis* (Curculionidae) to carbohydrates showed sucrose to be the strongest feeding stimulant, whereas fructose, glucose, galactose, mannose, and maltose also had a moderately stimulative effect (Akeson *et al.* 1970). Arabinose, on the other hand, acted as a feeding deterrent. A phagostimulatory effect of sucrose was also reported for four out of five weevil species by Shanks and Doss (1987). Receptor neurones sensitive to sucrose have been located on the mouthparts of the adult Colorado potato beetle *Leptinotarsa decemlineata* (Chrysomelidae) (Mitchell and Harrison 1984).

Amino acids and other stimulants

Little is known about taste perception in pollen-feeding Coleoptera. Best studied is the western corn rootworm *D. virgifera virgifera*. Laboratory (Hollister and Mullin 1999) and field studies (Moeser and Vidal 2001) have shown that the beetles generally prefer host over non-host pollen, but also preferences among non-host pollen have been reported. Further studies revealed that it is not sugars but amino acids (Mullin *et al.* 1994; Hollister and Mullin 1998, 1999) and probably certain lipids and secondary plant compounds (Lin and Mullin 1999) that are the primary phagostimulants for *D. virgifera virgifera*. Hollister and Mullin (1999) isolated and identified the primary metabolites from plant pollen that stimulate feeding in the adult beetle. Five amino acids (alanine, β-alanine, GABA, proline, serine) were dominant in all pollens and proved to be the active fraction responsible for the beetle's feeding response to crude pollen extracts. Hollister and Mullin (1998) furthermore investigated the dose-response profiles of adult beetles to these amino acids in no-choice feeding tests and made electrophysiological recordings from the maxillary (galea) contact-chemoreceptor sensilla. The behavioral and chemosensory dose-response curves corresponded closely to levels of free amino acids present in pollen from host plants. Interestingly, Kim and Mullin (1998) found that the strongest feeding response was elicited by certain non-essential amino acids, rather than by the nutritionally more important essential amino acids. Mitchell (1985) identified neurones in the sensilla of the galea of adult *L. decemlineata* that are specific for the amino acids L-alanine and GABA.

In addition to primary metabolites such as carbohydrates and amino acids, secondary metabolites also may be important feeding stimulants. In the case of *A. grandis*, several plant secondary compounds in extracts of cotton buds and anthers (i.e., esters of phytol and geranylgeraniol) have been identified as feeding elicitors (McKibben *et al.* 1985). Shanks and Doss (1987) found that feeding by various weevil species was not only elicited by sucrose but by certain sterols as well. Studies on *Zygogramma bicolorata* (Chrysomelidae) showed that parthenin (a sesquiterpene lactone) contained in pollen of the host plant *Parthenium hysterophorus* (Asteraceae) acts as an attractant and phagostimulant for adult beetles (Jayanth *et al.* 1993).

Nutritional requirements and food utilization

Pollen-feeding can impact various reproductive traits in Coleoptera. Consumption of pea pollen promotes the growth of the oocytes of the pea weevil (*B. pisorum*) compared to water-fed beetles (Pesho and Van Houten 1982; Annis and O'Keeffe 1984; but see Clement 1992). Also *Smicronyx fulvus* (Curculionidae) requires pollen for egg maturation, whereas sucrose-feeding does not appear to contribute to oocyte development (Rana and Charlet 1997). Pollen-feeding has positive effects on survival time and fat body development in *A. grandis* (Jones *et al.* 1993; Jones 1997). In the case of *Diabrotica* species, pollen-feeding positively affects longevity, fecundity, and oviposition period (Naranjo and Sawyer 1987; Elliott *et al.* 1990). Additionally, Pesho and Van Houten (1982) reported that access to pollen increases the frequency of copulation in *B. pisorum* and Tran and Huignard (1992) showed that pollen ingestion by *B. rufimanus* leads to the termination of reproductive diapause in both sexes.

Little is known about the effects of nectar compounds on beetle biology. A study by Clement (1992) indicates that in *B. pisorum* nectar may not only function as a source of energy but may also contribute to egg development. When *Acanthoscelides obtectus* (Bruchidae) fed on sucrose, glucose, or fructose, survival and egg production increased significantly but feeding lactose did not benefit the beetles (Leroi 1978). Honey-feeding increased longevity and fecundity of the bruchid *Callosobruchus chinensis* by a factor of 2.8 and 1.5, respectively (F. L. Wäckers, unpublished data). *Anthonomus grandis* lived significantly longer on a pollen diet when agar pellets containing 1% sucrose were added (Haynes and Smith 1992).

Conclusions and Implications

There is a vast body of information on the use of plant-provided food supplements by adult herbivorous insects. These plant-derived nutrients can have a profound impact on both herbivore longevity and fecundity. The presence and distribution of adult food sources may affect the population dynamics and population structure of Lepidoptera (Wiklund 1977; Wiklund and Åhrberg 1978; Ohsaki 1979; Murphy 1983; Murphy *et al.* 1983, 1984; Boggs 1987; Grossmueller and Lederhouse 1987; Shreeve 1992; Miller 1996; Karban 1997; Brommer and Fred 1999), Coleoptera (Cinereski and Chiang 1968), and Diptera (Malavasi *et al.* 1983; Hendrichs *et al.* 1991; Prokopy *et al.* 1996). It can also result in mate separation in cases where both sexes differ significantly in their requirements for plant-provided foods (Takakura 2004).

Flower-visiting herbivores can affect plant fitness both positively (pollination) and negatively (herbivory). These two capacities represent opposing selective forces in the evolution of floral traits, creating a potential dilemma for flowering plants. The extent of this dilemma depends on the costs and benefits represented by flower-visiting herbivores. As adults, a number of herbivorous insects obtain nectar from their larval host plants (Kästner 1929; Petherbridge *et al.* 1942), and herbivores may seek out these plants for feeding (Erhardt 1995; Peterson 1997). Oviposition by butterflies (Murphy *et al.* 1983, 1984; Grossmueller and Lederhouse 1987; Karban 1997) and Diptera (Averill and Prokopy 1993; Prokopy *et al.* 1996) is sometimes enhanced in the proximity of adult food sources. In some systems, feeding by the adult herbivore is followed by egg deposition on the same plant or even the same flower. The noctuid moth *Hadena bicruris* uses flowers of *Silene alba* (Caryophyllaceae) both as a source of nectar and as an oviposition substrate. Although the moth contributes to the pollination of *S. alba* flowers, this benefit is annihilated by the fact that the fruit-feeding larvae destroy approximately five capsules each (Brantjes 1976). The consumption of nectar is essential for egg maturation in this species; in fact, nectar drinking forms an integral part of the behavior leading to oviposition. When floral drinking is suppressed by satiating females with a sugar solution, floral ovipositions are replaced by (non-viable) random ovipositions (Brantjes 1976).

Also the presence of pollen can amplify herbivory. Both sexes of the cotton boll weevil (*A. grandis*) feed on cotton pollen, damaging or destroying several flowers and flower buds in the process (Isely 1928; Jones 1997). The female requires nutrients from pollen for egg maturation. Adding to the damage, matured eggs are subsequently deposited singly in buds or maturing fruits, often on the same plant where maturation feeding took place. As larval development usually results in abortion of the maturing fruits, pollen-feeding by *A. grandis* can have a dramatic impact on the reproductive output of the host. The same applies for pollen beetles (*Meligethes*). Females are attracted to flowering crucifers, including oilseed rape, where they feed on pollen (Free and Williams 1978). Oviposition then takes place on young flower buds (Fritzsche 1957). A study by Cook *et al.* (2004) revealed that pollen availability influenced adult distribution, with more pollen beetles being found on male-fertile plants than on male-sterile plants. It was not revealed whether this was due to an attraction of the beetles to visual or olfactory cues from the exposed pollen or an arrestment response. Pettersson (1992) observed an anthomyiid fly (*Delia flavifrons*) feeding on pollen of *Silene vulgaris* (Caryophyllaceae) before ovipositing in the flowers of the same plant. As this species does not act as pollinator, the interaction is strictly parasitic.

In other cases, the cost of flower use by herbivores can be outweighed by their contribution to pollination. Eventually this may lead to the co-evolution of an intimate mutualism between plant and herbivore. In such co-evolved systems, the plant may provide a feeding or oviposition substrate within the inflorescence, thus minimizing the impact of herbivory, while increasing the herbivore's flower constancy. Examples of such tight herbivory/pollination syndromes have been described involving Lepidoptera in the yucca–yucca moth system (Aker and Udovic 1981; Bronstein and Ziv 1997), Coleoptera in the Annonaceae (Thien 1980; Gottsberger 1989) and dwarf palm system (Anstett 1999), and Hymenoptera in the fig–fig wasp system (Janzen 1979; Bronstein and Hossaert-McKey 1996). The interdependency is strongest in the fig and yucca systems, in which the plant provides adult herbivores with oviposition sites and food for their herbivorous offspring. As the larval food substrate consists of plant fruits and seeds, the herbivore is made dependent on successful pollination.

The effects of nectar and pollen availability need not be limited to the food-providing plant, but may also carry over onto its neighbors. This applies particularly when neighbors are common host plants (Grossmueller and Lederhouse 1987; Karban 1997). In certain cases, the presence of adult food may also extend the host range of herbivorous insects. For example, following its introduction into India as part of a classical biological weed-control program the chrysomelid *Z. bicolorata* was found to attack sunflower, which had never been seen in previous host-specificity tests. Further studies showed that the beetles were only feeding on sunflower leaves that were dusted with pollen from the target weed (Jayanth *et al.* 1993). A similar effect, involving non-plant food, was reported for the Mediterranean fruit fly *C. capitata*: the odor of proteinaceous bird droppings lured flies to plants whose fruits emit little or no attractive odor, thus causing a (temporary) host range expansion (Prokopy *et al.* 1996). Although these plants did not serve as permanent hosts, they were nevertheless used for oviposition and produced viable offspring.

These examples show that the utilization of plant-provided food supplements by adult herbivores can have important implications for plant–insect interactions. Insight in nectar and pollen use by adult herbivores can also have obvious consequences for pest management. A prominent example of the influence on pest management based on knowledge of the pest's feeding behavior is the deployment of nectariless cotton varieties. Even though extrafloral nectar can be an important food source for beneficials (Rogers 1985; Schuster and Calderón 1986; Turlings and Wäckers 2004), many major cotton pests are known to feed on these extrafloral nectars as well (Lukefahr and Rhyne 1960; Beach *et al.* 1985). This has been a motivation to select for nectariless cotton varieties. In some studies these varieties indeed proved to be less sensitive to herbivory (Lukefahr and Rhyne 1960; Lukefahr *et al.* 1965; Schuster and Calderón 1986). Knowledge of herbivore food ecology should also be incorporated in programs on functional biodiversity. In part, these programs aim at tailoring agroecosystems to the needs of natural enemies. Whereas this approach may enhance the efficacy of biological control agents, herbivores may also benefit from the ecosystem services provided. By addressing the nutritional ecology of both adult herbivores and their natural enemies, it may become possible to identify "selective food sources," i.e., food sources that are suitable for predators or parasitoids, without providing a nutritional benefit to the herbivore (Rogers 1985; see also Van Rijn and Sabelis, Chapter 8). The feasibility of this concept was shown for the potato moth, *Phthorimaea operculella* (Lepidoptera: Gelechiidae) and its egg parasitoid *Copidosoma koehleri* (Hymenoptera: Encyrtidae) (Baggen and Gurr 1998; Baggen *et al.* 1999). Morphological characteristics prevent the moth from reaching the nectar of phacelia, nasturtium, and borage flowers, while the nectaries are accessible to the smaller parasitoid (Baggen *et al.* 1999). Another mechanism yielding sugar-source selectivity was described by Romeis and Wäckers (2000, 2002) and Wäckers (1999, 2001). They compared the gustatory response and sugar use of *P. brassicae* and its larval parasitoid *Cotesia glomerata* (Hymenoptera: Braconidae) and found that certain common sugars, such as glucose and melezitose, are not perceived by *P. brassicae*, while they elicit a strong feeding response in the parasitoid. In addition, the parasitoid is able to use melibiose, maltose, melezitose, and galactose, which do not provide a nutritional benefit to the pest insect. Sugar concentration might be a third mechanism conveying selectivity. Wäckers *et al.* (2001) proposed that the highly concentrated and viscous nature of extrafloral nectar might be adaptive in enabling plants to exclude nectar use by Lepidoptera, as their narrow tubular mouthparts restrict them to using nectar of low viscosity. Identification of sugar selectivity could be used to identify selective food sources, or be applied in developing selectivity in the production of artificial food sprays.

Acknowledgments

We thank Conrad C. Labandeira (National Museum of Natural History, Smithsonian Institution) and Micky D. Eubanks (Auburn University) for their comments on an earlier draft of this chapter.

References

Aker, C. L. and D. Udovic. 1981. Oviposition and pollination behaviour of the yucca moth, *Tegiticula maculata* (Lepidoptera: Proxididae), and its relation to the reproductive biology of *Yucca whipplei* (Agavaceae). *Oecologia* **459**: 96–101.

Akeson, W. R., F. A. Haskins, H. J. Gorz, and G. R. Manglitz. 1970. Feeding response of the sweetclover weevil to various sugars and related compounds. *Journal of Economic Entomology* **63**: 1079–1080.

Alm, J., T. E. Ohnmeiss, J. Lanza, and L. Vriesenga. 1990. Preference of cabbage white butterflies and honey bees for nectar that contains amino acids. *Oecologia* **84**: 53–57.

Aluja, M., F. Díaz-Fleischer, D. R. Papaj, G. Lagunes, and J. Sivinski. 2001. Effects of age, diet, female density, and the host resource on egg load in *Anastrepha ludens* and *Anastrepha obliqua* (Diptera: Tephritidae). *Journal of Insect Physiology* **47**: 975–988.

Aluja, M., I. Jácome, A. Birke, N. L. Lozada, and G. Quintero. 1993. Basic patterns of behavior in wild *Anastrepha striata* (Diptera: Tephritidae) flies under field-cage conditions. *Annals of the Entomological Society of America* **86**: 776–793.

Alzouma, I. and J. Huignard. 1981. Données préliminaires sur la biologie et le comportement de ponte dans la nature de *Bruchidius atrolineatus* (Pic) (Coléoptère Bruchidae) dans une zone sud-sahélienne au Niger. *Acta Oecologica, Oecologia Applicata* **2**: 391–400.

Andersson, S. 2003. Foraging responses in the butterflies *Inachis io*, *Aglais urticae*. (Nymphalidae), and *Gonepteryx ramni* (Pieridae) to floral scents. *Chemoecology* **13**: 1–11.

Angioy, A. M., A. Liscia, and P. Pietra. 1978a. Effects of salt and sugar solutions on taste hairs of *Ceratitis capitata* Wied. *Bollettino della Societá Italiana Biologia Sperimentale* **54**: 2108–2114.

1978b. The electrophysiological response of labellar and tarsal hairs of *Dacus oleae* Gmel. to salt and sugar stimulation. *Bollettino della Societá Italiana Biologia Sperimentale* **54**: 2115–2121.

Annis, B. and L. E. O'Keeffe. 1984. Effect of pollen source on oogenesis in the pea weevil, *Bruchus pisorum* L. (Coleoptera: Bruchidae). *Protection Ecology* **6**: 257–266.

Anstett, M. C. 1999. An experimental study of the interaction between the dwarf palm (*Chamaerops humilis*) and its floral visitor *Derelomus chamaeropsis* throughout the life cycle of the weevil. *Acta Oecologica* **20**: 551–558.

Averill, A. L. and R. J. Prokopy. 1993. Foraging of *Rhagoletis pomonella* flies in relation to interactive food and fruit resources. *Entomologia Experimentalis et Applicata* **66**: 179–185.

Baggen, L. R. and G. M. Gurr. 1998. The influence of food on *Copidosoma koehleri* (Hymenoptera: Encyrtidae), and the use of flowering plants as a habitat management tool to enhance biological control of potato moth, *Phthorimaea operculella* (Lepidoptera: Gelechiidae). *Biological Control* **11**: 9–17.

Baggen, L. R., G. M. Gurr, and A. Meats. 1999. Flowers in tri-trophic systems: mechanisms allowing selective exploitation by insect natural enemies for conservation biological control. *Entomologia Experimentalis et Applicata* **91**: 155–161.

Baker, H. G. and I. Baker. 1975. Studies of nectar-constitution and pollinator-plant coevolution. In L. E. Gilbert and P. H. Raven (eds.) *Coevolution of Animals and Plants*. Austin, TX: University of Texas Press, pp. 100–140.

1982. Chemical constituents of nectar in relation to pollination mechanisms and phylogeny. In M. H. Nitecki (ed.) *Biochemical Aspects of Evolutionary Biology*. Chicago, IL: University of Chicago Press, pp. 131–171.

1983. Floral nectar sugar constituents in relation to pollinator type. In C. E. Jones and R. J. Little (eds.) *Handbook of Experimental Pollination Biology*. New York: Scientific and Academic Editions, pp. 117–141.

Barbehenn, R. V., J. C. Reese, and K. S. Hagen. 1999. The food of insects. In C. B. Huffaker and A. P. Gutierrez (eds.) *Ecological Entomology*, 2nd edn. New York: John Wiley, pp. 83–121.

Bateman, M. A. 1972. The ecology of fruit flies. *Annual Review of Entomology* **17**: 493–518.

Beach, R. M., J. W. Todd, and S. H. Baker 1985. Nectaried and nectariless cotton cultivars as nectar sources for the adult soybean looper. *Journal of Entomological Sciences* **20**: 233–236.

Beenakkers, A. M. T., D. J. Van der Horst, and W. J. A. Van Marrewijk. 1984. Insect flight muscle metabolism. *Insect Biochemistry* **14**: 243–260.

Begon, M. and G. A. Parker. 1986. Should egg size and clutch size decrease with age? *Oikos* **47**: 293–302.

Benedict, J. H., D. A. Wolfenbarger, V. M. Bryant Jr., and D. M. George. 1991. Pollens ingested by boll weevils (Coleoptera: Curculionidae) in southern Texas and northeastern Mexico. *Journal of Economic Entomology* **84**: 126–131.

Blaney, W. M. and M. S. J. Simmonds. 1988. Food selection in adults and larvae of three species of Lepidoptera: a behavioural and electrophysiological study. *Entomologia Experimentalis et Applicata* **49**: 111–121.

1990. A behavioural and electrophysiological study of the role of tarsal chemoreceptors in feeding by adults of *Spodoptera*, *Heliothis virescens* and *Helicoverpa armigera*. *Journal of Insect Physiology* **36**: 743–756.

1994. Effect of age on the responsiveness of peripheral chemosensory sensilla of the turnip root fly (*Delia floralis*). *Entomologia Experimentalis et Applicata* **70**: 253–262.

Boggs, C. L. 1981. Nutritional and life-history determinants of resource allocation in holometabolous insects. *American Naturalist* **117**: 692–709.

1986. Reproductive strategies of female butterflies: variation in and constraints on fecundity. *Ecological Entomology* **11**: 7–15.

1987. Ecology of nectar and pollen feeding in Lepidoptera. In F. Slansky Jr. and J. G. Rodriguez (eds.) *Nutritional Ecology of Insects, Mites, Spiders, and Related Invertebrates*. New York: John Wiley, pp. 369–391.

1988. Rates of nectar feeding in butterflies: effects of sex, size, age and nectar concentration. *Functional Ecology* **2**: 289–295.

1990. A general model of the role of male-donated nutrients in female insects' reproduction. *American Naturalist* **136**: 598–617.

1995. Male nuptial gifts: phenotypic consequences and evolutionary implications. In S. R. Leather and J. Hardic (eds.) *Insect Reproduction*. Boca Raton, FL: CRC Press, pp. 215–242.

1997a. Reproductive allocation from reserves and income in butterfly species with differing adult diets. *Ecology* **78**: 181–191.

1997b. Dynamics of reproductive allocation from juvenile and adult feeding: radiotracer studies. *Ecology* **78**: 192–202.

Boggs, C. L., J. T. Smiley, and L. E. Gilbert. 1981. Patterns of pollen exploitation by *Heliconius* butterflies. *Oecologia* **48**: 284–289.

Bowdan, E. 1984. Electrophysiological responses of tarsal contact chemoreceptors of the apple maggot fly *Rhagoletis pomonella* to salt, sucrose and oviposition-deterrent pheromone. *Journal of Comparative Physiology A* **154**: 143–152.

Brantjes, N. B. M. 1976. Riddles around the pollination of *Melandrium album* (Mill.) Garcke (Caryophyllaceae) during the oviposition by *Hadena bicruris* Hufn. (Noctuidae, Lepidoptera). II. *Proceedings of the Royal Netherlands Academy of Sciences Series C* **1**: 127–141.

Brody, A. K. 1997. Effects of pollinators, herbivores, and seed predators on flowering phenology. *Ecology* **78**: 1624–1631.

Brommer, J. E. and M. S. Fred. 1999. Movement of the apollo butterfly *Parnassius apollo* related to host plant and nectar plant species. *Ecological Entomology* **24**: 125–131.

Bronstein, J. L. and M. Hossaert-McKey. 1996. Variation in reproductive success within a subtropical fig/pollinator mutualism. *Journal of Biogeography* **23**: 433–446.

Bronstein, J. L. and Y. Ziv. 1997. Costs of two non-mutualistic species in a yucca/yucca moth mutualism. *Oecologia* **112**: 379–385.

Canato, C. M. and F. S. Zucoloto. 1998. Feeding behaviour of *Ceratitis capitata* (Diptera, Tephritidae): influence of carbohydrate ingestion. *Journal of Insect Physiology* **44**: 149–155.

Cangussu, J. A. and F. S. Zucoloto. 1992. Nutritional value and selection of different diets by adult *Ceratitis capitata* flies (Diptera, Tephritidae). *Journal of Insect Physiology* **38**: 485–491.

Carroll, A. L. and D. T. Quiring. 1992. Sucrose ingestion by *Zeiraphera canadensis* Mut. and Free. (Lepidoptera: Tortricidae) increases longevity and lifetime fecundity but not oviposition rate. *Canadian Entomologist* **124**: 335–340.

Cate, J. R. and J. L. Skinner. 1978. Fate and identification of pollen in the alimentary canal of the boll weevil, *Anthonomus grandis*. *Southwestern Entomologist* **3**: 263–265.

Cheng, H. H. 1972. Oviposition and longevity of the dark-sided cutworm, *Euxoa messoria* (Lepidoptera: Noctuidae), in the laboratory. *Canadian Entomologist* **104**: 919–925.

Cinereski, J. E. and H. C. Chiang. 1968. The pattern of movements of adults of the northern corn rootworm inside and outside of corn fields. *Journal of Economic Entomology* **61**: 1531–1536.

Clement, S. L. 1992. On the function of pea flower feeding by *Bruchus pisorum*. *Entomologia Experimentalis et Applicata* **63**: 115–121.

Cook, S. M., D. A. Murray, and I. H. Williams. 2004. Do pollen beetles need pollen? The effect of pollen on oviposition, survival and development of a flower-feeding herbivore. *Ecological Entomology* **29**: 164–173.

Corbet, S. A. 2000. Butterfly nectaring flowers: butterfly morphology and flower form. *Entomologia Experimentalis et Applicata* **96**: 289–298.

Cresoni-Pereira, C. and F. S. Zucoloto. 2001. Dietary self-selection and discrimination threshold in wild *Anastrepha obliqua* females (Diptera: Tephritidae). *Journal of Insect Physiology* **47**: 1127–1132.

Crowson, R. A. 1981. *The Biology of the Coleoptera*. London: Academic Press.

Cunningham, J. P., S. A. West, and D. J. Wright. 1998. Learning in the nectar foraging behaviour of *Helicoverpa armigera*. *Ecological Entomology* **23**: 363–369.

David, W. A. L. and B. O. C. Gardiner. 1962. Oviposition and the hatching of the eggs of *Pieris brassicae* (L.) in a laboratory culture. *Bulletin of Entomological Research* **53**: 91–109.

Delisle, J., J. N. McNeil, E. W. Underhill, and D. Barton. 1989. *Helianthus annuus* pollen, an oviposition stimulant for the sunflower moth, *Homoeosoma electellum*. *Entomologia Experimentalis et Applicata* **50**: 53–60.

Dethier, V. G. 1976. *The Hungry Fly*. Cambridge, MA: Harvard University Press.

Dethier, V. G. and F. E. Hanson. 1965. Taste papillae of blowfly. *Journal of Cellular and Comparative Physiology* **65**: 93–100.

DeVries, P. J. 1979. Pollen-feeding rainforest *Parides* and *Battus* butterflies in Costa Rica. *Biotropica* **11**: 237–238.

Deyrup, M. A. 1988. Pollen-feeding in *Poecilognathus punctipennis* (Diptera: Bombyliidae). *Florida Entomologist* **71**: 597–605.

Dobson, H. E. M. 1994. Floral volatiles in insect biology. In E. A. Bernays (ed.) *Insect–Plant Interactions*. Boca Raton, FL: CRC Press, pp. 47–81.

Doucette, C. F. and P. M. Eide. 1955. Influence of sugars on oviposition of narcissus bulb fly. *Annals of the Entomological Society of America* **48**: 343–344.

Dover, J. W. 1997. Conservation headlands: effects on butterfly distribution and behaviour. *Agriculture, Ecosystems and Environment* **63**: 31–49.

Downes, J. A. 1955. The food habits and description of *Atrichopogon pollinivorus* sp. n. (Diptera: Ceratopogonidae). *Transactions of the Royal Entomological Society (London)* **106**: 439–453.

Downes, W. L. and G. A. Dahlem. 1987. Keys to the evolution of Diptera: role of Homoptera. *Environmental Entomology* **16**: 847–854.

Drew, R. A. I. and B. Yuval. 2000. The evolution of fruit fly feeding behavior. In M. Aluja and A. L. Norrbom (eds.) *Fruit Flies (Tephritidae): Phylogeny and Evolution of Behavior*. Boca Raton, FL: CRC Press, pp. 731–749.

Duan, J. J. and R. J. Prokopy. 1993. Toward developing pesticide-treated spheres for controlling apple maggot flies, *Rhagoletis pomonella* (Walsh) (Dipt., Tephritidae). *Journal of Applied Entomology* **115**: 176–184.

Elliott, N. C., R. D. Gustin, and S. L. Hanson. 1990. Influence of adult diet on the reproductive biology and survival of the western corn rootworm, *Diabrotica virgifera virgifera*. *Entomologia Experimentalis et Applicata* **56**: 15–21.

Engelmann, F. 1970. *The Physiology of Insect Reproduction*. Oxford, UK: Pergamon Press.

Erhardt, A. 1991. Nectar sugar and amino acid preferences of *Battus philenor* (Lepidoptera, Papilionidae). *Ecological Entomology* **16**: 425–434.

1992. Preferences and non-preferences for nectar constituents in *Ornithoptera priamus poseidon* (Lepidoptera, Papilionidae). *Oecologia* **90**: 581–585.

1995. Ecology and conservation of alpine Lepidoptera. In A. S. Pullin (ed.) *Ecology and Conservation of Butterflies*. London: Chapman and Hall, pp. 258–276.

Erhardt, A. and I. Baker. 1990. Pollen amino acids: an additional diet for a nectar feeding butterfly? *Plant Systematics and Evolution* **169**: 111–121.

Erhardt, A. and H. P. Rusterholz. 1998. Do peacock butterflies (*Inachis io* L.) detect and prefer nectar amino acids and other nitrogenous compounds? *Oecologia* **117**: 536–542.

Erhardt, A. and J. A. Thomas. 1991. Lepidoptera as indicators of change in the semi-natural grasslands of lowland and upland Europe. In M. Collins and J. A. Thomas (eds.) *The Conservation of Insects and Their Habitats*. London: Academic Press, pp. 213–236.

Esquivel, J. F. and P. D. Lingren. 2002. Citrus pollen retention by adult *Helicoverpa zea* (Lepidoptera: Noctuidae) after exposure to citrus blooms. *Journal of Economic Entomology* **95**: 1174–1178.

Faegri, K. and L. Van der Pijl. 1979. *The Principles of Pollination Ecology*, 3rd edn. Oxford, UK: Pergamon Press.

Finch, S. 1971. The fecundity of the cabbage root fly *Erioischia brassicae* under field conditions. *Entomologia Experimentalis et Applicata* **14**: 147–160.

1974. Sugars available from flowers visited by the adult cabbage root fly, *Erioischia brassicae* (Bch.) (Diptera, Anthomyiidae). *Bulletin of Entomological Research* **64**: 257–263.

Finch, S. and T. H. Coaker. 1969. Comparison of the nutritive values of carbohydrates and related compounds to *Erioischia brassicae*. *Entomologia Experimentalis et Applicata* **12**: 441–453.

Finch, S., C. J. Eckenrode, and M. E. Cadoux. 1986. Behaviour of onion maggot (Diptera: Anthomyiidae) in commercial onion fields treated regularly with parathion sprays. *Journal of Economic Entomology* **79**: 107–113.

Fox, C. W. 1993. Multiple mating, lifetime fecundity and female mortality of the bruchid beetle *Callosobruchus maculatus* (Coleoptera: Bruchidae). *Functional Ecology* **7**: 203–208.

Free, J. B. and I. H. Williams. 1978. The responses of the pollen beetle, *Meligethes aeneus*, and the seed weevil, *Ceuthorhynchus assimilis*, to oilseed rape, *Brassicae napus*, and other plants. *Journal of Applied Ecology* **15**: 761–774.

Frings, H. and M. Frings. 1956. The loci of contact chemoreceptors involved in feeding reactions in certain Lepidoptera. *Biological Bulletin* **110**: 291–299.

Fritzsche, R. 1957. Zur Biologie und Ökologie der Rapsschädlinge aus der Gattung *Meligethes. Zeitschrift für angewandte Entomologie* **40**: 222–280.

Galun, R. 1989. Phagostimulation of the mediterranean fruit fly, *Ceratitis capitata* by ribonucleotides and related compounds. *Entomologia Experimentalis et Applicata* **50**: 133–139.

Galun, R., S. Gothilf, S. Blondheim, and A. Lachman. 1980. Responses of the Mediterranean fruit fly *Ceratitis capitata* to amino acids. Proc. 16th Int. Congr. of Entomology, Japan, pp. 55–63.

Gilbert, L. E. 1972. Pollen feeding and reproductive biology of *Heliconius* butterflies. *Proceedings of the National Academy of Sciences of the USA* **69**: 1403–1407.

Gilbert, L. E. and M. C. Singer. 1975. Butterfly ecology. *Annual Review of Ecology and Systematics* **6**: 365–397.

Gothilf, S., R. Galun, and M. Bar-Zeev. 1971. Taste reception in the Mediterranean fruit fly: electrophysiological and behavioural studies. *Journal of Insect Physiology* **17**: 1371–1384.

Gottsberger, G. 1977. Some aspects of beetle pollination in the evolution of flowering plants. *Plant Systematics and Evolution* Suppl. **1**: 211–216.

1989. Comments on flower evolution and beetle pollination in the genera *Annona* and *Rollinia* (Annonaceae). *Plant Systematics and Evolution* **167**: 189–194.

Grimaldi, D. 1999. The co-radiations of pollinating insects and angiosperms in the Cretaceous. *Annals of the Missouri Botanical Garden* **86**: 373–406.

Grossmueller, D. W. and R. C. Lederhouse. 1987. The role of nectar source distribution in the habitat use and oviposition by the tiger swallowtail butterfly. *Journal of the Lepidopterists' Society* **41**: 159–165.

Gu, H. and W. Danthanarayana. 1990. The role of availability of food and water to the adult *Epiphyas postvittana*, the light brown apple moth, in its reproductive performance. *Entomologia Experimentalis et Applicata* **54**: 101–108.

Gunn, A. and A. G. Gatehouse. 1985. Effects of the availability of food and water on reproduction in the African armyworm, *Spodoptera exempta. Physiological Entomology* **10**: 53–63.

Gurr, G. M., H. F. Van Emden, and S. D. Wratten. 1998. Habitat manipulation and natural enemy efficacy: implications for the control of pests. In P. Barbosa (ed.) *Conservation Biological Control*. San Diego, CA: Academic Press, pp. 155–183.

Hardee, D. D., G. D. Jones, and L. C. Adams. 1999. Emergence, movement, and host plants of boll weevils (Coleoptera: Curculionidae) in the Delta of Mississippi. *Journal of Economic Entomology* **92**: 130–139.

Haslett, J. R. 1983. A photographic account of pollen digestion by adult hoverflies. *Physiological Entomology* **8**: 167–171.

Haynes, J. W. and J. W. Smith. 1992. Longevity of laboratory-reared boll weevils (Coleoptera: Curculionidae) offered honey bee-collected pollen and plants unrelated to cotton. *Journal of Entomological Sciences* **27**: 366–374.

Hendrichs, J. and J. Reyes. 1987. Reproductive behaviour and post-mating female guarding in the monophagous multivoltine *Dacus longistylus* (Diptera: Tephritidae) in southern Egypt. In A. P. Economopoulos (ed.) *Fruit Flies*. Amsterdam, the Netherlands: Elsevier Science, pp. 303–313.

Hendrichs, J., B. I. Katsoyannos, D. R. Papaj, and R. J. Prokopy. 1991. Sex differences in movement between natural feeding and mating sites and tradeoffs between food consumption, mating success and predator evasion in Mediterranean fruit flies (Diptera: Tephritidae). *Oecologia* **86**: 223–231.

Hendrichs, J., B. S. Fletcher, and R. J. Prokopy. 1993. Feeding behaviour of *Rhagoletis pomonella* flies (Diptera, Tephritidae): effect of initial food quantity and quality on food foraging, handling costs, and bubbling. *Journal of Insect Behaviour* **6**: 43–64.

Hill, C. J. 1989. The effect of adult diet on the biology of butterflies. II. The common crow butterfly, *Euploea core corinna*. *Oecologia* **81**: 258–266.

Hill, C. J. and N. E. Pierce. 1989. The effect of adult diet on the biology of butterflies. I. The common imperial blue, *Jalmenus evagoras*. *Oecologia* **81**: 249–257.

Hill, R. E. and Z. B. Mayo. 1980. Distribution and abundance of corn rootworm species as influenced by topography and crop rotation in eastern Nebraska. *Environmental Entomology* **9**: 122–127.

Hocking, B. 1953. The intrinsic rate and speed of flight of insects. *Transactions of the Royal Entomological Society London* **104**: 223–346.

Hollister, B. and C. A. Mullin. 1998. Behavioral and electrophysiological dose-response relationships in adult western corn rootworm (*Diabrotica virgifera virgifera* LeConte) for host pollen amino acids. *Journal of Insect Physiology* **44**: 463–470.

1999. Isolation and identification of primary metabolite feeding stimulants for adult western corn rootworm, *Diabrotica virgifera virgifera* LeConte, from host pollens. *Journal of Chemical Ecology* **25**: 1263–1280.

Howell, J. F. 1981. Codling moth: the effect of adult diet on longevity, fecundity, fertility, and mating. *Journal of Economic Entomology* **74**: 13–18.

Inouye, D. W. 1980. The terminology of floral larceny. *Ecology* **61**: 1251–1253.

Inouye, D. W. and G. D. Waller. 1984. Responses of honey bees (*Apis mellifera*) to amino acid solutions mimicking floral nectars. *Ecology* **65**: 618–625.

Isely, D. 1928. Oviposition of the boll weevil in relation to food. *Journal of Economic Entomology* **21**: 152–155.

Jácome, I., M. Aluja, and P. Liedo. 1999. Impact of adult diet on demographic and population parameters of the tropical fruit fly *Anastrepha serpentina* (Diptera: Tephritidae). *Bulletin of Entomological Research* **89**: 165–175.

Janzen, D. 1979. How to be a fig. *Annual Review of Ecology and Systematics* **10**: 13–51.

Jayanth, K. P., S. Mohandas, R. Asokan, and P. N. G. Visalakshy. 1993. Parthenium pollen induced feeding by *Zygogramma bicolorata* (Coleoptera: Chrysomelidae) on sunflower (*Helianthus annuus*) (Compositae). *Bulletin of Entomological Research* **83**: 595–598.

Jennersten, O. 1984. Flower visitation and pollination efficiency of some European butterflies. *Oecologia* **63**: 80–89.

Jensen, R. L., L. D. Newsom, and J. Gibbens. 1974. The soybean looper: effects of adult nutrition on oviposition, mating frequency, and longevity. *Journal of Economic Entomology* **67**: 467–470.

Johnson, J. B. and M. P. Stafford. 1985. Adult Noctuidae feeding on aphid honeydew and a discussion of honeydew feeding by adult Lepidoptera. *Journal of the Lepidopterists' Society* **39**: 321–327.

Jones, G. D. and J. R. Coppedge. 1999. Foraging resources of boll weevils (Coleoptera: Curculionidae). *Journal of Economic Entomology* **92**: 860–869.

2000. Foraging resources of adult Mexican corn rootworm (Coleoptera: Chrysomelidae) in Bell County, Texas. *Journal of Economic Entomology* **93**: 636–643.

Jones, R. W. 1997. Pollen feeding by the boll weevil (Coleoptera: Curculionidae) following cotton harvest in east central Texas. *Southwestern Entomologist* **22**: 419–429.

Jones, R. W., J. R. Cate, E. Martinez Hernandez, and E. Salgado Sosa. 1993. Pollen feeding and survival of the boll weevil (Coleoptera: Curculionidae) on selected plant species in northeastern Mexico. *Environmental Entomology* **22**: 99–108.

Josens, R. B. and W. M. Farina. 2001. Nectar feeding by the hovering hawk moth *Macroglossum stellatarum*: intake rate as a function of viscosity and concentration of sucrose solutions. *Journal of Comparative Physiology A* **187**: 661–665.

Judd, G. J. R. and J. H. Borden. 1991. Sensory interaction during trap-finding by female onion flies: implications for ovipositional host-plant finding. *Entomologia Experimentalis et Applicata* **58**: 239–249.

Kaib, M. 1974. Die Fleisch- und Blumenduftrezeptoren auf der Antenne der Schmeissfliege *Calliphora vicina*. *Journal of Comparative Physiology* **95**: 105–121.

Karban, R. 1997. Neighbourhood affects a plant's risk of herbivory and subsequent success. *Ecological Entomology* **22**: 433–439.

Karlsson, B. 1994. Feeding habits and change in body composition with age in three nymphalid butterfy species. *Oikos* **69**: 224–230.

Kästner, A. 1929. Untersuchungen zur Lebensweise und Bekämpfung der Zwiebelfliege (*Hylemyia antiqua* Meigen). II. Morphologie und Biologie. *Zeitschrift für Morphologie und Ökologie der Tiere* **15**: 363–422.

Katsoyannos, B. I. 1989. Response to shape, size and colour. In A. S. Robinson and G. Hooper (eds.) *Fruit Flies, Their Biology, Natural Enemies and Control*, vol. 3A. Amsterdam, the Netherlands: Elsevier, pp. 307–324.

Kazi, A. S. 1976. Studies on the field habits of adult melon fruit fly *Dacus (Strumeta) cucurbitae*, Coquillet. *Pakistan Journal of Scientific and Industrial Research* **19**: 71–76.

Keiser, I. and E. L. Schneider. 1969. Need for immediate sugar and ability to withstand thirst by newly emerged oriental fruit flies, melon flies, and mediterranean fruit flies untreated or sexually sterilized with gamma radiation. *Journal of Economic Entomology* **62**: 539–540.

Kelber, A. and M. Pfaff. 1997. Spontaneous and learned preferences for visual flower features in a diurnal hawkmoth. *Israel Journal of Plant Sciences* **45**: 235–245.

Kevan, P. G. and H. G. Baker. 1999. Insects on flowers. In C. B. Huffaker and A. P. Gutierrez (eds.) *Ecological Entomology*, 2nd edn. New York: John Wiley, pp. 553–584.

Kim, J. H. and C. A. Mullin. 1998. Structure–phagostimulatory relationships for amino acids in adult western corn rootworm, *Diabrotica virgifera virgifera*. *Journal of Chemical Ecology* **24**: 1499–1511.

Kingsolver, J. G. and T. L. Daniel. 1979. On the mechanics and energetics of nectar feeding in butterflies. *Journal of Theoretical Biology* **76**: 167–179.

Kirk, W. D. J. 1984. Pollen-feeding in thrips (Insecta: Thysanoptera). *Journal of Zoology* **204**: 107–117.

1997. Feeding. In T. Lewis (ed.) *Thrips as Crop Pests*. Wallingford, UK: CAB International, pp. 119–174.

Koptur, S. and J. H. Lawton. 1988. Interactions among vetches bearing extrafloral nectaries, their biotic protective agents, and herbivores. *Ecology* **69**: 278–283.

Körting, A. 1940. Zur Biologie und Bekämpfung der Möhrenfliege. *Arbeiten über physiologische und angewandte Entomologie aus Berlin-Dahlem* **7**: 209–232.

Kozhantshikov, I. W. 1938. Carbohydrate and fat metabolism in adult Lepidoptera. *Bulletin of Entomological Research* **29**: 103–114.

Krenn, H. W. and C. M. Penz. 1998. Mouthparts of *Heliconius* butterflies (Lepidoptera: Nymphalidae): a search for anatomical adaptations to pollen-feeding behavior. *International Journal of Insect Morphology and Embryology* **27**: 301–309.

Krupnick, G. A. and A. E. Weis. 1999. The effect of floral herbivory on male and female reproductive success in *Isomeris arborea*. *Ecology* **80**: 135–149.

Krupnick, G. A., A. E. Weis, and D. R. Campbell. 1999. The consequences of floral herbivory for pollinator service to *Isomeris arborea*. *Ecology* **80**: 125–134.

Kusano, T. and H. Adachi. 1969. Proboscis extending time on distilled water, sugars and salts and their nutritive value in the cabbage butterfly (*Pieris rapae crucivora*). *Kontyû* **36**: 427–436.

Kusano, T. and K. Nishide. 1978. Digestion and utilization of carbohydrates in the cabbage butterfly, *Pieris rapae crucivora* Boisduval. *Kontyû* **46**: 302–311.

Kusano, T. and H. Sato. 1980. The sensitivity of tarsal chemoreceptors for sugars in the cabbage butterfly, *Pieris rapae crucivora* Boisduval. *Applied Entomology and Zoology* **15**: 385–391.

Labandeira, C. C. 1998. How old is the flower and the fly? *Science* **280**: 57–59.

Lance, D. R., N. C. Elliott, and G. L. Hein. 1989. Flight activity of *Diabrotica* spp. at the borders of cornfields and its relation to ovarian stage in *D. barberi*. *Entomologia Experimentalis et Applicata* **50**: 61–67.

Landis, D. A., S. D. Wratten, and G. M. Gurr. 2000. Habitat management to conserve natural enemies of arthropod pests in agriculture. *Annual Review of Entomology* **45**: 175–201.

Lanza, J. and B. R. Krauss. 1984. Detection of amino acids in artificial nectars by two tropical ants *Leptothorax* and *Monomorium*. *Oecologia* **63**: 423–425.

Larson, B. M. H., P. G. Kevan, and D. W. Inouye. 2001. Flies and flowers: taxonomic diversity of anthophiles and pollinators. *Canadian Entomologist* **133**: 439–465.

Leahy, T. C. and D. A. Andow. 1994. Egg weight, fecundity, and longevity are increased by adult feeding in *Ostrinia nubilalis* (Lepidoptera: Pyralidae). *Annals of the Entomological Society of America* **87**: 342–349.

Lederhouse, R. C., M. P. Ayres, and J. M. Scriber. 1990. Adult nutrition affects male virility in *Papilio glaucus* L. *Functional Ecology* **4**: 743–751.

Le Metayer, M., M.-H. Pham-Delegue, D. Thiery, and C. Masson. 1993. Influence of host- and non-host plant pollen on the calling and oviposition behaviour of the European sunflower moth *Homoeosoma nebulellum* (Lepidoptera: Pyralidae). *Acta Oecologica* **14**: 619–626.

Leroi, B. 1978. Alimentation des adultes d'*Acanthoscelides obtectus* Say (Coléoptère, Bruchidae): influence sur la longévité et la production ovarienne des individus vierges. *Annales de Zoologie–Écologie Animale* **10**: 559–567.

Lewis, A. C. 1993. Learning and the evolution of resources: pollinators and flower morphology. In D. R. Papaj and A. C. Lewis (eds.) *Insect Learning*. New York: Chapman and Hall, pp. 219–242.

Lin, S. and C. A. Mullin. 1999. Lipid, polyamide, and flavonol phagostimulants for adult western corn rootworm from sunflower (*Helianthus annuus* L.) pollen. *Journal of Agricultural and Food Chemistry* **47**: 1223–1229.

Lopez, J. D. Jr., T. N. Shaver, and P. D. Lingren. 1994. Evaluation of feeding stimulants for adult *Helicoverpa zea*. Proc. Beltwide Cotton Production Res. Conf., National Cotton Council, Memphis, TN, pp. 920–924.

Ludwig, K. A. and R. E. Hill. 1975. Comparison of gut contents of adult western and northern corn rootworms in northeast Nebraska. *Environmental Entomology* **4**: 435–438.

Lukefahr, M. J. and D. F. Martin. 1964. The effects of various larval and adult diets on the fecundity and longevity of the bollworm, tobacco budworm, and cotton leafwom. *Journal of Economic Entomology* **57**: 233–235.

Lukefahr, M. J. and C. Rhyne. 1960. Effects of nectariless cottons on populations of three lepidopterous species. *Journal of Economic Entomology* **53**: 242–244.

Lukefahr, M. J., D. F. Martin, and J. R. Meyer. 1965. Plant resistance to five Lepidoptera attacking cotton. *Journal of Economic Entomology* **58**: 516–518.

Malavasi, A., J. S. Morgante, and R. J. Prokopy. 1983. Distribution and activities of *Anastrepha fraterculus* (Diptera: Tephritidae) flies on host and non-host trees. *Annals of the Entomological Society of America* **76**: 286–292.

Masters, A. R. 1991. Dual role of pyrrolizide alkaloids in nectar. *Journal of Chemical Ecology* **17**: 195–205.

May, P. G. 1985. Nectar uptake rates and optimal nectar concentrations of two butterfly species. *Oecologia* **66**: 381–386.

1992. Flower selection and the dynamics of lipid reserves in two nectarivorous butterflies. *Ecology* **73**: 2181–2191.

McAlpine, J. F. 1981. Morphology and terminology: adults. In J. F. McAlpine, B. V. Peterson, G. E. Shewell, *et al.* (eds.) *Manual of Nearctic Diptera*, vol. 1, Monograph no. 27. Ottawa: Research Branch Agriculture Canada, pp. 9–63.

McDonald, R. S. and J. H. Borden. 1996. Dietary constraints on sexual activity, mating success, and survivorship of male *Delia antiqua*. *Entomologia Experimentalis et Applicata* **81**: 243–250.

McEwen, P. K. and H. Liber. 1995. The effect of adult nutrition on the fecundity and longevity of the olive moth *Prays oleae* (Bern.). *Journal of Applied Entomology* **119**: 291–294.

McKibben, G. H., M. J. Thompson, W. L. Parrott, A. C. Thompson, and W. R. Lusby. 1985. Identification of feeding stimulants for boll weevils from cotton buds and anthers. *Journal of Chemical Ecology* **11**: 1229–1238.

McLeod, D. G. R. 1964. Nutrition and feeding behavior of the adult onion maggot, *Hylemya antiqua*. *Journal of Economic Entomology* **57**: 845–847.

McQuate, G. T., G. D. Jones, and C. D. Sylva. 2003. Assessment of corn pollen as a food source for two tephritid fruit fly species. *Environmental Entomology* **32**: 141–150.

Metcalf, R. L., R. L. Lampman, and P. A. Lewis. 1998. Comparative kairomonal chemical ecology of diabroticite beetles (Coleoptera: Chrysomelidae: Galerucinae: Luperini: Diabroticina) in a reconstituted tallgrass prairie ecosystem. *Journal of Economic Entomology* **91**: 881–890.

Mevi-Schütz, J. and A. Erhardt. 2002. Can *Inachis io* detect nectar amino acids at low concentrations? *Physiological Entomology* **27**: 256–260.

2003a. Larval nutrition affects female nectar amino acid preference in the map butterfly (*Araschnia levana*). *Ecology* **84**: 2788–2794.

2003b. Mating frequency influences nectar amino acid preference of *Pieris napi*. *Proceedings of the Royal Society of London Series B* **271**: 153–158.

2003c. Effects of nectar amino acids on fecundity of the wall brown butterfly (*Lasiommata megera* L.). *Basic and Applied Ecology* **4**: 413–421.

Mevi-Schütz, J., M. Goverde, and A. Erhardt. 2003. Effects of fertilization and elevated CO_2 on larval food and butterfly nectar amino acid preference in *Coenonympha pamphilus* L. *Behavioral Ecology and Sociobiology* **54**: 36–43.

Miller, W. E. 1987. Spruce budworm (Lepidoptera: Tortricidae): role of adult imbibing in reproduction. *Environmental Entomology* **16**: 1291–1295.

1988. European corn borer reproduction: effects of honey in imbibed water. *Journal of the Lepidopterists' Society* **42**: 138–143.

1989. Reproductive enhancement by adult feeding: effects of honeydew in imbibed water of spruce budworm. *Journal of the Lepidopterists' Society* **43**: 167–177.

1996. Population behaviour and adult feeding capability in Lepidoptera. *Environmental Entomology* **25**: 213–226.

Mitchell, B. K. 1985. Specificity of an amino acid-sensitive cell in the adult Colorado beetle, *Leptinotarsa decemlineata*. *Physiological Entomology* **10**: 421–429.

Mitchell, B. K. and G. D. Harrison. 1984. Characterization of galeal chemosensilla in the adult Colorado potato beetle, *Leptinotarsa decemlineata*. *Physiological Entomology* **9**: 49–56.

Moeser, J. and S. Vidal. 2001. Alternative food resources for adult *Diabrotica virgifera virgifera* in southern Hungary. Proc. 21st. IOBC/IWGO Conference and 7th *Diabrotica* subgroup meeting, Venice, Italy, pp. 19–24.

Moore, R. A. and M. C. Singer. 1987. Effects of maternal age and adult diet on egg weight in the butterfly *Euphydryas editha*. *Ecological Entomology* **12**: 401–408.

Mullin, C. A., S. Chyb, H. Eichenseer, B. Hollister, and J. L. Frazier. 1994. Neuroreceptor mechanisms in insect gustation: a pharmacological approach. *Journal of Insect Physiology* **40**: 913–931.

Murphy, D. D. 1983. Nectar sources as constraints on the distribution of egg masses by the checkerspot butterfly, *Euphydryas chalcedona* (Lepidoptera: Nymphalidae). *Environmental Entomology* **12**: 463–466.

Murphy, D. D., W. E. Launer, and P. R. Ehrlich. 1983. The role of adult feeding in egg production and population dynamics of the checkerspot butterfly *Euphydryas editha*. *Oecologia* **56**: 257–263.

Murphy, D. D., M. S. Menninger, and P. R. Ehrlich. 1984. Nectar source distribution as a determinant of oviposition host species in *Euphydryas chalcedona*. *Oecologia* **62**: 269–271.

Naranjo, S. E. 1991. Movement of corn rootworm beetles, *Diabrotica* spp. (Coleoptera: Chrysomelidae), at cornfield boundaries in relation to sex, reproductive status, and crop phenology. *Environmental Entomology* **20**: 230–240.

Naranjo, S. E. and A. J. Sawyer. 1987. Reproductive biology and survival of *Diabrotica barberi* (Coleoptera: Chrysomelidae): effect of temperature, food, and seasonal time of emergence. *Annals of the Entomological Society of America* **80**: 841–848.

Nicolson, S. W. 1994. Pollen feeding in the eucalypt nectar fly, *Drosophila flavohirta*. *Physiological Entomology* **19**: 58–60.

Nishida, T. 1958. Extrafloral glandular secretions, a food source for certain insects. *Proceedings of the Hawaiian Entomological Society* **16**: 379–386.

Nolte, H.-W. 1959. Untersuchungen zum Farbensehen des Rapsglanzkäfers (*Meligethes aeneus* F.). I. Die Reaktion des Rapsglanzkäfers auf Farben und die ökologische Bedeutung des Farbensehens. *Biologisches Zentralblatt* **78**: 63–107.

Norris, M. J. 1936. The feeding habits of the adult Lepidoptera Heteroneura. *Transactions of the Royal Entomological Society (London)* **85**: 61–90.

O'Brian, D. M., C. L. Boggs, and M. L. Fogel. 2003. Pollen feeding in the butterfly *Heliconius charitonia*: isotopic evidence for essential amino acid tranfer from pollen to eggs. *Proceedings of the Royal Society of London Series B* **270**: 2631–2636.

O'Brian, D. M., M. L. Fogel, and C. L. Boggs. 2002. Renewable and nonrenewable resources: amino acid turnover and allocation to reproduction in Lepidoptera. *Proceedings of the National Academy of Sciences of the USA* **99**: 4413–4418.

O'Brian, D. M., D. P. Schrag, and C. Martínez del Rio. 2000. Allocation to reproduction in a hawkmoth: a quantitative analysis using stable carbon isotopes. *Ecology* **81**: 2822–2831.

Ohsaki, N. 1979. Comparative population studies of three *Pieris* butterflies, *P. rapae*, *P. melete* and *P. napi*, living in the same area. I. Ecological requirements for habitat resources in the adults. *Researches on Population Ecology* **20**: 278–296.

Ômura, H. and K. Honda. 2003. Feeding responses of adult butterflies, *Nymphalis xanthomelas*, *Kanisca canace*, and *Vanessa indica*, to components in tree sap and rotting fruits: synergistic effects of ethanol and acetic acid on sugar responsiveness. *Journal of Insect Physiology* **49**: 1031–1038.

Penz, C. M. and H. W. Krenn. 2000. Behavioural adaptions to pollen-feeding in *Heliconius* butterflies (Nymphalidae, Heliconiinae): an experiment using *Lantana* flowers. *Journal of Insect Behavior* **13**: 865–880.

Pesho, G. R. and R. J. Van Houten. 1982. Pollen and sexual maturation of the pea weevil (Coleoptera: Bruchidae). *Annals of the Entomological Society of America* **75**: 439–443.

Peterson, M. A. 1997. Host plant phenology and butterfly dispersal: causes and consequences of uphill movement. *Ecology* **78**: 167–180.

Petherbridge, F. R., D. W. Wright, and P. G. Davies. 1942. Investigations on the biology and control of the carrot fly. *Annals of Applied Biology* **29**: 380–392.

Pettersson, M. W. 1992. Taking a chance on moths: oviposition by *Delia flavifrons* (Diptera: Anthomyiidae) on the flowers of bladder campion, *Silene vulgaris* (Caryophyllaceae). *Ecological Entomology* **17**: 57–62.

Pivnick, K. A. and J. N. McNeil. 1987. Diel patterns of activity of *Thymelicus lineola* adults (Lepidoptera: Hesperiidae) in relation to weather. *Ecological Entomology* **12**: 197–207.

Porter, K., C. A. Steel, and J. A. Thomas. 1992. Butterflies and communities. In R. L. H. Dennis (ed.) *The Ecology of Butterflies in Britain*. Oxford, UK: Oxford University Press, pp. 139–177.

Proctor, M., P. Yeo, and A. Lack. 1996. *The Natural History of Pollination*. London: HarperCollins.

Prokopy, R. J., J. J. Duan, and R. I. Vargas. 1996. Potential for host range expansion in *Ceratitis capitata* flies: impact of proximity of adult food to egg-laying sites. *Ecological Entomology* **21**: 295–299.

Ramaswamy, S. B. 1987. Behavioural responses of *Heliothis virescens* (Lepidoptera: Noctuidae) to stimulation with sugars. *Journal of Insect Physiology* **33**: 755–760.

Rana, R. L. and L. D. Charlet. 1997. Feeding behavior and egg maturation of the red and gray sunflower seed weevils (Coleoptera: Curculionidae) on cultivated sunflower. *Annals of the Entomological Society of America* **90**: 693–699.

Raubenheimer, D. and S. J. Simpson. 1999. Integrating nutrition: a geometrical approach. *Entomologia Experimentalis et Applicata* **91**: 67–82.

Raulston, J. R., S. D. Pair, P. D. Lingren, W. H. Hendrix III, and T. N. Shaver. 1998. The role of population dynamics in the development of control strategies for adult *Helicoverpa zea* and other Noctuidae. *Southwestern Entomologist* (suppl.) **21**: 25–35.

Rickson, F. R., M. Cresti, and J. H. Beach. 1990. Plant cells which aid in pollen digestion within a beetle's gut. *Oecologia* **82**: 424–426.

Rockstein, M. and J. Miquel. 1973. Aging in insects. In M. Rockstein (ed.) *The Physiology of Insecta*, vol. I, 2nd edn. New York: Academic Press, pp. 371–478.

Rogers, C. E. 1985. Extrafloral nectar: entomological implications. *Bulletin of the Entomological Society of America* **31**: 15–20.

Romeis, J. and F. L. Wäckers. 2000. Feeding responses by female *Pieris brassicae* butterflies to carbohydrates and amino acids. *Physiological Entomology* **25**: 247–253.

2002. Nutritional suitability of individual carbohydrates and amino acids for adult *Pieris brassicae*. *Physiological Entomology* **27**: 148–156.

Ross, D. W., D. J. Pree, and D. P. Toews. 1977. Digestive enzyme activity in the gut of the adult apple maggot, *Rhagoletis pomonella*. *Annals of the Entomological Society of America* **70**: 417–422.

Roulston, T. H. and J. H. Cane. 2000. Pollen nutritional content and digestibility for animals. *Plant Systematics and Evolution* **222**: 187–209.

Rusterholz, H. P. and A. Erhardt. 1997. Preferences for nectar sugars in the peacock butterfly, *Inachis io*. *Ecological Entomology* **22**: 220–224.

2000. Can nectar properties explain sex-specific flower preferences in the adonis blue butterfly *Lysandra bellargus*? *Ecological Entomology* **25**: 81–90.

Ruther, J. and K. Thiemann. 1997. Response of the pollen beetle *Meligethes aenus* to volatiles emitted by intact plants and conspecifics. *Entomologia Experimentalis et Applicata* **84**: 183–188.

Salama, H. S., A. Khalifa, N. Azmy, and A. Sharaby. 1984. Gustation in the lepidopterous moth *Spodoptera littoralis* (Boisd.). *Zoologische Jahrbücher, Abteilung für allgemeine Zoologie und Physiologie der Tiere* **88**: 165–178.

Samuelson, G. A. 1994. Pollen consumption and digestion by leaf beetles. In P. H. Jolivet, M. L. Cox, and E. Petitpierre (eds.) *Novel Aspects of the Biology of Chrysomelidae*. Dordrecht, the Netherlands: Kluwer Academic Publishers, pp. 179–183.

Scherer, C. and G. Kolb. 1987. Behavioural experiments on the visual processing of colour stimuli in *Pieris brassicae* L. (Lepidoptera). *Journal of Comparative Physiology A* **160**: 645–656.

Schmidt, J. M. and W. G. Friend. 1991. Ingestion and diet destination in the mosquito *Culiseta inornata*: effects of carbohydrate configuration. *Journal of Insect Physiology* **37**: 817–828.

Scholz, J. and H. J. Dörner. 1976. Untersuchungen über das Auftreten, die Schadwirkung und die Bekämpfung des Ackerbohnenkäfers in Ackerbohnenbeständen der Bezirke Halle und Leipzig. *Nachrichtenblatt für den Pflanzenschutz in der DDR* **30**: 212–216.

Schoonhoven, L. M., T. Jermy, and J. J. A. van Loon. 1998. *Insect–Plant Biology: From Physiology to Evolution*. London: Chapman and Hall.

Schuster, M. F. and M. Calderón. 1986. Interactions of host plant resistant genotypes and beneficial insects in cotton ecosystems. In D. J. Boethel and R. D. Eikenbarry (eds.) *Interactions of Plant Resistance and Parasitoids and Predators of Insects*. Chichester, UK: Ellis Horwood, pp. 84–97.

Schuster, M. F., M. J. Lukefahr, and F. G. Maxwell. 1976. Impact of nectariless cotton on plant bugs and natural enemies. *Journal of Economic Entomology* **69**: 400–402.

Shanks, C. H. and R. P. Doss. 1987. Feeding responses by adults of five species of weevils (Coleoptera, Curculionidae) to sucrose and sterols. *Annals of the Entomological Society of America* **80**: 41–46.

Sharp, J. L. and D. L. Chambers. 1984. Consumption of carbohydrates, proteins, and amino acids by *Anastrepha suspensa* (Loew) (Diptera: Tephritidae) in the laboratory. *Environmental Entomology* **13**: 768–773.

Shelly, T. E., S. S. Kennelly, and D. O. McInnis. 2002. Effect of adult diet on signaling activity, mate attraction, and mating success in male Mediterranean fruit flies (Diptera: Tephritidae). *Florida Entomologist* **85**: 150–155.

Shorey, H. H. 1963. The biology of *Trichoplusia ni* (Lepidoptera: Noctuidae). II. Factors affecting adult fecundity and longevity. *Annals of the Entomological Society of America* **56**: 476–480.

Shreeve, T. G. 1992. Adult behaviour. In R. L. H. Dennis (ed.) *The Ecology of Butterflies in Britain.* Oxford, UK: Oxford Universitry Press, pp. 22–45.

Siddappaji, C. and T. K. S. Gowda. 1980. Rhizobial nodules eating insect - *Rivellia* sp. – a new pest of pulse crops in India. *Current Research* **9**: 122–123.

Siegfried, B. D. and C. A. Mullin. 1990. Effects of alternative host plants on longevity, oviposition, and emergence of western and northern corn rootworms (Coleoptera: Chrysomelidae). *Environmental Entomology* **19**: 474–480.

Skevington, J. H. and P. T. Dang. 2002. Exploring the diversity of flies (Diptera). *Biodiversity* **3:** 3–27.

Städler, E. 1971. An improved mass-rearing method of the carrot rust fly, *Psila rosae* (Diptera: Psilidae). *Canadian Entomologist* **103**: 1033–1038.

1978. Chemoreception of host plant chemicals by ovipositing females of *Delia (Hylemya) brassicae. Entomologia Experimentalis et Applicata* **24**: 711–720.

Städler, E. and R. Schöni. 1991. High sensitivity to sodium in the sugar chemoreceptor of the cherry fruit fly after emergence. *Physiological Entomology* **16**: 117–129.

Städler, E. and W. D. Seabrook. 1975. Chemoreceptors on the proboscis of the female eastern spruce budworm: electrophysiological study. *Entomologia Experimentalis et Applicata* **18**: 153–160.

Stensmyr, M. C., I. Urru, I. Collu, *et al.* 2002. Rotting smell of dead-horse arum flowers. *Nature* **420**: 625.

Stoffolano, J. G. Jr. 1995. Regulation of a carbohydrate meal in the adult Diptera, Lepidoptera, and Hymenoptera. In R. F. Chapman and G. De Boer (eds.) *Regulatory Mechanisms in Insect Feeding.* New York: Chapman and Hall, pp. 210–247.

Swirski, E., Y. Izhar, M. Wysoki, E. Gurevitz, and S. Greenberg. 1980. Integrated control of the long-tailed mealybug, *Pseudococcus longispinus* (Hom, Pseudococcidae), in avocado plantations in Israel. *Entomophaga* **25**: 415–426.

Tahhan, O. and H. F. Van Emden. 1989. Biology of *Bruchus dentipes* Baudi (Coleoptera: Bruchidae) on *Vicia faba* and a method to obtain gravid females during the imaginal quiescence period. *Bulletin of Entomological Research* **79**: 201–210.

Takakura, K. 1999. Active female courtship behaviour and male nutritional contribution to female fecundity in *Bruchidius dorsalis* (Fahraeus) (Coleoptera: Bruchidae). *Researches on Population Ecology* **41**: 269–273.

2004. The nutritional contribution of males affects the feeding behaviour and spatial distribution of females in a bruchid beetle, *Bruchidius dorsalis. Journal of Ethology* **22**: 37–42.

Thien, L. B. 1980. Patterns of pollination in the primitive angiosperms. *Biotropica* **12**: 1–13.

Thien, L. B., P. Bernhardt, G. W. Gibbs, *et al.* 1985. The pollination of *Zygogynum* (Winteraceae) by a moth, *Sabatinca* (Micropterigidae): an ancient association. *Science* **227**: 540–543.

Tingle, F. C. and E. R. Mitchell. 1992. Attraction of *Heliothis virescens* (F.) (Lepidoptera: Noctuidae) to volatiles from extracts of cotton flowers. *Journal of Chemical Ecology* **18**: 907–914.

Tooker, J. F., P. F. Reagel, and L. M. Hanks. 2002. Nectar sources of day-flying Lepidoptera of central Illinois. *Annals of the Entomological Society of America* **95**: 84–96.

Topper, C. P. 1987. Nocturnal behaviour of adults of *Heliothis armigera* (Hübner) (Lepidoptera: Noctuidae) in the Sudan Gezira and pest control implications. *Bulletin of Entomological Research* **77**: 541–554.

Tran, B. and J. Huignard. 1992. Interactions between photoperiod and food affect the termination of reproductive diapause in *Bruchus rufimanus* (Boh.), (Coleoptera, Bruchidae). *Journal of Insect Physiology* **38**: 633–642.

Tsiropoulos, G. J. 1977. Reproduction and survival of the adult *Dacus oleae* feeding on pollens and honeydew. *Environmental Entomology* **6**: 390–392.

1978. Holidic diets and nutritional requirements for survival and reproduction of the adult walnut husk fly. *Journal of Insect Physiology* **24**: 239–242.

1980a. Major nutritional requirements of adult *Dacus oleae*. *Annals of the Entomological Society of America* **73**: 251–253.

1980b. Carbohydrate utilization by normal and γ-sterilized *Dacus oleae*. *Journal of Insect Physiology* **26**: 633–637.

1983. The importance of dietary amino acids on the reproduction and longevity of adult *Dacus oleae* (Gmelin) (Diptera Tephritidae). *Archives Internationales de Physiologie et de Biochimie* **91**: 159–164.

1984. Effect of specific phagostimulants on adult *Dacus oleae* feeding behaviour. In: *Fruit Flies of Economic Importance* 84, Proc. CEC/IOBC *ad-hoc* meeting, Hamburg, pp. 95–98.

Tsitsipis, J. A. 1989. Nutrition: requirements. In A. S. Robinson and G. Hooper (eds.) *Fruit Flies, Their Biology, Natural Enemies and Control*, vol. 3A. Amsterdam, the Netherlands: Elsevier, pp. 103–119.

Turlings, T. C. J. and F. L. Wäckers. 2004. Recruitment of predators and parasitoids by herbivore-injured plants. In R. T. Cardé and J. Millar (eds.) *Advances in Insect Chemical Ecology*. Cambridge, UK: Cambridge University Press, pp. 21–75.

Vijaysegaran, S., G. H. Walter, and R. A. I. Drew. 1997. Mouthpart structure, feeding mechanisms, and natural food sources of adult *Bactrocera* (Diptera: Tephritidae). *Annals of the Entomological Society of America* **90**: 184–201.

Wacht, S., K. Lunau, and K. Hansen. 2000. Chemosensory control of pollen ingestion in the hoverfly *Eristalis tenax* by labellar taste hairs. *Journal of Comparative Physiology A* **186**: 193–203.

Wäckers, F. L. 1999. Gustatory response by the hymenopteran parasitoid *Cotesia glomerata* to a range of nectar and honeydew sugars. *Journal of Chemical Ecology* **25**: 2863–2877.

2001. A comparison of nectar- and honeydew sugars with respect to their utilization by the hymenopteran parasitoid *Cotesia glomerata*. *Journal of Insect Physiology* **47**:1077–1084.

Wäckers, F. L., D. Zuber, R. Wunderlin, and F. Keller. 2001. The effect of herbivory on the temporal and spatial dynamics of extrafloral nectar production. *Annals of Botany* **87**: 365–370.

Wada, A., Y. Isobe, S. Yamaguchi, R. Yamaoka, and M. Ozaki. 2001. Taste-enhancing effects of glycine on the sweetness of glucose: a gustatory aspect of symbiosis between the ant, *Camponotus japonicus*, and the larvae of the lycaenid butterfly, *Niphanda fusca*. *Chemical Senses* **26**: 983–992.

Waldbauer, G. P., A. P. Marciano, and P. K. Pathak. 1980. Life-span and fecundity of adult rice leaf folders, *Cnaphalocrocis medinalis* (Guenée) (Lepidoptera: Pyralidae), on sugar sources, including honeydew from the brown plant hopper, *Nilaparvata lugens* (Stal) (Hemiptera: Delpacidae). *Bulletin of Entomological Research* **70**: 65–71.

Watt, W. B., P. C. Hoch, and S. G. Mills. 1974. Nectar resource use by *Colias* butterflies. *Oecologia* **14**: 353–374.

Weis, I. 1930. Versuche über die Geschmacksrezeption durch die Tarsen des Admirals, *Pyrameis atalanta* L. *Zeitschrift für vergleichende Physiologie* **12**: 206–248.

Wheeler, D. 1996. The role of nourishment in oogenesis. *Annual Review of Entomology* **41**: 407–431.

Wickman, P.-O. and B. Karlsson. 1987. Changes in egg colour, egg weight and oviposition rate with the number of eggs laid by wild females of the small heath butterfly, *Coenonympha pamphilus*. *Ecological Entomology* **12**: 109–114.

Wiesenborn, W. D. and T. C. Baker. 1990. Upwind flight to cotton flowers by *Pectinophora gossypiella* (Lepidoptera: Gelechiidae). *Environmental Entomology* **19**: 490–493.

Wiklund, C. 1977. Oviposition, feeding and spatial separation of breeding and foraging habitats in a population of *Leptidea sinapis* (Lepidoptera). *Oikos* **28**: 56–68.

Wiklund, C. and C. Åhrberg. 1978. Host plants, nectar source plants and habitat selection of males and females of *Anthocharis cardamines* (Lepidoptera). *Oikos* **31**: 169–183.

Wiklund, C. and B. Karlsson. 1984. Egg size variation in satyrid butterflies: adaptive vs. historical, 'Bauplan', and mechanistic explanations. *Oikos* **43**: 391–400.

Wiklund, C. and A. Persson. 1983. Fecundity, and the relation of egg weight variation to offspring fitness in the speckled wood butterfly *Pararge aegeria*, or why don't butterfly females lay more eggs? *Oikos* **40**: 53–63.

Wiklund, C., T. Eriksson, and H. Lundberg. 1979. The wood white butterfly *Leptidea sinapis* and its nectar plants: a case of mutualism or parasitism? *Oikos* **33**: 358–362.

Willers, J. L., J. C. Schneider, and S. B. Ramaswamy. 1987. Fecundity, longevity and caloric patterns in female *Heliothis virescens*: changes with age due to flight and supplemental carbohydrate. *Journal of Insect Physiology* **33**: 803–808.

Part III PLANT-PROVIDED FOOD AND BIOLOGICAL CONTROL

8

Impact of plant-provided food on herbivore–carnivore dynamics

PAUL C. J. VAN RIJN AND MAURICE W. SABELIS

Introduction

Arthropod predators and parasitoids play an important role in reducing herbivore damage to plants. Although most of these arthropods are mainly carnivorous, they also use plant-provided food (PPF) as a source of nutrients during at least part of their life cycle. These foods affect longevity, fecundity, and the distribution of carnivores (Olson *et al.*, Chapter 5 and Eubanks and Styrsky, Chapter 6), and thus also the population dynamics of herbivore–carnivore systems.

Despite the importance of this type of omnivory for herbivore–carnivore interactions in general and biological control in particular, its population-dynamical consequences are not fully understood. Relatively few population studies have addressed the topic (Bakker and Klein 1992; Alomar and Wiedemann 1996; Stapel *et al.* 1997; Eubanks and Denno 2000; Van Rijn *et al.* 2002; Wäckers 2003), and even fewer theoretical studies (Krivan and Sirot 1997; Van Baalen *et al.* 2001; Van Rijn *et al.* 2002; Kean *et al.* 2003). Omnivory in general, however, has gained much attention since Polis' seminal papers (Polis *et al.* 1989; Polis and Holt 1992), both among empirical (Diehl 1995; Holyoak and Sachdev 1998; Pringle and Hamazaki 1998; Gillespie and McGregor 2000; Coll and Guershon 2002) and theoretical ecologists (Pimm and Lawton 1978; Holt and Polis 1997; McCann and Hastings 1997; Polis 1998; Mylius *et al.* 2001). These studies have mainly focussed on the consequences of omnivory for population persistence and community stability, and much less on the consequences for herbivory. It is our aim to fill this gap, by the application of general ecological theory and the development of simple mathematical models.

Plant-Provided Food for Carnivorous Insects, ed. F. L. Wäckers, P. C. J. van Rijn, and J. Bruin.
Published by Cambridge University Press.

In the first section we discuss the mechanism by which PPF, through its effect on carnivore behavior and life history, may affect herbivore abundance. We focus on simple consumer-resource models that ignore stage or spatial structure. In the second section we use stage-structured models to investigate how the impact of PPF on herbivory may depend on the life history and feeding requirements of carnivorous arthropods. We compare parasitoids and predators, as well as predators with different types of omnivory. In the third section we explicitly consider the impact of spatial structure. In the fourth section, we discuss the impact of more complex food-web structure, due to, for example, top carnivores, competitors, and intraguild predators. Finally, we discuss how the theoretical exercise in this chapter can be applied to reduce the impact of herbivores in crops.

General theory

Impact on equilibrium: apparent competition

To understand how PPF, through its effect on the carnivore, will affect herbivore abundance, we consider a system where the predator population directly controls the herbivore population. When this system is at equilibrium (showing neither increase nor decline), the addition of PPF will initially result in an increase of the predator population, simply because more food is available. This increase will only halt when a sufficient decrease of the herbivore population has compensated the added PPF. Thus, adding food will lead to a decrease in the herbivore population via the consumers they share. As the impact on population level is similar to competition, Holt labeled this principle *apparent* (i.e., *predator-mediated*) competition, to contrast it with resource and interference competition (Holt 1977). Although originally formulated in terms of two (self-reproducing) prey populations sharing the same predator, the principle also holds when one of the two prey species is replaced by a non-reproducing food source (Van Baalen *et al.* 2001; Van Rijn *et al.* 2002).

The principle and its conditions can be illustrated by modifying the Lotka–Volterra predator–prey model:

$$\frac{dN}{dt} = rN(1 - N/K) - F(N,A)P$$
$$\frac{dP}{dt} = G(N,A)P - \mu P. \qquad (8.1a, b)$$

This model represents the change in prey (N) and predator (P) population sizes resulting from the difference between reproduction and mortality. The amount of PPF is assumed to be constant (A). The parameters r and K represent the growth rate and the carrying capacity of the prey population, F the per capita predation rate, G the predator reproduction rate, and μ the predator mortality rate. In absence of an alternative food source ($A = 0$) the ratio G/F indicates the assimilation or conversion efficiency of the predators.

If the PPF and the prey are *substitutable* foods (Tilman 1982), the overall food level can be represented by $N + \phi A$, where ϕ indicates the food value of A relative to that of N. When the per capita predation rate (F), and consequently the predator's reproduction rate (G), increases linearly with prey or food densities, the functional and numerical responses can simply be written as:

$$F(N, A) = fN \text{ and } G(N, A) = g(N + \phi A). \tag{8.2a, b}$$

The equilibrium density of the herbivore, solved from (Eqn 8.1b), now becomes:

$$N^* = \frac{\mu}{g} - \phi A \text{ and } P^* = \frac{r}{f}. \tag{8.3}$$

This expression directly shows that for positive ϕ (i.e., PPF has a nutritional value for the predator), increasing the PPF density (A) will decrease the equilibrium prey density (N^*).

It may, however, be more realistic to assume that the per capita predation rate is a decelerating function of prey and food density. This yields saturating, type II functional and numerical responses of the following forms:

$$F(N, A) = f\frac{N}{N + \phi A + h} \text{ and } G(N, A) = g\frac{N + \phi A}{N + \phi A + h}, \tag{8.4a, b}$$

where h is the prey density at which the predation and reproduction rates are half their maximum, and where food preference (incorporated in ϕ) is assumed to be constant (Van Rijn *et al.* 2004, unpublished data). With this saturating functional response the equilibrium densities become

$$N^* = \frac{h\mu}{g - \mu} - \phi A, \text{ and } P^* = \frac{r}{f}h\left(1 + \frac{\mu}{g - \mu}\right). \tag{8.5}$$

This expression for N^* is only different from (Eqn 8.3) in the first term. Thus, the shape of the functional response does not change the qualitative (negative) impact of PPF on the herbivore equilibrium.

Different predictions emerge when food and prey are not substitutable but *complementary*. Two food types will, for example, be complementary when they affect different life-history aspects, for example when prey affects reproduction and PPF affects mortality. Now the term ϕA should be eliminated from (Eqns 8.4a,b) for the functional and numerical responses ($F(.)$ and $G(.)$), whereas predator mortality, μ, should be written as a decreasing function of A, e.g.,

$$\mu(A) = \frac{\mu_0}{1 + \phi A}, \tag{8.6}$$

where μ_0 is the (background) mortality in absence of PPF. Assuming a linear functional response, the equilibrium densities now become:

$$N^* = \frac{\mu_0}{g(1 + \phi A)} \text{ and } P^* = \frac{r}{f}. \tag{8.7a, b}$$

For the case of a saturating functional response:

$$N^* = \frac{h\mu_0}{g(1 + \phi A) - \mu_0}, \text{ and } P^* = \frac{r}{f}(h + N^*) = \frac{r}{f}h\left[1 + \frac{\mu_0}{g(1 + \phi A) - \mu_0}\right]. \tag{8.8a, b}$$

Whereas for substitutable resources the equilibrium prey density can be brought to zero by adding sufficient amounts of PPF, for complementary resources the equilibrium herbivore density will *asymptotically* decrease to a minimum level defined by the minimum level of mortality (e.g., due to predation or senescence) (compare drawn and dashed lines in Fig. 8.2–C2 below).

What can we learn from these equations concerning the impact of PPF on equilibrium herbivore–carnivore interactions?

Carnivore reproduction and survival

By enhancing reproduction or survival of the carnivore, PPF will indirectly decrease equilibrium prey density. When PPF and prey are substitutable foods there may be a critical amount of PPF at which the equilibrium prey density (N^*) becomes zero, indicating that the herbivore will go extinct. At that point, the predator population is maintained by PPF only (Van Rijn *et al.* 2002). When PPF and prey are complementary rather than substitutable, PPF will also result in lower equilibrium prey levels (N^*), but not to the point of prey extinction.

Predation rate

When predators feed on pollen or plant tissue this often goes at the expense of their consumption of prey, possibly due to limitation of time or gut capacity (Eubanks and Styrsky, Chapter 6). For the case of a saturating functional response (Eqn 8.4) this impact is incorporated by putting A in the denominator. This impact has, however, no effect on the equilibrium prey density, N^* (Eqn 8.5), which results from the numerical response only. Even a stronger inhibition of prey consumption, for example due to food-type switching (Van Baalen *et al.* 2001) or asymmetric satiation (Van Rijn *et al.*, unpublished data), will have no effect on prey equilibrium, as long as carnivore reproduction and survival (resulting from the assimilation of both prey and PPF) are not negatively affected by the alternative food. This condition is likely to be met, since feeding on PPF would clearly be maladaptive if it were to result in reduced reproduction and survival (Holt 1983). To understand the insensitivity of the herbivore equilibrium to the per capita predation rate, one should realize that a reduction in the per capita predation rate (F) is compensated by an increase in the predator population (P^*), leaving the overall predation on the herbivore population (FP^*) unaffected.

Herbivore performance

In special cases PPF may not only be used by the carnivore but by the herbivore as well (Romeis *et al.*, Chapter 7). For example, pollen used by predatory mites can also be used by the thrips they feed on (Van Rijn *et al.* 2002). The PPF could in that case enhance the population growth rate of the herbivore (r). In our model this has no effect on equilibrium herbivore density, as this parameter does not show up in the equations for N^* (Eqns. 8.3 and 8.5). This result is well known from food-chain dynamics literature (Oksanen *et al.* 1981). At equilibrium, the enhanced herbivore performance due to PPF will (again) be compensated by predation (FP^*) from a larger predator population.

However, when the different species are assigned more flexible responses (such as predator avoidance behavior) the herbivore is not only regulated top-down, but also bottom–up (Polis 1994, 1999; Abrams and Vos 2003). In this case, PPF-related changes in predation rate and herbivore performance *will* translate in increased herbivore abundance at equilibrium as well.

Impact on transient dynamics: a single-generation model

The equilibrium approach holds when environmental conditions, such as climate or the availability of PPF, remain unchanged over a sufficiently long period. How long the conditions need to be constant to approximate the equilibrium depends on the initial densities of the interacting populations, their generation times, and the stability of the equilibrium. For carnivorous mites and herbivorous thrips with generation times of about 3 weeks, populations were already within the 10% range of their equilibrium level after 12 weeks following their introduction in a cucumber crop (Van Rijn *et al.* 2002). After this period the impact of a regularly supplied food source on mite and thrips populations can adequately be predicted from equilibrium equations only.

Larger arthropods generally have longer generation times and their populations require more time to settle around some equilibrium (Sabelis 1992). For insects with only one or two generations per year and with food sources available only during part of the year, an equilibrium approach is unlikely to hold. In that case, one should rather focus on the dynamics displayed before the system approaches its equilibrium state (the so-called transient dynamics). Models of such systems require proper representation of delays, such as the juvenile period, by incorporating age or stage structure (Caswell 1989; Nisbet 1997; De Roos and Persson 2001), and can usually be solved by numerical techniques only.

The higher complexity of such models, as well as the dependency on numerical solutions, makes it more difficult to derive transparent results. It is therefore useful to start with a simple model that can be solved analytically and that yields qualitative insight. (see Box 8.1)

BOX 8.1. Single-generation model

Consider the impact of PPF on the first generation of carnivores and herbivores only. Yet unaffected by reproduction, the population size of the first generation is determined only by mortality and migration. Assuming that after an initial colonization period further immigration is negligible, and that the per capita mortality and emigration rates are constant in time, the carnivore population (P) will decline exponentially with time (Fig. 8.1A):

$$P(t) = P_0 e^{-\mu t}.$$

Here P_0 is the initial carnivore density relative to that of the herbivores, and μ is the sum of mortality and emigration rates, each of which may depend on the availability of PPF. To allow explicit solutions, we further assume that herbivore density does not affect the decline in the number of carnivores, as will be realistic for life-history omnivores such as parasitoids and lacewings.

The herbivore population will, in addition to background mortality and emigration (ν), suffer from a constant per capita attack rate by the carnivores (f). The herbivore population will then (at every time step) decline at a rate equal to $\nu + fP(t)$, and the number of herbivores surviving the time period from 0 to T can then be calculated as:

$$N(T) = \exp\left[-\int_0^T (\nu + fP(t))dt\right].$$

Note that when P represents parasitoids, N indicates non-parasitized herbivores only.

If we now insert the carnivore equation into the herbivore equation and solve the integral, we obtain an expression that links predation rate and carnivore mortality to herbivore density (Fig. 8.1B):

$$N(T) = \exp\left[-\int_0^T (\nu + fP_0 e^{-\mu t})dt\right] = \exp\left[-\nu T - fP_0 \frac{1 - e^{-\mu T}}{\mu}\right].$$

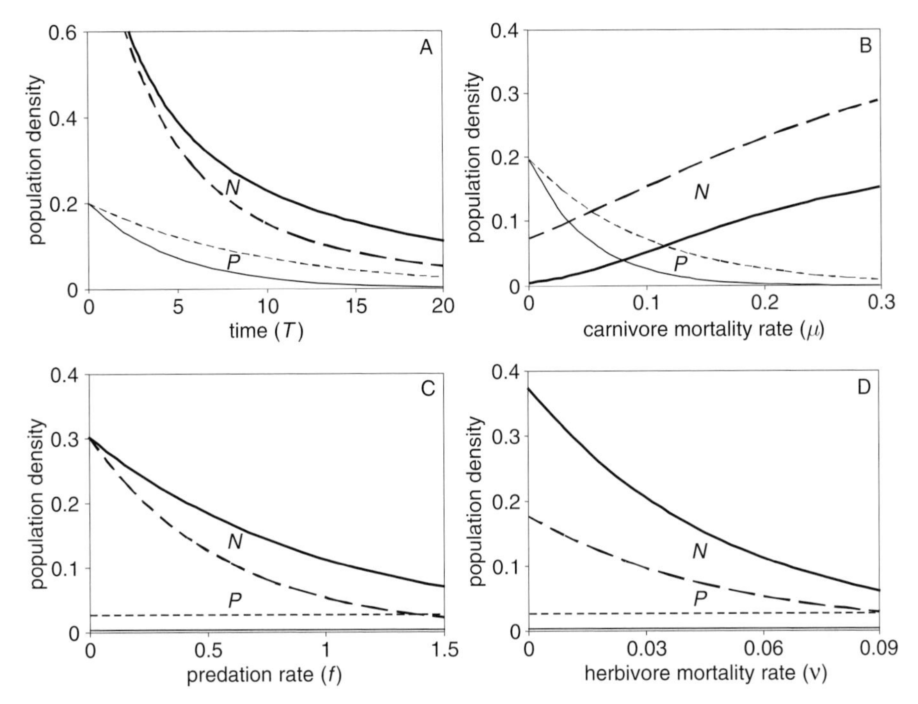

Figure 8.1 Results of single-generation herbivore–carnivore model (Box 8.1). (A) Decline in the numbers of carnivores (P, thin lines) and herbivores (N, heavy lines) for two levels of carnivore mortality: $\mu = 0.2$/day (drawn line) and 0.1/day (dashed line). (B) Effect of carnivore mortality (μ) on carnivore and herbivore density after interaction period of $T = 10$ (dashed line) and 20 days (drawn line). (C,D) Effect of predation rate (f) and herbivore mortality (ν) on carnivore and herbivore density after $T = 20$ days for two levels of carnivore mortality: $\mu = 0.2$/day ("without PPF", drawn lines) and 0.1/day ("with PPF", dashed lines). PPF is expected to *reduce* the parameter on the x-axes of the latter three panels ($P_0 = 0.2$, $\mu = 0.2$, $f = 1$/day, $\nu = 0.06$/day).

The Single-generation model of Box 8.1 leads to the following conclusions (see also Fig. 8.1):

Carnivore mortality

When PPF reduces carnivore mortality but affects none of the other parameters (as can be expected for parasitoids), the enhanced survival of the carnivores will, in due time, result in reduced herbivore numbers (Fig. 8.1A). For a given interaction period, herbivore number shows a positive relationship with carnivore mortality (Fig. 8.1B).

Predation rate

In other cases it can be expected that PPF reduces the carnivore's predation rate as well, for example in predators or host-feeding parasitoids where foraging is time or satiation-limited (Kidd and Jervis 1996; Van Rijn *et al.* 2004). A reduced predation rate now results in increased herbivore numbers. Figure 8.1C indicates that when food supply would result in a twofold reduction of the predation rate, this can only be compensated when at the same time it would result in a more than twofold reduction in carnivore mortality rate.

Herbivore mortality

When PPF is not only used by the carnivore but by the herbivore as well (Romeis *et al.*, Chapter 7), this may result in reduced herbivore mortality (ν), and consequently, in higher herbivore numbers. Figure 8.1D shows that, except when herbivore mortality (ν) is low, the impact of a twofold reduction in herbivore mortality due to PPF consumption will only be compensated when carnivore mortality (μ) is reduced more than twofold by PPF.

Section summary

The equilibrium and the single-generation approximations both predict that when PPF increases carnivore performance this will result in enhanced herbivore suppression. However, when PPF (also) reduces predation rate or enhances reproduction of the herbivore, this will have a negative impact on herbivore suppression in the short run, as indicated by the first-generation model, but not (or to a much lesser extent) in the long run, as indicated by the equilibrium approach. The explanation is that a positive effect of PPF on survival, arrestment or oviposition will result in enhanced growth of the local carnivore population, which, on a longer time scale, will (largely) compensate for a reduced per capita predation rate, and even for enhanced herbivore reproduction.

The impact of PPF is not only determined by the timescale of the interactions, but also by the level of complementarity of PPF and prey for the dominant carnivore. If PPF and prey are substitutable, increasing PPF abundance potentially reduces the equilibrium herbivore prey density to zero, but not if PPF and prey are complementary. Conversely, when PPF is complementary to prey, it may even be essential for a carnivore's existence, and consequently for herbivore control.

Impact of stage-related food requirements

In the introductory chapter of this book we discussed how feeding habits may change during the ontogeny of the carnivore, and how different types of omnivory result from it (Wäckers and van Rijn, Chapter 1). For our understanding of ecological interactions, such as predator–prey and parasitoid–host interactions, it is especially important to consider which life stage, and consequently which life-history characteristic (development, reproduction, or survival), is affected by prey (or host) density and PPF. In parasitoids, for instance, the larvae feed on herbivores, but it is the adult female that actively searches for them. By her foraging decisions she takes care that her offspring are provided with sufficient food to develop and mature. As a result herbivore density affects the oviposition rate of the female parasitoid, but not the survival or development of the larvae. Depending on the stages of the arthropod carnivore that actively searches for herbivores, there are three scenarios for the effect of herbivore density, describing three carnivore types:

(A) Herbivore density affects only adult performance, such as adult survival, attack rate and oviposition rate.
(B) Herbivore density affects only juvenile performance, such as juvenile survival and development.
(C) Herbivore density affects both adult and juvenile performance.

Also for the impact of PPF various scenarios are possible, depending on the type of food and the insect stage affected (see Olson *et al.*, Chapter 5 and Eubanks and Styrsky, Chapter 6):

(1) PPF allows the carnivorous arthropods feeding on it to survive longer and (when mature) to reproduce longer.
(2) PPF affects the rate at which the carnivore searches for the herbivores; positively when PPF is essential or when searching requires much energy, or negatively when time budget or gut capacity are constrained.
(3) PPF (especially when rich in proteins) enhances assimilation processes and increases the oviposition rate of adult carnivore females or the development of juveniles.

The various assumptions (scenarios) on herbivore-density dependence (A, B, and C) result in three different stage-structured herbivore–carnivore models (see Box 8.2). For each of these models the impact of the different PPF scenarios (1, 2, and 3) is analyzed and discussed separately. We focus on equilibrium equations and discuss transient effects only briefly at the end of this section. Stability of the equilibrium is a reasonable starting point for our analysis because we assume that only one herbivore stage is vulnerable for attack by the carnivore (Murdoch *et al.* 1987), and because we choose appropriate parameter values (Table 8.1).

BOX 8.2. Model with stage-related food requirements

To study the impact of stage-related food requirements in carnivores on population dynamics and herbivore suppression we subdivided the life cycles of herbivores and carnivores in three stages: two juvenile (non-reproducing) and one adult (reproducing) phase. The often strongly stage-specific vulnerability of hosts and prey is incorporated by restricting the attack to the first juvenile stage only (N_1). Stage-specific host or prey vulnerability is important as it increases the herbivore level and the stability of the equilibrium (Van den Bosch and Diekmann 1986; Murdoch 1987). Two other stages (N_2 and N_3) represent the remaining juvenile period and the adult phase. The first carnivore stage (P_1, typically including the egg stage) is unable to attack herbivores. The remaining juveniles (P_2) and the adults (P_3) may or may not. In general, the system can be represented by the following set of ordinary differential equations (ODEs):

$$\frac{dN_1}{dt} = rN_3 - F(N_1, A)P_C - d_1N_1 \qquad \frac{dP_1}{dt} = G(N_1, A)P_3 - e_1P_1$$

$$\frac{dN_2}{dt} = d_1N_1 - d_2N_2 \qquad \frac{dP_2}{dt} = e_1P_1 - \mu_2(N_1, A)P_2 - e_2(N_1, A)P_2$$

$$\frac{dN_3}{dt} = d_2N_2 - \nu N_3 \qquad \frac{dP_3}{dt} = e_2(N_1, A)P_2 - \mu_3(N_1, A)P_3$$

For the herbivores (N), r stands for the (net) reproduction rate of adults, $d_{1,2}$ for the development rate of stage 1 and 2, respectively, and ν for the adult mortality rate. $F(.)$ indicates the daily number of herbivores attacked per carnivore (functional response). For the carnivore (P), $G(.)$ indicates the per adult (net) reproduction rate, $e_{1,2}$, the development rates of juvenile stages 1 and 2, and $\mu_{2,3}$ the mortality rates of juveniles (stage 2) and adults (stage 3) (Table 8.1). As indicated, many parameters of carnivore life history (development, reproduction, survival) are potentially affected both by herbivore density (N_1) and PPF availability (A).

Three types of carnivore are distinguished, based on the stage that attacks the herbivores: (A) the adults, (B) the juveniles, or (C) both. It is assumed that only the stages that attack the herbivores are affected by herbivore density. For each carnivore type specific simplifications are sought that yield model equilibria that can be solved analytically (see Table 8.2 and main text).

Also regarding the impact of PPF three scenarios are studied. PPF is assumed to affect either: (1) mortality, (2) host/prey searching rate, or (3) (maximum) reproduction rate. Table 8.2 shows the functions describing these effects of PPF, as well as the solutions for the equilibrium densities of herbivore and carnivore for all the (3 × 3) combinations, and it provides the criteria for PPF to be essential.

Here, it is assumed that feeding on PPF does not go at the expense of foraging for prey. If, however, there is a trade-off between these two activities, this can be incorporated by adding a term with A in the denominator of the functional and numerical response function (Van Baalen *et al.* 2001), as in:

$$F(N, A) = f\frac{N}{N + \phi A + h}.$$

This modification does not affect the herbivore density at equilibrium, but it does increase the carnivore population at equilibrium (C3 in Table 8.2). In our simple models, no reduction of the per capita predation rate affects the equilibrium herbivore density, as long as the effect on reproduction is compensated by PPF. The equilibrium carnivore density will then further increase with PPF, such that the *population* predation rate stays unaffected (see above, pp. 224–230).

(A) Herbivore density affects adult performance

The first type of carnivore, in which only adult performance is affected by herbivore density, is most typically represented by parasitoid flies and wasps. As parasitoids oviposit only in or near their arthropod host, the rate of oviposition will be strongly linked to host density. Parasitoid females usually avoid oviposition on already parasitized hosts (avoidance of superparisitism), hence the juvenile performance in terms of development and survival is usually independent of host density. Only at high parasitoid/host ratios, when fewer unparasitized hosts are available, females can decide to superparasitize, despite the lower chance for offspring survival.

The juveniles of parasitoids are purely carnivorous, but the adults can often only meet their energy demands by feeding on nectar or honeydew. In agreement with the three PPF scenarios proposed above (Table 8.2), the adult parasitoid may, by feeding on these carbohydrate sources, enhance one or more of the following functions: (1) adult survival, (2) host searching rate, and (3) egg maturation rate (see Olson *et al.*, Chapter 5). With respect to the second scenario, we envision two ways in which PPF density can have a positive effect on host searching rate: (a) PPF can relieve the energy constraints on the energy-demanding host searching process, and (b) increasing availability of PPF reduces the time required to search for it, leaving more time for the parasitoids to search for hosts (Lewis *et al.* 1998).

Table 8.1 *Descriptions and values of parameters used in stage-structured carnivore–herbivore models. Herbivore parameter values are chosen to approximate the life history of some thrips and aphids pest species; carnivore parameter values are chosen to be valid for parasitoids as well as small predators, such as predatory mites, both at 16–20 °C*

Symbol	Parameter description	Value		Unit
		Default	If PPF essential	
	Herbivore biology			
$1/d_1$	developmental time vulnerable prey phase	12		days
$1/d_2$	developmental time invulnerable prey phase	24		days
r	net reproduction rate	0.5		offspring·adult^{-1}·day^{-1}
ν	instantaneous decline in net reproduction rate (including mortality)	0.04		/day
	Carnivore biology and response			
f	maximum rate of predation	2	1	prey·adult^{-1}·day^{-1}
h	prey density at which predation is half it maximum	1		prey·dm^{-2}
j	predation rate of juveniles relative to adults	0.5		[ratio]
$1/e_1$	developmental time non-feeding juveniles (e.g., eggs and pupae)	10		days
$1/e_2$	developmental time feeding juveniles	20		days
μ_2	maximum mortality rate feeding juveniles (scenario B only)	0.02		/day
c	conversion efficiency: female offspring per attacked herbivore	0.25		[ratio]
$g = cf$	maximum rate of net reproduction	0.5	0.25	offspring·adult^{-1}·day^{-1}
w	maintenance costs (scenario C only)	0.1		offspring·adult^{-1}·day^{-1}
μ_3	maximum decline in adult net reproduction rate	0.25		/day

Table 8.2 *Assumptions on the effect of herbivore density (N) and plant-provided food (PPF) density (A) on carnivore traits, and the resulting equilibrium state of the herbivore–carnivore system*

	Type of carnivore		
	A (parasitoids)	B ("hoverflies")	C (real omnivores)
Herbivore searching stage	$P_C = P_3$ (adult)	$P_C = P_2$ (juvenile)	$P_C = jP_2 + P_3$ (both adult and juvenile)
Herbivore density (*N*) affects	Oviposition rate	Juvenile development and/or mortality	Net reproduction rate
Numerical response function(s)	$G(N, A) = cF(N, A)$	$e_2(N) = e_{2m} \frac{N}{N+h_n}$, $\mu_2(N) = \mu_{2m} \left[\frac{N}{N+h_n}\right]^{-1}$	$G(N, A) = cF(N, A) - w$
PPF (*A*) affects	Resulting equilibrium herbivore and carnivore densities (N_1^* and P_C^*)*		
(1) Mortality $\mu_3(A) = \frac{\mu_{30}}{1+A}$ $F(N) = f\frac{N}{N+h}$	$N_1^* = h\frac{\mu_{30}}{cf(1+A)-\mu_{30}}$ $P_c^*\left(\frac{r}{\nu}-1\right)\frac{d_1}{f}(N_1^*+h)$	$h_n\left[\left[\frac{e_{2m}}{\mu_{2m}}\left((1+A)\frac{g}{\mu_{30}}-1\right)\right]^p-1\right]^{-1}$ $\left(\frac{r}{\nu}-1\right)\frac{d_1}{f}(N_1^*+h)$	$h\frac{\mu_{30}(1+A)^{-1}+w}{cf-\mu_{30}(1+A)^{-1}-w}$ $\left(\frac{r}{\nu}-1\right)\frac{d_1}{f}(N_1^*+h)$
(2) Searching rate $F(N, A) = f\frac{N(1+A)}{N(1+A)+h}$	$\frac{h}{(1+A)}\cdot\frac{\mu_3}{(cf-\mu_3)}$ $\left(\frac{r}{\nu}-1\right)\frac{d_1}{f}\left(N_1^*+\frac{h}{1+A}\right)$	--- ---	$\frac{h}{(1+A)}\cdot\frac{\mu_3+w}{cf-\mu_3-w}$ $\left(\frac{r}{\nu}-1\right)\frac{d_1}{f}\left(N_1^*+\frac{h}{1+A}\right)$
(3) Reproduction rate	$\frac{h(1+A)\mu_3}{(1+A)cf-\mu_3}$ $\left(\frac{r}{\nu}-1\right)\frac{d_1}{f}\left(\frac{N_1^*}{1+A}+h\right)$	$h_n\left[\left[\frac{e_{2m}}{\mu_{2m}}\left((1+A)\frac{g}{\mu_{30}}-1\right)\right]^p-1\right]^{-1}$ $\left(\frac{r}{\nu}-1\right)\frac{d_1}{f}(N_1^*+h)$	$h\frac{\mu_3+w}{cf-\mu_3-w}-\phi A$ $\left(\frac{r}{\nu}-1\right)\frac{d_1}{f}(N_1^*+\phi A+h)$
function $G(.)$ varies with carnivore type)	$G(N, A) = cf\frac{N(1+A)}{N+h(1+A)}$	$G(A) = g(1+A)$	$G(N, A) = cf\frac{N+\phi A}{N+\phi A+h}$
PPF essential when	$cf < \mu_3$	$g < \mu_3\left(1+\frac{\mu_{2m}}{e_{2m}}\right)$	$cf - w < \mu_3$

* Total equilibrium population densities can be obtained from: $\sum N^* = N_1^* + N_2^* + N_3^* = \left(\frac{d_1^{-1}+d_2^{-1}+\nu^{-1}}{d_1^{-1}}\right)N_1^*$ and $\sum P^* = P_1^* + P_2^* + P_3^* = \left(\frac{e_1^{-1}+e_2^{-1}+\mu_3^{-1}}{j_2 e_2^{-1}+j_3\mu_3^{-1}}\right)P_C^*$, where the values of (j_2, j_3) for the carnivore types A, B, and C are (0,1), (1,0), and (*j*,1), respectively.

To understand the population-dynamical consequences it is important to realize that in contrast to the other carnivore types discussed, most parasitoids have a numerical response that results directly from its functional response, as long as the number of offspring resulting from each parasitized host (*c*, in Tables 8.1 and 8.2) is independent of host density (a condition generally assumed in parasitoid–host models). The plateau of these responses can therefore both be limited by host-handling time and by egg-maturation rate. Egg maturation, a PPF-amendable process, will not affect the steepness of the rising part of the functional response as in this part parasitization is limited by searching rate only.

When the host–parasitoid system approaches an equilibrium state, the densities of both populations relate to PPF as represented in Table 8.2 (column A) and illustrated in Fig. 8.2A. Since an equilibrium will only occur at host densities at which the numerical response is increasing, PPF is expected to have a stronger impact on herbivore density when affecting searching rate (scenario 2) than when affecting maturation rate (scenario 3), assuming that PPF has quantitatively similar effects on these rates. The impact of PPF on adult life span (the inverse of mortality rate, scenario 1) will directly affect lifetime reproduction, independent of host density, and can have the strongest impact on herbivore suppression (Fig. 8.2A). When PPF affects other life-history traits as well, the herbivore reduction factors of the various PPF scenarios should be multiplied to gain an estimate of the overall effect (Fig. 8.2A).

In suitable habitats certain parasitoids may persist despite the absence of PPF. In other habitats or for other parasitoid species a minimum level of PPF is needed for population persistence. PPF is *essential* when in absence of it the maximum reproduction (at high host densities) is insufficient to compensate mortality (or $cf < \mu_3$) (see Table 8.2). Rephrased at the individual level: in absence of carbohydrate feeding, the number of surviving female offspring per female parasitoid is less that unity ($R_0 < 1$) and therefore insufficient to replace a female when she dies. Clearly, when PPF is essential a much more pronounced effect on herbivore suppression is to be expected (Fig. 8.2–A1).

Host-feeding parasitoids

The females of some groups of parasitoid wasps (Jervis 1998) and flies (Gilbert and Jervis 1998) can also feed on the hemolymph of their hosts (host feeding). As hemolymph contains all the essential nutrients, it may enhance all three functions mentioned for PPF (survival, searching rate, and egg maturation). The sugars available from the host's hemolymph are responsible for enhanced survival of the parasitoids (Giron *et al.* 2002). Not all host-feeding parasitoids are able to absorb the typical hemolymph sugars (such as trehalose) from their gut, and these may still benefit from additional nectar-feeding (McAuslane and

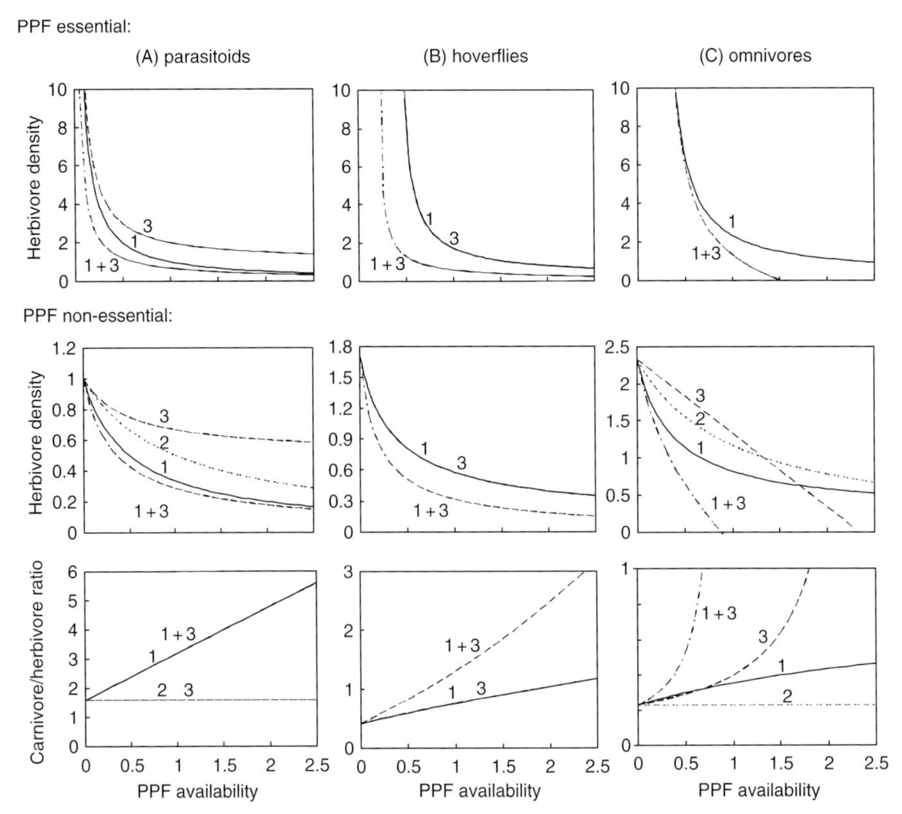

Figure 8.2 The impact of plant-provided food (PPF) on carnivore–herbivore systems at equilibrium for different scenarios and parameter combinations, based on models and solutions presented in Table 8.2. Each column represents another carnivore type defined by the effect of herbivore density: (A) only adult carnivore's performance affected (typical examples are parasitoids), (B) only juvenile performance (typical examples are hoverflies), (C) both adult and juvenile performance. The upper row presents equilibrium herbivore densities when PPF is essential to obtain a positive equilibrium carnivore density ($g = 0.25$ offspring per female per day). The middle and lower rows present equilibrium herbivore densities and carnivore/herbivore ratios, when PPF is not essential ($g = 0.5$ offspring per female per day). Different lines represent different scenarios for the effect of PPF on carnivore performance: PPF affects (1) mortality (drawn line), (2) prey/host searching rate (dotted line), (3) net reproduction rate (dashed line), or (1 + 3) both mortality and reproduction (dash-dotted line). Note that when PPF is essential, scenario 2 offers no solution. Parameter values are according to Table 8.1.

Nguyen 1996; Heimpel *et al.* 1997). Searching for hosts to be parasitized may be enhanced by host feeding due to higher energy reserves; in the short term it may go down due to the time spend on host-feeding. Although egg maturation is sometimes enhanced by nectar-feeding (Tylianakis *et al.* 2004), it is actually more commonly affected by host-feeding (Chan and Godfray 1993; Collier 1995; Heimpel and Collier 1996; Ueno 1999a).

When host-feeding and feeding on PPF serve similar functions, host feeding is expected to lower the need for (and impact of) PPF. Conversely, ample availability of PPF may in some species reduce the need to host-feed. As host-feeding contributes to host killing, this may result - on a short timescale - in reduced herbivore mortality. However, when nectar production invokes a gradual, yet ultimately adaptive, shift from host-feeding to nectar-feeding, it should have no negative impact on the reproduction (i.e., parasitization) rate, and consequently no negative impact on the equilibrium host density (as shown in the first section).

(B) Herbivore density affects juvenile performance

Also for this type of carnivore, where the immature stages are affected by herbivore density, representatives should be found among life-history omnivores (see Table 1.1 in Wäckers and van Rijn, Chapter 1), with carnivorous juveniles and herbivorous/nectarivorous adults. Now, the juveniles are predators rather than parasites, and consequently have to search for prey themselves. Examples of this lifestyle may be found among hoverflies (Syrphidae) and lacewings (Neuroptera). Although it is likely that the oviposition rate of females shows some dependence on prey density as well, we focus here on the special case that only aspects of juvenile performance, such as development and juvenile survival, are affected by herbivore density.

PPF is especially important for the adults of these insects. Feeding on carbohydrate sources, such as nectar and honeydew, will generally enhance adult survival (scenario 1) (McEwen *et al.* 1996; Limburg and Rosenheim 2001), whereas pollen-feeding (Stelzl 1991; Pilcher *et al.* 1997) will also enhance egg maturation (scenario 3, Table 8.2, column B). As the oviposition rate is not limited by prey density, no impact of PPF via the adult's searching rate is to be expected (hence, not scenario 2!). Juveniles may now profit from PPF sources as well (McEwen *et al.* 1993), but because their mobility is limited, only when close to their prey. This special case is not further elaborated here.

When assuming that juvenile mortality rate and developmental time are similarly affected by prey density, a simple expression for the prey population equilibrium is obtained (Box 8.1, Table 8.2, column B). If only one of these rates is affected by prey density ($p = 1$), PPF has a stronger impact on prey equilibrium, than if both parameters are affected ($p = 0.5$, shown in Fig. 8.2B). The two PPF scenarios (PPF affects survival or egg maturation) have identical consequences for herbivore and carnivore equilibrium densities (Table 8.2, column B), but if both parameters are affected, the impact of PPF on herbivore suppression is much stronger (Fig. 8.2B). If PPF is essential (Fig. 8.2–B1) also the critical

PPF level is lower when both survival and egg maturation are affected. At higher levels of PPF the equilibrium herbivore density decreases hyperbolically, and full eradication of herbivores is not possible under equilibrium conditions.

(C) Herbivore density affects both adult and juvenile performance

In the third type of carnivore, prey density affects both juveniles and adults. This type includes arthropods that only feed on prey in the juvenile phase, but have an oviposition rate that depends strongly on prey density (e.g., some lacewings, hoverflies, and aphid-eating midges). The most typical representatives, however, are those arthropods that feed on prey during the juvenile and adult phase, such as some lacewings (e.g., Hemerobiidae), predatory bugs (e.g., Anthocoridae), predatory beetles (e.g., Coccinellidae), earwigs (e.g., Forficulidae), and predatory mites (e.g., Phytoseiidae). For these arthropods prey density will affect both development and oviposition rate (Alomar and Wiedemann 1996; Van Rijn *et al.* 2002; Xia *et al.* 2003), as well as juvenile and adult mortality. In our modeling approach it is possible to make each of these parameters a function of prey density, but this would make the model too complex for explicit solutions. Instead, we assembled all prey dependencies into one function representing the net reproduction rate, i.e., oviposition rate corrected for the proportion of offspring surviving from egg to adult (cf. Van Rijn *et al.* 2002).

Typical representatives of this type of carnivore not only feed on PPF in the adult phase but often in the juvenile phase as well, and should be considered as true (temporal or permanent) omnivores (Table 1.1 in Wäckers and van Rijn, Chapter 1). This means that they can feed on prey as well as PPF, and that they can change their diet composition in response to food availability. PPF may not only involve nectar and pollen, but (for heteropteran bugs) also plant juices that are extracted from leaf, stem, or developing fruit (Eubanks and Styrsky, Chapter 6). PPF that provides mainly carbohydrates (nectar) is likely to affect (1) adult mortality, and (2) prey searching rate (Table 8.1). PPF that is richer in proteins (pollen) is likely to affect the net reproduction rate as well, either by enhancing egg maturation rate or by promoting development and juvenile survival (scenario 3). Whereas in the first case prey and PPF are complementary food sources by definition (as they affect different processes), in the latter case they may be complementary, substitutable, or both. To include this aspect in our model, we assume that prey and PPF are fully substitutable with respect to their effect on the net reproduction rate (cf. Eqn 8.4).

The population-dynamical consequences of complementary PPF (scenarios 1 and 2) are comparable to those described for the other types: with increasing PPF availability, the equilibrium herbivore density decreases asymptotically to some lower value, whereas the carnivore / herbivore ratio is unaffected (scenario 2), or increases slightly (scenario 1). With substitutable PPF (scenario 3) the predictions are very different: with increasing PPF availability, the herbivore density decreases linearly until the herbivore population is exterminated locally (Fig. 8.2–C2). A zero herbivore population is possible because the carnivore population can now persist on PPF alone. Below the critical PPF density, the total food (PPF + prey) availability remains unchanged, and so does the equilibrium carnivore density, resulting in a strong effect of PPF on the carnivore/herbivore ratio (Fig. 8.2–C3). When PPF affects both adult survival and net reproduction rate (combining scenarios 1 and 3), even stronger herbivore suppression is attained (Fig. 8.2–C2). Under different model assumptions, inducible defenses or flexible avoidance behavior may prevent herbivore extermination (Van Rijn *et al.* 2002; Vos *et al.* 2004).

When PPF is fully substitutable with prey, it can never be essential for carnivore persistence. When PPF is complementary with respect to at least one process, it can be essential and its availability may result in a transition from high (unchecked) herbivore levels to much lower herbivore levels checked by the carnivore (Box 8.1, Fig. 8.2–C1).

Section summary

In this section we have analyzed the equilibria of three types of stage-structured carnivore–herbivore models, which differ in the carnivore's life-history components (mortality, development, reproduction) affected by herbivore density. For each model the impact of PPF is studied for various scenarios that differ in the parameters affected by PPF (adult mortality, searching rate and reproduction rate), leading to the following conclusions:

- The impact of PPF on the efficacy of carnivores in terms of herbivore suppression differs not necessarily between carnivores with different life histories.
- When PPF and prey affect similar processes (only possible for true omnivores) and are substitutable food sources, there may be a critical supply of PPF above which the herbivore goes extinct.
- When, in contrast, PPF and prey affect different processes (and are necessarily complementary), PPF availability asymptotically reduces herbivore density to some positive value, at which the PPF-affected processes are limited by other factors, such as top predation or aging.
- An important special case of complementary foods occurs when, either by life-history features (such as synovigeny) or by adverse environmental conditions (high mortality rate), the carnivore population cannot persist

in absence of PPF. PPF can now be called *essential* and a minimum amount of PPF is required for herbivore control.

- When PPF in not essential but merely speeds up the foraging process (scenario 2), it has a stronger impact on equilibrium herbivore levels than when PPF limits egg maturation rate (scenario 3). PPF has the strongest impact through its effect on adult mortality (scenario 1). The explanation is that an equilibrium can only occur in the rising part of the numerical response, which is more affected by searching rate than by maturation rate, whereas the effect of mortality is direct and not mediated by the numerical response.
- When PPF affects different life-history processes, the overall impact on herbivore suppression will even be stronger; the impact (in terms of herbivore reduction) of the different scenarios is approximately multiplicative.
- When PPF enhances the searching rate (scenario 2) this has no effect on the carnivore / herbivore ratio, but in the other scenarios PPF results in increased carnivore / herbivore ratios (or proportion parasitized hosts in case the carnivores are parasitoids).

Impact of distribution and spatial structure

So far, we have ignored the spatial component of ecosystems, and have assumed implicitly that carnivores, herbivores, and food plants are well mixed. This assumption is valid as long as the grain size of the spatial heterogeneity is well below the average foraging range of the consumers. In many cases, however, this assumption does not hold. Spatial heterogeneity at higher spatial scales affects the probability of interactions and this may have important consequences for population dynamics and community structure.

When herbivore prey and food plants co-occur only at spatial scales that are beyond the search range of individual carnivores, new mechanisms may come into play that are absent in fine-grained environments: carnivore aggregation, spatial subsidies, and metapopulation dynamics.

Local carnivore aggregation: short-term apparent competition

In a heterogeneous environment, PPF may enhance biological control not only by affecting life-history components of the carnivore, but also by affecting *dispersal* characteristics of the carnivore, such as dispersal rate and directionality. In fact, many of the experiments in which food supplements are sprayed ("artificial honeydew") (Evans and Swallow 1993; Evans and Richards

1997; Jacob and Evans 1998; Rogers and Potter 2004) lasted so short a time (2–14 days) that the observed increase in (adult) predators can only result from an aggregation of predators in the sprayed area, drawing on the pool of predators outside the sprayed area. Such aggregation of predators can very well result from extended *retention* (arrestment) due to the sprayed food. The differences in retention time inside and outside the sprayed area can create a net influx of predators into the treated area, and result in a locally higher predator density (Box 8.3). A similar aggregation may result from *attraction*, due to distant perception of visual or olfactory stimuli associated with the food source (Drukker *et al.* 1995; Lambin *et al.* 1996; Vet *et al.* 1998). The spatial range at which these stimuli can be effective, however, is limited (Berec and Krivan 2000).

BOX 8.3 Local carnivore aggregation results in short-term apparent competition

An elegant way to model the impact of behavioral responses to locally provided food can be found in Van Rijn *et al.* (2002). The proportion of the herbivore-inhabited area that is provided with PPF (area 1) is represented by α. The distribution of the carnivore is assumed to result from arrestment (enhanced retention time) that is proportional to the local food availability (prey and PPF). When the carnivores are allowed to distribute themselves freely while food availability remains fixed, their distribution will approach an equilibrium. At this equilibrium the proportion of carnivores (γ) in the area with PPF is equal to the proportion of the food (PPF and prey) present in this area (Van Rijn *et al.* 2002) or

$$\gamma = \frac{\alpha N + \phi A}{N + \phi A},$$

where N represents the size of the herbivore population, αN the herbivore population present in the area with PPF, and ϕA the amount of PPF valued relative to the nutritional value of the prey. This carnivore distribution results in a predation pressure in areas with and without PPF that is proportional to

$$q_1 = \frac{\gamma}{\alpha} = \frac{N + \phi A \alpha^{-1}}{N + \phi A} \text{ and } q_0 = \frac{1-\gamma}{1-\alpha} = \frac{N}{N + \phi A}, \text{ respectively.}$$

If we assume that the herbivore distribution is unaffected by the alternative food source, and that the A/N ratio does not change over time (implying non-selective feeding by the carnivore), the numbers of herbivores in both areas ($i = 1,0$) will decline according to:

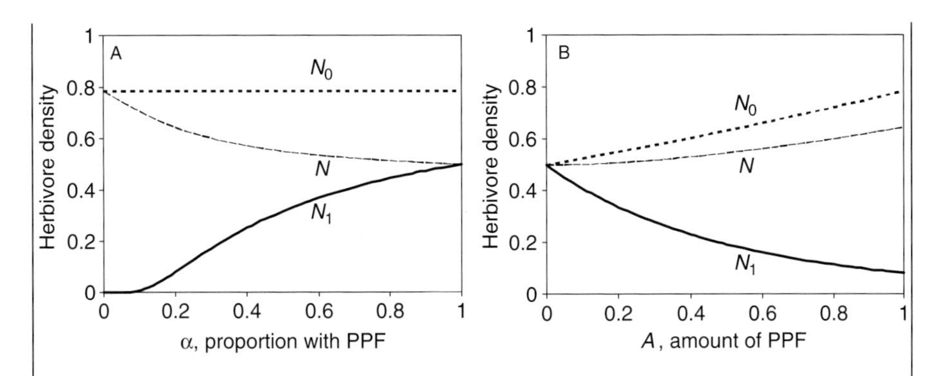

Impact of plant-provided food (PPF) distribution (A) and abundance (B) on herbivore density (N) after (T=) 10 days, in the area with PPF (N_1, drawn line) and in the connected area without PPF (N_0, dashed line), as a result of carnivore aggregative response (see text and Box 8.3). The overall mean density is indicated by N and a thin dashed line. ($N(0) = 10$, $A = 1$, $P = 1$, $f = 0.5$/day, $r = 0.2$/day, $\alpha = 0.2$).

$$N_i(t) = N_i(0)\exp(r - q_i fP),$$

provided that the autonomous growth rate of the herbivore $r < q_i fP$.

Higher levels of PPF (A) will result in a more asymmetric distribution of carnivores, a stronger reduction of herbivores in the target area, and a weaker reduction in the other area (B). A further concentration of PPF (by lowering α) intensifies the impact in the target area, but leaves the *density* in the non-target area unaffected (A).

Irrespective of the mechanism, the aggregation of predators will usually increase the predation pressure on the local herbivore population, which eventually results in a lower herbivore density (Kean *et al.* 2003). This indirect effect of PPF is comparable with "apparent (predator-mediated) competition" (see first section), but as it results from behavioral rather than life-history responses, it acts already on timescales shorter than one generation, and has therefore been labeled "short-term apparent competition" (Holt and Kotler 1987; Evans and England 1996). This principle has already been useful in explaining spatial "avoidance" in plants due to increased seed predation in the neighborhood of other seed-producing plants (Veech 2001). As a method to enhance biological control, however, it has clear limitations. The local increase in predator density goes at the expense of predator densities in the surrounding fields (Box 8.2). Only when the surrounding "source" area is sufficiently large compared to the "target" area, the impact may be negligible (Kean *et al.* 2003). Moreover, if the target herbivores occur in the source area as well, they may profit from the reduced predation there, and diffusion may ultimately cancel out the herbivore reduction in the target area. Clearly, this short-term effect of PPF will work most effectively when the agricultural fields are surrounded by larger areas of (semi-)natural habitat with low numbers of potential pest organisms and with some redundancy in the carnivorous species feeding on them.

An empirical example where the principle worked nicely is provided by Harmon *et al.* (2000): alfalfa plots with many pollen-producing dandelions contained higher densities of the predatory beetle, *Coleomegilla maculata* and lower densities of pea aphids, *Acyrthosiphon pisum*, than nearby plots with few dandelions. Laboratory studies showed that the predators aggregate in plots with dandelions and that this results in increased predation on aphids. Another example may be provided by Tylianakis *et al.* (2004) who observed that the parasitism rate of aphids (*Metopolophium dirhodum*) on wheat declined exponentially with increasing distance from flowering buckwheat patches, which provided the parasitoids (*Aphidius rhopalosiphi*) with nectar. Unfortunately, in this case the overall impact on aphid numbers has not been studied.

Spatial subsidies: source–sink dynamics

Many crop plants are unable to provide the supplementary food that can support the natural enemies of their pests. When PPF is essential for these natural enemies, they can only be maintained in these crops if neighboring plants provide the essential resources. Such food plants could be available in the surrounding habitats, at the crop edges, or even in rows within the crop (see Wilkinson and Landis, Chapter 10). To have an impact on herbivore suppression in the crop the carnivore should either (1) commute regularly between food plant and crop, or (2) be maintained in sufficient numbers on the food plants allowing a surplus to disperse into the crop, thus creating a spatial subsidy (Polis *et al.* 1997; Fagan *et al.* 1999). The first situation comes close to the well-mixed situation addressed in the first part of this chapter. The second situation requires a separate, spatially explicit, approach. The spatial relationships will differ depending on the level of substitutability of prey and PPF, as will be discussed below (Dunning *et al.* 1992).

Complementary resources: PPF essential

When the PPF is not present in a certain habitat (e.g., a crop) yet essential for a carnivore to attain sufficient reproduction ($g < \mu_0$ in Eqn 8.8), the carnivore population can only persist in this habitat by immigration (Box

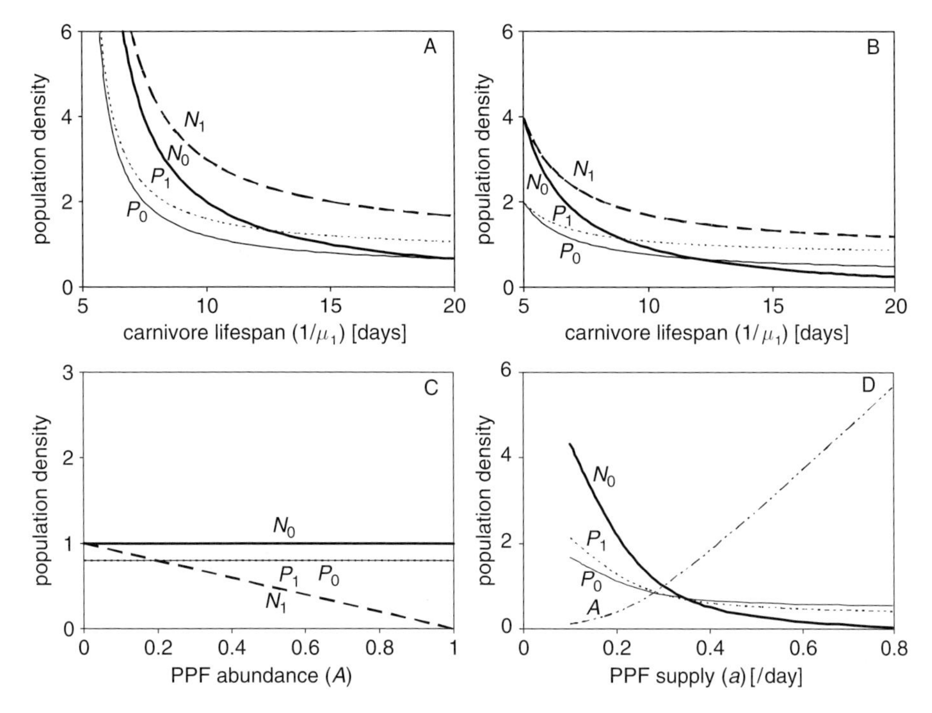

Figure 8.3 Impact of plant-provided food (PPF) present in one habitat (1), on carnivore (P) and herbivore (N) densities in a connected habitat (0, the crop) that is deficient in PPF (see text and Box 8.4). (A,B) By reducing mortality (μ_1) when PPF and prey are *complementary* resources: (A) Essential PPF ($g = 0.2$); (B) non-essential PPF ($g = 0.25$) ($f = 0.5$/day, $r = 0.2$/day, $\delta = 0.2$, $h = 1$, $\mu_0 = 0.2$). (C,D) When PPF and prey are *substitutable* resources: (C) with prey in habitat 1; (D) without prey in habitat 1. ($f = 0.5$/day, $r = 0.2$/day, $\delta = 0.2$, $h = 1$, $\mu_0 = \mu_1 = 0.1$, $\phi = 1$).

8.3A). When a nearby habitat provides all resources, including PPF and prey, and allows the carnivore population to grow, this habitat can act as a population source from where individuals disperse into the sink habitat in search for new prey. At the landscape scale the system can thus be described as a *source–sink system* (Pulliam 1988; Dunning *et al.* 1992). The PPF in the source habitat is now essential for the persistence and abundance of the carnivore in the sink habitat (crop), and thus contributes to the suppression of herbivores in the sink habitat. This principle is analyzed in Box 8.4A and illustrated in Fig. 8.3A. Although the carnivore's reproduction within the sink habitat is (by definition) insufficient to cancel out mortality, it still contributes to the population at the landscape level (Box 8.4A). In addition, it can increase the stability of the source–sink system, both in ecological (Jansen and Yoshimura 1998) and in evolutionary terms (Holt 1997).

BOX 8.4 Source–sink dynamics

(A) Complementary resources

When a carnivore needs both herbivore prey/host and PPF to obtain reproduction that exceeds mortality, and a habitat does not have sufficient PPF, this habitat is a sink for the carnivore population. The local carnivore–herbivore dynamics could be described by the same Lotka–Volterra type model used in the first section (Eqns 8.1 and 8.3), with the special condition that carnivore mortality rate that is higher than the maximum reproduction rate ($\mu_0 > g$). When this sink habitat ("0") is connected through diffusive carnivore dispersal (δ) with a ("source") habitat ("1") that provides the resources to maintain a local carnivore population ($\mu_1 < g$), the system is described by:

$$\frac{dN_0}{dt} = rN_0(1 - kN_0) - f\frac{N_0}{N_0+h}P_0 \qquad \frac{dN_1}{dt} = rN_1(1 - kN_1) - f\frac{N_1}{N_1+h}P_1$$

$$\frac{dP_0}{dt} = g\frac{N_0}{N_0+h}P_0 - \mu_0 P_0 + \delta(P_1 - P_0) \qquad \frac{dP_1}{dt} = g\frac{N_1}{N_1+h}P_1 - \mu_1 P_1 + \delta(P_0 - P_1)$$

(Eqn Box 8.4.1)

The net immigration from the source habitat is big enough to maintain a carnivore population in the sink habitat ($P_0^* > 0$), when

$$2\delta + \mu_1 > g.$$

In that case, the equilibrium state of the system (for $k = 0$) can be described by:

$$N_0^* = h\left(\frac{g}{\mu_0 + \delta(1-p)} - 1\right)^{-1} \quad N_1^* = h\left(\frac{g}{\mu_1 + \delta(1-p^{-1})} - 1\right)^{-1},$$
$$P_0^* = \frac{r}{f}\left(h + N_0^*\right) \qquad P_1^* = \frac{r}{f}\left(h + N_1^*\right)$$

(Eqn Box 8.4.2)

where

$$p \equiv \frac{P_1^*}{P_0^*} = \frac{2\delta + \mu_0 - g}{2\delta + \mu_1 - g}.$$

(Eqn Box 8.4.3)

These results show that in a habitat where the herbivore would remain unchecked due to insufficient resources for the carnivore, the top-down control of the herbivore can be restored by an adequate connection with a habitat that has sufficient resources for the carnivore. A further increase of these (plant-provided) resources, here modeled as a further reduction in carnivore mortality (μ_1), gives a further reduction in equilibrium herbivore density in both source and sink habitats (see also Figure 8.3B). A further increase in dispersal rate (δ) will, however, increase herbivore density.

(B) Substitutable resources

When herbivore prey and PPF are substitutable resources for a carnivore, its population persistence does not require both food sources to be present. Under these conditions, a habitat without PPF is not necessarily a sink, as increased herbivore levels can compensate its absence. To model a system in which a PPF-free habitat is connected with a habitat that does contain PPF, we simply adjust the source part of the model (Eqn Box 8.4.1) by adding up the density of the PPF to that of the prey (N_1 becomes $N_1 + \phi A$) in the functional and numerical response terms (see Eqns 8.4a,b).
At PPF levels that are insufficient to support the carnivore population in absence of prey, i.e.,

$$\phi A < h \frac{\mu}{g - \mu},$$

this two-habitat system has a non-zero equilibrium. We make the simplifying assumption that the parameters are equal in both habitats ($\mu_0 = \mu_1$), which does not affect the conclusions. As a result, the equilibrium ratio between the carnivore populations in both habitats, p, becomes unity, and the equilibrium becomes independent of the dispersal rate (δ):

$$\begin{aligned} N_0^* &= h\tfrac{\mu}{g-\mu} & N_1^* &= h\tfrac{\mu}{g-\mu} - \phi A \\ P_0^* &= \tfrac{r}{f}\left(h + N_0^*\right) & P_1^* &= \tfrac{r}{f}\left(h + N_1^* + \phi A\right) \end{aligned} \qquad \text{(Eqn Box 8.4.4)}$$

These equations show that under strict substitutability, PPF (A) reduces the herbivore level in habitat 1 in such a way that the overall food level in habitat 1 ($N_1 + \phi A$) remains the same. Consequently the carnivore levels in both habitats are unaffected, and so is the herbivore level in habitat 0 (Fig. 8.3C).

At higher PPF levels the herbivore can no longer survive in the source habitat (apparent competition). In the absence of a feedback with its food source the carnivore would continue to grow exponentially. This defect can be corrected by including the dynamics of (and feedback with) the PPF, as in the following model:

$$\begin{aligned} \tfrac{dN_0}{dt} &= rN_0(1 - kN_0) - f\tfrac{N_0}{N_0+h}P_0 & \tfrac{dA_1}{dt} &= a - bA_1 - f\tfrac{A_1}{A_1+h}P_1 \\ \tfrac{dP_0}{dt} &= g_0\tfrac{N_0}{N_0+h}P_0 - \mu_0 P_0 + \delta(P_1 - P_0) & \tfrac{dP_1}{dt} &= g_0\tfrac{A_1}{A_1+h}P_1 - \mu_1 P_1 + \delta(P_0 - P_1) \end{aligned}, \qquad \text{(Eqn Box 8.4.5)}$$

where a represents the production rate of PPF and b its natural decay rate (see Van Rijn *et al.* 2002). The equilibrium of this system is only slightly different from the equation set above (Eqn Box 8.4.2):

$$A^* = h\left(\frac{g}{\mu_0+\delta(1-p)} - 1\right)^{-1} \quad N_1^* = h\left(\frac{g}{\mu_1+\delta(1-p^{-1})} - 1\right)^{-1},$$
$$P_0^* = \tfrac{1}{f}\left(h + N_0^*\right)\left(\tfrac{a}{A^*} - b\right) \quad P_1^* = \tfrac{r}{f}\left(h + N_1^*\right) \qquad \text{(Eqn Box 8.4.6)}$$

The solution for p, however, is now a function of a:

$$p = \frac{h + A^*}{h + N_0^*}\left(\frac{a}{A^*} - b\right)\frac{1}{r}. \qquad \text{(Eqn Box 8.4.7)}$$

Since N_0^* and A^* are functions of p as well (and do not cancel out), p can now only be solved implicitly by a numerical root finding algorithm. The calculations show that increased PPF-production levels (a) result in increased carnivore migration from habitat 1 to habitat 0, and consequently decreased herbivore numbers in habitat 0 (Fig. 8.3D).

Complementary resources: PPF not essential

When the complementary food is not essential for sustained reproduction, a PPF-free crop will not be a sink *by definition* ($g > \mu_0$). The carnivore populations in both habitats can now persist independently and impose some top-down control on the local herbivore populations. When, however, one habitat contains PPF, this not only results in lower herbivore numbers in this habitat, but also in higher carnivore numbers. This results in an increase in carnivores dispersing into the connected (crop) habitat, and causes a reduction in herbivore number there as well (Box 8.4A). The impact on the herbivores, however, is less dramatic than when PPF is essential (Fig. 8.3B).

Mixed substitutable resources

When PPF and prey are substitutable, and are both present in the non-crop habitat, the PPF will first result in reduced herbivore numbers in the non-crop habitat. At equilibrium (and under the assumptions of our simple model: see Box 8.4B) the decrease in local prey density exactly cancels out the increase of PPF and leaves the local carnivore population unaffected. In that case, the PPF will not enhance the number of carnivores migrating to other habitats, and will not enhance herbivore suppression in these habitats (Fig. 8.3C). However, more realistic models, allowing for a mix of bottom-up and top-down control (Abrams and Vos 2003; Vos *et al.* 2004), would predict a moderate trans-habitat effect.

Separate substitutable resources

A very different situation occurs when in the non-crop habitat the PPF availability is such that it can support the carnivore population even in absence of prey. Now (due to apparent competition) the herbivore population can no longer exist in this habitat (Holt and Lawton 1994). As a result, a further increase in PPF availability will no longer be compensated by a decrease in prey density, but will translate directly into a larger carnivore population (Box 8.4B). Diffusion of carnivores into the connected (crop) habitat now leads to lower herbivore numbers (Fig. 8.3D).

Temporal subsidies

Conditions may not be such that a dynamic equilibrium will be approached for the system as a whole, as was assumed so far. For several empirical systems, it has been shown that seasonal patterns may turn sources into sinks and vice versa (Corbett and Plant 1993), or that the direction of the main subsidy flow may reverse (Nakano and Murakami 2001). Such patterns may promote the stability of food-web dynamics at the landscape scale (Takimoto *et al.* 2002). For certain predatory mite species the availability of pollen early in spring seems to be critical for their abundance and subsequent suppression of herbivorous mites later in the year. In orchards this food may be subsidized by stands of wind-pollinated trees, such as birches and pines (Addison *et al.* 2000). In other environments it may be available in the surrounding vegetation such as hedgerows (Kreiter *et al.* 2002; Duso *et al.* 2004). The pollen allows the predators to build up a population in these non-crop habitats, from where they can migrate into the crop later in the season when the herbivorous mite population starts to build up.

Metapopulation dynamics

Whether or not a population persists may depend on the spatial scale under consideration. In many species, local populations go extinct. On a larger spatial scale the aggregate of local populations (the "metapopulation") can still persist due to regular recolonizations of empty habitats from nearby surviving populations. This spatial process of local extinctions and recolonizations is called metapopulation dynamics (Hanski 1998).

Metapopulation dynamics is especially well studied and described for predatory mite – spider mite interactions (Huffaker 1958; Sabelis and Van der Meer 1986; Ellner *et al.* 2001; Pels *et al.* 2002; Nachman and Zemek 2003). At larger spatial and temporal scales, also some insect herbivore–carnivore interactions should be characterized as metapopulation dynamics (Ives and Settle 1997), as documented for host–parasitoid interactions (Lei and Hanski 1997; Eber 2001; Kean and Barlow 2001; Steffan-Dewenter and Tscharntke 2002; Van Nouhuys and Hanski 2002).

Since the rise of metapopulation theory (Levins 1969), a range of models has been developed to describe and understand metapopulation dynamics (Hanski

1998). Complex simulation models describe both the dynamics of local populations and their connectivity resulting from dispersal between populations. However, there are also simpler models of the patch-occupance type. For example, the model by Levins (1969) does not describe the densities of predator and prey, but only the number of patches occupied by prey or by predator and prey, resulting from local extinctions and recolonizations. In many instances this simple model already captures many of the basic characteristics of metapopulation dynamics (Bascompte and Sole 1998; Holyoak 2000; Pels *et al.* 2002). But even when the dynamics of the local (within-patch) populations are considered explicitly, tractable and insightful results are still feasible under certain simplifying assumptions (Diekmann *et al.* 1988; Sabelis *et al.*, Chapter 4).

Between-patch dispersal

In a metapopulation environment, where herbivore patches are transient, the dynamics are importantly affected by the ability of carnivores to find new herbivore patches. In Levins' model this ability is represented by the predator's dispersal (or colonization) rate, defined as the number of predators per patch that are able to colonize a new patch. This dispersal rate is likely to be affected by the availability of PPF in the landscape to be explored, as PPF may allow the predators to "refuel", and increase time and energy available for searching. Increasing the predator dispersal rate in Levins' model will decrease the number of herbivore-occupied patches (Fig. 8.4) and therefore contributes to herbivore control.

Within-patch dynamics

When additional food sources are (also) affecting the local within-patch dynamics, for example by affecting predator reproduction or predation rate, the predictions are less straightforward. In this case we need a population model to describe the local within-patch dynamics. To link this with metapopulation processes we may either (1) take a large number of these patches and connect them through dispersal (see for example Pels *et al.* 2002), or (2) use the local population model to predict how food may affect the parameters of Levins' model (such as turnover rate and dispersal rate), and use Levins' model to predict the effects at the metapopulation level. Here we use the latter two-step approach, which may be less accurate, yet more transparent, than the modeling of metapopulation dynamics from local dynamical processes. To describe the unstable within-patch dynamics we use a simple, explicitly solvable, predator–prey model with a constant per capita predation rate (the so-called "pancake-predation" model) (Sabelis 1992; Janssen and Sabelis 1992; Van Baalen and Sabelis 1995; Sabelis and van Rijn 1997). Below, the impact on local and

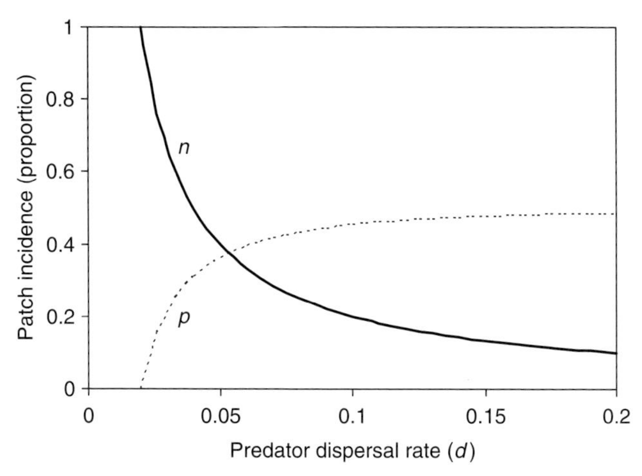

Figure 8.4 The impact of predator dispersal rate (d), positively affected by PPF availability in the environment, on the proportion of patches occupied by herbivores (n), or herbivores and carnivores (p) in a Levins' metapopulation model ($c = 0.04$, $e = 0.02$).

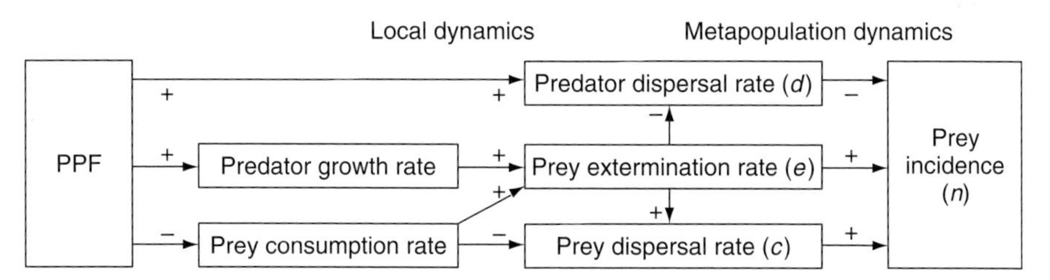

Figure 8.5 The possible causal pathways between plant-provided food (PPF) and herbivore prey incidence though metapopulation dynamics. The effect of local population-dynamical parameters (second column) on metapopulation parameters (third column) is predicted by the local "pancake-predation" model. The effect of metapopulation parameters on prey incidence is predicted by a Levins' metapopulation model (see text).

metapopulation dynamics is discussed for two parameters that may be affected by PPF: predator growth rate, and prey consumption rate (Fig. 8.5).

Predator growth rate. – When PPF increases the population growth rate of the predator without affecting its prey consumption rate, the local model predicts faster prey extermination (e). Since the predators grow exponentially and are assumed to disperse only after local prey extermination, the number of dispersing predators (d) is strongly affected by the time to prey extermination. Consequently, the predator dispersal rate generally declines with predator growth rate. Faster prey extermination and lower predator dispersal both *increase* (according to Levins' model) the proportion of patches occupied by herbivore prey. The increase in prey extermination rate, however, also reduces the number of dispersing prey (c), which ultimately *decreases* prey incidence (Fig. 8.5). The overall result of enhanced predator growth rate on prey incidence can therefore both be positive and negative, depending on the precise parameter setting.

Prey consumption rate. – In some cases PPF may reduce the prey consumption rate. According to the local dynamics model, a reduction in prey consumption extends the time to prey extermination ($1/e$). Therefore it basically has the opposite effect on all three metapopulation parameters, as compared to an increase in predator growth rate (Fig. 8.5). By extending the time to prey extermination, a reduction in prey consumption may therefore even facilitate the reduction in herbivore incidence.

These results clearly indicate that conclusions based on local dynamics do not necessarily match the conclusions based on metapopulation dynamics, since at different spatial scales the types of feedback may differ.

Section summary

The spatial separation or proximity of plants hosting herbivores and plants providing food for carnivores may strongly affect the impact of PPF on biological control. In this section the various possibilities are illustrated by novel model approaches.

- When the two plant types are in close proximity, or even merged into one plant species, the net influx of carnivores from the surrounding vegetation to the PPF plants may locally increase the predation pressure on the target herbivores and decrease local herbivore numbers.
- When the two plant types are further apart, the PPF plants may subsidize the carnivore population on the other plants, and help to suppress herbivore numbers there. The impact of PPF on herbivores in the other habitat will however vary with the complementarity of PPF and prey, and with the distribution of the herbivores. Based on our simple trophic cascade models the impact of PPF may vary from none (when PPF and prey are mixed and substitutable resources) to strong (when PPF is essential, or when PPF is substitutable but not co-occurring with herbivore prey).
- When local population extinctions are likely to occur, recolonizations from nearby non-extinct populations may help to maintain the species

at a larger metapopulation scale. PPF may then contribute to the maintenance of a carnivore population by refueling the dispersing individuals. When PPF affects the local herbivore–carnivore dynamics (by affecting population growth rate or predation rate), its impact on herbivore incidence may by very different from its impact in more homogeneous systems.

Impact of food web structure

We have seen how PPF affects herbivore suppression, and how this impact may be modified by stage- and space-related interactions. With few exceptions, the overall pattern is that PPF somehow enhances herbivore suppression. We thereby consistently assumed that the herbivore and carnivore are part of a simple tritrophic food chain with plants at the lowest trophic level. In reality, herbivore and carnivore are part of a much complexer food web, interacting with many more species in their habitat (Polis and Strong 1996). How will the interactions with other species affect our conclusions on the impact of PPF?

Top-carnivores

Carnivores may have their own suite of (hyper-)predators, (hyper-)parasitoids, and pathogens (Rosenheim 1998; Sullivan and Volkl 1999). When we again assume simple trophic cascade models, and assume that the primary productivity of our system is large enough to support a top-carnivore, the top-carnivore will at equilibrium control the primary carnivore. As a result the herbivore is released from top–down control (Oksanen *et al.* 1981). The equilibrium density of the primary carnivore would be determined by the traits of the top-carnivore, whereas the herbivore would grow to a density where it is limited from bottom up. This implies that at equilibrium, food provided to the primary carnivore would no longer affect its own density or that of the herbivore!

However, when the different species are assigned more flexible responses (such as predator avoidance behavior) the carnivore is not only regulated from top down, but also from bottom up (Polis 1994, 1999; Abrams and Vos 2003), and plant feeding by the primary carnivore may have some effect on carnivore and herbivore abundance. This will especially occur when feeding on plant material makes the primary carnivore less vulnerable to attack by the top-carnivore, for example by outgrowing their vulnerable stage more quickly (Van Rijn *et al.* 2002). Such effects of PPF on the vulnerability to predation or parasitization have, to our knowledge, not yet been studied experimentally.

In some cases, PPF can (also) be used by the top-predator or hyperparasitoid (Chang *et al.* 1994). PPF will now likely reduce the density of the primary carnivore, and consequently have a negative rather than a positive impact on herbivore biological control.

Competition

Often, herbivore species have multiple enemies. This may result in competition among those enemies for the same resource. If the dominant competitor utilizes the PPF, the system behaves effectively like one with a single carnivore species. If, however, a subdominant competitor utilizes the PPF more effectively than the dominant carnivore, a small increase in PPF will initially have no impact on the herbivore level. But at a higher, critical PPF level the initial subdominant species may become dominant, and a further increase in PPF will now enhance herbivore suppression. This means that, in theory, PPF can reverse the outcome of competitive exclusion (Tilman 1982; Holt *et al.* 1994), and that a reduction of the former keystone species does not necessarily indicate that biological control fails.

Intraguild predation

The food web becomes even more complex when some species are intraguild predators, i.e., top-carnivore and competitor at the same time (Polis *et al.* 1989; Polis and Holt 1992).

Among arthropod predators, intraguild predation (IGP) often depends on the relative body sizes of the interacting individuals and arises from the larger instars attacking smaller instars of competing predators (Lucas *et al.* 1998; Hindayana *et al.* 2001). At the species level, intraguild predation can consequently be reciprocal (Lucas *et al.* 1998; Hindayana *et al.* 2001) as well as one-sided (Rosenheim 2001; Snyder and Ives 2001).

Among parasitoids, IGP can occur by host-feeding on already parasitized hosts (Ueno 1999b; Collier and Hunter 2001), by multiparasitism followed by larval killing (Hagvar 1988; Mackauer *et al.* 1992; Pijls *et al.* 1995; Collier and Hunter 2001), and by autoparasitism (Williams and Polaszek 1996; Hunter and Woolley 2001; Hunter *et al.* 2002). In the latter case the parasitoids parasitize primary hosts as well as (con- or heterospecific) parasitoid larvae inside primary hosts.

IGP can also result from interactions between predators and parasitoids. When parasitoids and predators attack the same herbivore population, the predators will often feed on parasitized herbivores as well (Brodeur and Rosenheim 2000; Snyder and Ives 2001). Very few parasitoids, however, will attack predators as well as herbivores (but see Mansfield and Mills 2002).

For its impact on herbivore suppression the allocation of costs and benefits of IGP is crucial. In the interaction between parasitoids and predators, the parasitoids will usually suffer from IGP, whereas the predator encounters little costs and no benefits. In asymmetric IGP among predators, the dominant intraguild predator may directly benefit from the interaction by having an additional food source.

An assemblage of two or more carnivores may temporarily have an additive impact on herbivore suppression (Bogran *et al.* 2002). On a longer timescale, the carnivore species that can survive at the lowest herbivore level will outcompete the other carnivores (competitive exclusion). When, however, a weaker competitor is a better intraguild predator, it may reverse this process, now ending up with a less effective enemy and a higher herbivore abundance (Rosenheim *et al.* 1995; Briggs and Collier 2001; Rosenheim 2001; Snyder and Ives 2001; Hunter *et al.* 2002).

PPF may change the outcome of IGP by promoting one species more than others (Evans and England 1996). If PPF supports a species (*b*) that (without PPF) is a worse competitor but a good intraguild predator, a small increase in PPF should result in lower numbers of the better competitor (species *a*) and in higher herbivore numbers (Holt and Polis 1997; Mylius *et al.* 2001). However, beyond the PPF level that results in exclusion of species *a*, species *b* may continue to respond positively and this will cause herbivore numbers to decrease again.

Epilogue

Throughout this chapter, we either assumed stationary conditions or transient dynamics. To analyze these two cases, we focussed on two extremes of a continuum: the equilibrium and the first generation. In reality, the time-frame (typically one season) is often beyond one generation, but also too short to approximate an equilibrium state. Which of the two extremes gives the best approximation for real situations will depend on the number of generations, the stage-dependent interaction patterns, and the initial numbers of carnivore and herbivore. For timescales intermediate between a single generation and many generations (when the equilibrium approach suffices), computer simulations are the only way to obtain answers, next to experiments. At this timescale, the transient dynamics are strongly affected by stage structure and foraging behavior. This timescale problem may complicate the application of theory to empirical and practical situations. To illustrate this, we briefly discuss two experimental studies among the few that are able to link PPF, through its impact on carnivores, with herbivore population dynamics.

Bean pods, predatory bugs, and herbivores

Eubanks and Denno (1999, 2000; see also Eubanks and Styrsky, Chapter 6) studied the impact of PPF in a system of lima bean plants, several herbivore species such as pea aphids and corn earworm (*Helicoverpa zea*), and the predatory big-eyed bug (*Geocoris punctipes*). They showed that bean pods are a food source for the bugs which extends their lifespan after prey-free periods (Eubanks and Denno 1999), but also reduces their feeding on aphids and earworm eggs (Eubanks and Denno 2000). A field experiment over a 4-week period showed that, with some delay, the predator population benefits from the presence of bean pods (Fig. 8.6A). On a short timescale (within 1 week) the herbivore numbers were similar or even higher in the presence of bean pods; on a longer timescale, however, their numbers remained much lower on plants with bean pods (Fig. 8.6B). The latter result is precisely what our basic theory predicts: initially, when predator numbers are still unaffected, the lower per capita predation rate in the presence of bean pods will result in lower predation pressure. Later, this less than twofold lower per capita predation rate will be more that compensated by a twofold and fourfold increase in the number of juvenile and adult predators.

Pollen, predatory mites, and thrips

Van Rijn *et al.* (2002) studied the impact of PPF in a system of cucumber plants, western flower thrips (*Frankliniella occidentalis*), and predatory mites (*Iphiseius degenerans*). The pollen-sterile cucumber plants were artificially supplied with cattail pollen. Laboratory experiments had shown that pollen is a good food source for the predators which allows full development and reproduction, even in absence of prey (Van Rijn and Tanigoshi 1999), but that it also reduces their feeding on thrips as prey (Van Rijn *et al.* 2004b). A replicated greenhouse experiment during a 13-week period showed that the growth of the predator population is strongly enhanced by the presence of pollen. This resulted in strongly inhibited growth of the thrips population. That the thrips did not, not even for a short period, benefit from the pollen, whereas thrips can use pollen as a food source as well, is explained by the spatial arrangement of the pollen that resulted in aggregation of the predators near pollen sites (see Box 8.2) and avoidance of predators at pollen sites by the thrips (see also Sabelis *et al.*, Chapter 4). A stage-structured population model (similar to the one used in Box 8.1), extended with a spatial component and with foraging and avoidance behavior, yielded accurate predictions of the experimental results (Fig. 4.1 in Sabelis *et al.*, Chapter 4).

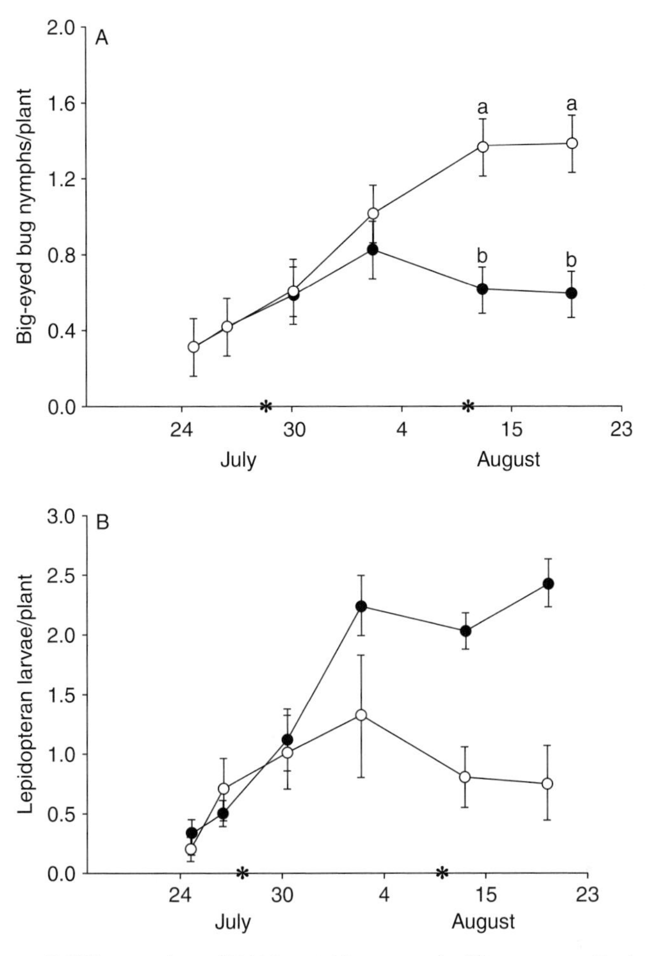

Figure 8.6 The number of (A) big-eyed bug nymphs (*Geocorus punctipes*) and (B) lepidopteran larvae (*Helicoverpa zea*) per plant in plots of lima beans with many pods (open circles) and with few pods (solid circles). Means with different letters are significantly different (Bonferroni means separation test, $P < 0.05$). Asterisks indicate applications of fruit thinner. From Eubanks and Denno (2000).

Conclusions

Theoretical considerations, inspired and checked by a series of modeling exercises, provide the following list of conditions that are important for the impact of PPF on herbivore suppression.

Short term vs. long term

PPF often enhances the persistence of the carnivore population, by increasing carnivore survival or reproduction. Such effects however require time (to build up carnivore numbers) to translate into increased herbivore suppression. Other effects of PPF, such as reduced predation rates, will translate directly into reduced herbivore suppression. As a result, on the short timescale, PPF may lead to higher herbivore numbers. On the long timescale, however, it likely leads to reduced herbivore numbers, provided PPF indeed increases survival or reproduction of the carnivores despite reduced predation rate.

Substitutable vs. complementary food sources

When PPF and herbivore prey are substitutable food sources for the carnivore, supply of PPF can ultimately lead to herbivore extermination, even under equilibrium conditions. This is not possible when the food sources are complementary.

Essential vs. non-essential PPF

When (for a certain habitat) the predator or parasitoid cannot obtain sufficient offspring to persist in absence of PPF (not even at high herbivore numbers), a minimum amount of PPF is needed before carnivore persistence and herbivore control will become manifest. When PPF is not essential, herbivore control is possible also without PPF, and any amount of PPF may improve herbivore control, albeit less dramatic than when PPF is essential.

Life-history scenarios

Carnivore species that differ in the stages that are affected by herbivore density and PPF availability are not necessarily affected differently by PPF at equilibrium. Only for true omnivores, of which single stages are affected by both food sources, the food sources can be substitutable. PPF may affect different functions, with different consequences for herbivore suppression, and the different effects can be multiplicative when acting in concert.

Impact via aggregation of carnivores

When the two plant types are in close proximity, or even merged into one plant species, the net influx of carnivores from the surrounding vegetation to the PPF plants may locally increase the predation pressure on the target herbivores and decrease local herbivore numbers.

Spatial configuration of source and sink habitats

When the two plant types are further apart, the PPF plants may subsidize the carnivore population on the other plants, and help to suppress herbivore numbers there. The impact of PPF on herbivores in the other habitat will vary with the complementarity of PPF and prey, and with the distribution of the herbivores. The impact of PPF may range from none (when prey and PPF are mixed and substitutable) to strong (when PPF is essential or separate from and substitutable with herbivore prey).

Metapopulation structure

When at the specific spatial scale population extinctions are likely to occur, recolonizations from nearby non-extinct populations may help maintain the species at a larger metapopulation scale. PPF may then contribute to the maintenance of a carnivore population by refueling the dispersing individuals. When PPF affects the local herbivore–carnivore dynamics (by affecting population growth rate or predation rate), its impact on herbivore incidence may be very different from its impact in more homogeneous systems.

Top–down vs. bottom–up effects

Our (Lotka–Volterra-type) models describing herbivore–carnivore interactions predict that anything, including PPF, that enhances *carnivore* reproduction or survival reduces equilibrium herbivore levels (top–down effect), whereas anything that enhances *herbivore* reproduction or survival, including PPF or increased plant productivity, has no (bottom–up) effect on herbivore numbers. More realistic models, which include more flexible attack and defense responses, predict that PPF will also have a bottom–up effect when used by the herbivores as well.

Top-carnivores and intraguild predators

Top-predators, hyperparasitoids, and intraguild predators may undermine the top–down effect of the primary carnivores, and impair the impact of PPF on herbivore control. This impact may even be reversed when these organisms at the fourth trophic level can benefit from the PPF as well.

Acknowledgments

We thank Matthijs Vos, Felix Wäckers, and Jan Bruin for discussions and comments.

References

Abrams, P. A. and M. Vos. 2003. Adaptation, density dependence and the responses of trophic level abundances to mortality. *Evolutionary Ecology Research* **5**: 1113–1132.

Addison, J. A., J. M. Hardman, and S. J. Walde. 2000. Pollen availability for predaceous mites on apple: spatial and temporal heterogeneity. *Experimental and Applied Acarology* **24**: 1–18.

Alomar, O. and R. N. Wiedemann (eds.). 1996. *Zoophytophagous Heteroptera: Implications for Life History and Integrated Pest Management.* Lanham, MD: Entomological Society of America.

Bakker, F. M. and M. E. Klein. 1992. Transtrophic interactions in cassava. *Experimental and Applied Acarology* **14**: 299–311.

Bascompte, J. and R. V. Sole. 1998. Effects of habitat destruction in a prey–predator metapopulation model. *Journal of Theoretical Biology* **195**: 383–393.

Berec, L. and V. Krivan. 2000. A mechanistic model for partial preferences. *Theoretical Population Biology* **58**: 279–289.

Bogran, C. E., K. M. Heinz, and M. A. Ciomperlik. 2002. Interspecific competition among insect parasitoids: field experiments with whiteflies as hosts in cotton. *Ecology* **83**: 653–668.

Briggs, C. J. and T. R. Collier. 2001. Autoparasitism, interference and parasitoid–pest population dynamics. *Theoretical Population Biology* **60**: 33–57.

Brodeur, J. and J. A. Rosenheim. 2000. Intraguild interactions in aphid parasitoids. *Entomologia Experimentalis et Applicata* **97**: 93–108.

Caswell, H. 1989. *Matrix Population Models.* Sunderland, MA: Sinauer Associates.

Chan, M. S. and H. C. J. Godfray. 1993. Host-feeding strategies of parasitoid wasps. *Evolutionary Ecology* **7**: 593–604.

Chang, Y. D., J. Y. Lee, and Y. N. Youn. 1994. Primary parasitoids and hyperparasitoids of the soybean aphid, *Aphis glycines* Matsumura (Homoptera: Aphididae). *Korean Journal of Applied Entomology* **33**(2): 51–55.

Coll, M. and M. Guershon. 2002. Omnivory in terrestrial arthropods: mixing plant and prey diets. *Annual Review of Entomology* **47**: 267–297.

Collier, T. R. 1995. Host feeding, egg maturation, resorption and longevity in the parasitoid *Aphytis melinus* (Hymenoptera, Aphelinidae). *Annals of the Entomological Society of America* **88**: 206–214.

Collier, T. R. and M. S. Hunter. 2001. Lethal interference competition in the whitefly parasitoids *Eretmocerus eremicus* and *Encarsia sophia. Oecologia* **129**: 147–154.

Corbett, A. and R. E. Plant. 1993. Role of movement in the response of natural enemies to agroecosystem diversification: a theoretical evaluation. *Environmental Entomology* **22**: 519–531.

De Roos, A. M. and L. Persson. 2001. Physiologically structured models: from versatile technique to ecological theory. *Oikos* **94**: 51–71.

Diehl, S. 1995. Direct and indirect effects of omnivory in a littoral lake community. *Ecology* **76**: 1727–1740.

Diekmann, O., J. A. J. Metz, and M. W. Sabelis. 1988. Mathematical models of predator–prey–plant interactions in a patchy environment. *Experimental and Applied Acarology* **5**: 319–342.

Drukker, B., P. Scutareanu, and M. W. Sabelis. 1995. Do anthocorid predators respond to synomones from *Psylla*-infested pear trees under field conditions? *Entomologia Experimentalis et Applicata* **77**: 193–203.

Dunning, J. B., B. J. Danielson, and H. R. Pulliam. 1992. Ecological processes that affect populations in complex landscapes. *Oikos* **65**: 169–175.

Duso, C., V. Malagnini, A. Paganelli, *et al.* 2004. Pollen availability and abundance of predatory phytoseiid mites on natural and secondary hedgerows. *BioControl* **49**: 397–415.

Eber, S. 2001. Multitrophic interactions: the population dynamics of spatially structured plant–herbivore–parasitoid systems. *Basic and Applied Ecology* **2**: 27–33.

Ellner, S. P., E. McCauley, B. E. Kendall, *et al.* 2001. Habitat structure and population persistence in an experimental community. *Nature* **412**: 538–543.

Eubanks, M. D. and R. F. Denno. 1999. The ecological consequences of variation in plants and prey for an omnivorous insect. *Ecology* **80**: 1253–1266.

2000. Host plants mediate omnivore–herbivore interactions and influence prey suppression. *Ecology* **81**: 936–947.

Evans, E. W. and S. England. 1996. Indirect interactions in biological control of insects: pests and natural enemies in alfalfa. *Ecological Applications* **6**: 920–930.

Evans, E. W. and D. R. Richards. 1997. Managing the dispersal of ladybird beetles (Col.: Coccinellidae): use of artificial honeydew to manipulate spatial distributions. *Entomophaga* **42**: 93–102.

Evans, E. W. and J. G. Swallow. 1993. Numerical responses of natural enemies to artificial honeydew in Utah alfalfa. *Environmental Entomology* **22**: 1392–1401.

Fagan, W. E., R. S. Cantrell, and C. Cosner. 1999. How habitat edges change species interactions. *American Naturalist* **153**: 165–182.

Gilbert, F. and M. Jervis. 1998. Functional, evolutionary and ecological aspects of feeding-related mouthpart specializations in parasitoid flies. *Biological Journal of the Linnean Society* **63**: 495–535.

Gillespie, D. R. and R. R. McGregor. 2000. The functions of plant feeding in the omnivorous predator *Dicyphus hesperus*: water places limits on predation. *Ecological Entomology* **25**: 380–386.

Giron, D., A. Rivero, N. Mandon, E. Darrouzet, and J. Casas. 2002. The physiology of host feeding in parasitic wasps: implications for survival. *Functional Ecology* **16**: 750–757.

Hagvar, E. B. 1988. Multiparasitism of the green peach aphid, *Myzus persicae*: competition in the egg stage between *Aphidius matricariae* and *Ephedrus cerasicola*. *Entomologia Experimentalis et Applicata* **47**: 275–282.

Hanski, I. 1998. Metapopulation dynamics. *Nature* **396**: 41–49.

Harmon, J. P., A. R. Ives, J. E. Losey, A. C. Olson, and K. S. Rauwald. 2000. *Coleomegilla maculata* (Coleoptera: Coccinellidae) predation on pea aphids promoted by proximity to dandelions. *Oecologia* **125**: 543–548.

Heimpel, G. E. and T. R. Collier. 1996. The evolution of host-feeding behaviour in insect parasitoids. *Biological Reviews of the Cambridge Philosophical Society* **71**: 373–400.

Heimpel, G. E., J. A. Rosenheim, and D. Kattari. 1997. Adult feeding and lifetime reproductive success in the parasitoid *Aphytis melinus*. *Entomologia Experimentalis et Applicata* **83**: 305–315.

Hindayana, D., R. Meyhofer, D. Scholz, and H. M. Poehling. 2001. Intraguild predation among the hoverfly *Episyrphus balteatus* de Geer (Diptera: Syrphidae) and other aphidophagous predators. *Biological Control* **20**: 236–246.

Holt, R. D. 1977. Predation, apparent competition and the structure of predator–prey communities. *Theoretical Population Biology* **12**: 197–229.

1983. Optimal foraging and the form of the predator isocline. *American Naturalist* **122**: 521–541.

1997. On the evolutionary stability of sink populations. *Evolutionary Ecology* **11**: 723–731.

Holt, R. D. and B. P. Kotler. 1987. Short-term apparent competition. *American Naturalist* **130**: 412–430.

Holt, R. D. and J. H. Lawton. 1994. The ecological consequences of shared natural enemies. *Annual Review of Ecology and Systematics* **25**: 495–520.

Holt, R. D. and G. A. Polis. 1997. A theoretical framework for intraguild predation. *American Naturalist* **149**: 745–764.

Holt, R. D., J. Grover, and D. Tilman. 1994. Simple rules for interspecific dominance in systems with exploitative and apparent competition. *American Naturalist* **144**: 741–771.

Holyoak, M. 2000. Habitat patch arrangement and metapopulation persistence of predators and prey. *American Naturalist* **156**: 378–389.

Holyoak, M. and S. Sachdev. 1998. Omnivory and the stability of simple food webs. *Oecologia* **117**: 413–419.

Huffaker, C. B. 1958. Experimental studies on predation: dispersion factors and predator–prey oscillations. *Hilgardia* **27**: 343–383.

Hunter, M. S. and J. B. Woolley. 2001. Evolution and behavioral ecology of heteronomous aphelinid parasitoids. *Annual Review of Entomology* **46**: 251–290.

Hunter, M. S., T. R. Collier, and S. E. Kelly. 2002. Does an autoparasitoid disrupt host suppression provided by a primary parasitoid? *Ecology* **83**: 1459–1469.

Ives, A. R. and W. H. Settle. 1997. Metapopulation dynamics and pest control in agricultural systems. *American Naturalist* **149**: 220–246.

Jacob, H. S. and E. W. Evans. 1998. Effects of sugar spray and aphid honeydew on field populations of the parasitoid *Bathyplectes curculionis* (Hymenoptera: Ichneumonidae). *Environmental Entomology* **27**: 1563–1568.

Jansen, V. A. A. and J. Yoshimura. 1998. Populations can persist in an environment consisting of sink habitats only. *Proceedings of the National Academy of Sciences of the USA* **95**: 3696–3698.

Janssen, A. and M. W. Sabelis. 1992. Phytoseiid life-histories, local predator–prey dynamics and strategies for control of tetranychid mites. *Experimental and Applied Acarology* **14**: 233–250.

Jervis, M. 1998. Functional and evolutionary aspects of mouthpart structure in parasitoid wasps. *Biological Journal of the Linnean Society* **63**: 461–493.

Kean, J., S. Wratten, J. Tylianakis, and N. Barlow. 2003. The population consequences of natural enemy enhancement and implications for conservation biological control. *Ecology Letters* **6**: 604–612.

Kean, J. M. and N. D. Barlow. 2001. A spatial model for the successful biological control of *Sitona discoideus* by *Microctonus aethiopoides*. *Journal of Applied Ecology* **38**: 162–169.

Kidd, N. A. C. and M. A. Jervis. 1996. Population dynamics. In M. A. Jervis and N. A. C. Kidd (eds.) *Insect Natural Enemies*. London: Chapman and Hall, pp. 293–379.

Kreiter, S., M. S. Tixier, B. A. Croft, P. Auger, and D. Barret. 2002. Plants and leaf characteristics influencing the predaceous mite *Kampimodromus aberrans* (Acari: Phytoseiidae) in habitats surrounding vineyards. *Environmental Entomology* **31**: 648–660.

Krivan, V. and E. Sirot. 1997. Searching for food or hosts: the influence of parasitoid behavior on host–parasitoid dynamics. *Theoretical Population Biology* **51**: 201–209.

Lambin, M., A. Ferran, and K. Maugan. 1996. Perception of visual information in the ladybird *Harmonia axyridis* Pallas. *Entomologia Experimentalis et Applicata* **79**: 121–130.

Lei, G. C. and I. Hanski. 1997. Metapopulation structure of *Cotesia melitaearum*, a specialist parasitoid of the butterfly *Melitaea cinxia*. *Oikos* **78**: 91–100.

Levins, R. 1969. Some demographic and genetic consequences of environmental heterogeneity for biological control. *Bulletin of the Entomological Society of America* **15**: 237–240.

Lewis, W., J. Stapel, A. Cortesero, and K. Takasu. 1998. Understanding how parasitoids balance food and host needs: importance to biological control. *Biological Control* **11**: 175–183.

Limburg, D. D. and J. A. Rosenheim. 2001. Extrafloral nectar consumption and its influence on survival and development of an omnivorous predator, larval *Chrysoperla plorabunda* (Neuroptera: Chrysopidae). *Environmental Entomology* **30**: 595–604.

Lucas, E., D. Coderre, and J. Brodeur. 1998. Intraguild predation among aphid predators: characterization and influence of extraguild prey density. *Ecology* **79**: 1084–1092.

Mackauer, M., B. Bai, A. Chow, and T. Danyk. 1992. Asymmetric larval competition between two species of solitary parasitoid wasps: the influence of superparasitism. *Ecological Entomology* **17**: 233–236.

Mansfield, S. and N. J. Mills. 2002. Host egg characteristics, physiological host range and parasitism following inundative releases of *Trichogramma platneri* (Hymenoptera: Trichogrammatidae) in walnut orchards. *Environmental Entomology* **31**: 723–731.

McAuslane, H. J. and R. Nguyen. 1996. Reproductive biology and behavior of a thelytokous species of *Eretmocerus* (Hymenoptera: Aphelinidae) parasitizing *Bemisia argentifolii* (Homoptera: Aleyrodidae). *Annals of the Entomological Society of America* **89**: 686–693.

McCann, K. and A. Hastings. 1997. Re-evaluating the omnivory-stability relationship in food webs. *Proceedings of the Royal Society of London Series B* **264**: 1249–1254.

McEwen, P. K., M. A. Jervis, and N. A. C. Kidd. 1993. Influence of artificial honeydew on larval development and survival in *Chrysoperla carnea* (Neur., Chrysopidae). *Entomophaga* **38**: 241–244.

1996. The influence of an artificial food supplement on larval and adult performance in the green lacewing *Chrysoperla carnea* (Stephens). *International Journal of Pest Management* **42**: 25–27.

Murdoch, W. W., R. M. Nisbet, S. M. Blythe, W. S. C. Gurney, and J. D. Reeve. 1987. An invulnerable age class and stability in delay-differential parasitoid–host models. *American Naturalist* **129**: 263–282.

Mylius, S. D., K. Klumpers, A. M. de Roos, and L. Persson. 2001. Impact of intraguild predation and stage structure on simple communities along a productivity gradient. *American Naturalist* **158**: 259–276.

Nachman, G. and R. Zemek. 2003. Interactions in a tritrophic acarine predator–prey metapopulation system. V. Within-plant dynamics of *Phytoseiulus persimilis* and *Tetranychus urticae* (Acari: Phytoseiidae, Tetranychidae). *Experimental and Applied Acarology* **29**: 35–68.

Nakano, S. and M. Murakami. 2001. Reciprocal subsidies: dynamic interdependence between terrestrial and aquatic food webs. *Proceedings of the National Academy of Sciences of the USA* **98**: 166–170.

Nisbet, R. M. 1997. Delay-differential equations for structured populations. In S. Tuljapurkar and H. Caswell (eds.) *Structured-Population Models in Marine, Terrestrial and Freshwater Systems*. New York: Chapman and Hall, pp. 89–118.

Oksanen, L., S. D. Fretwell, J. Arruda, and P. Niemela. 1981. Exploitation ecosystems in gradients of primary productivity. *American Naturalist* **118**: 240–261.

Pels, B., A. M. de Roos, and M. W. Sabelis. 2002. Evolutionary dynamics of prey exploitation in a metapopulation of predators. *American Naturalist* **159**: 172–189.

Pijls, J., K. D. Hofker, M. J. Van Staalduinen, and J. J. M. Van Alphen. 1995. Interspecific host discrimination and competition in *Apoanagyrus (Epidinocarsis) lopezi* and *A. (E.) diversicornis*, parasitoids of the cassava mealybug *Phenacoccus manihoti*. *Ecological Entomology* **20**: 326–332.

Pilcher, C. D., J. J. Obrycki, M. E. Rice, and L. C. Lewis. 1997. Preimaginal development, survival and field abundance of insect predators on transgenic *Bacillus thuringiensis* corn. *Environmental Entomology* **26**: 446–454.

Pimm, S. L. and J. H. Lawton. 1978. On feeding on more than one trophic level. *Nature* **27**: 542–544.

Polis, G. A. 1994. Food webs, trophic cascades and community structure. *Australian Journal of Ecology* **19**: 121–136.

1998. Ecology: stability is woven by complex webs. *Nature* **395**: 744–745.

1999. Why are parts of the world green? Multiple factors control productivity and the distribution of biomass. *Oikos* **86**: 3–15.

Polis, G. A. and R. D. Holt. 1992. Intraguild predation: the dynamics of complex trophic interactions. *Trends in Ecology and Evolution* **7**: 151–154.

Polis, G. A. and D. R. Strong. 1996. Food web complexity and community dynamics. *American Naturalist* **147**: 813–846.

Polis, G. A., W. B. Anderson, and R. D. Holt. 1997. Toward an integration of landscape and food web ecology: the dynamics of spatially subsidized food webs. *Annual Review of Ecology and Systematics* **28**: 289–316.

Polis, G. A., C. A. Myers, and R. D. Holt. 1989. The ecology and evolution of intraguild predation: potential competitors that eat each other. *Annual Review of Ecology and Systematics* **20**: 297–330.

Pringle, C. M. and T. Hamazaki. 1998. The role of omnivory in a neotropical stream: separating diurnal and nocturnal effects. *Ecology* **79**: 269–280.

Pulliam, H. R. 1988. Sources, sinks and population regulation. *American Naturalist* **132**: 652–661.

Rogers, M. E. and D. A, Potter. 2004. Potential for sugar sprays and flowering plants to increase parasitism of white grubs (Coleoptera: Scarabeidae) by tiphiid wasps (Hymenoptera: Tiphiidae). *Environmental Entomology* **33**: 619–626.

Rosenheim, J. A. 1998. Higher-order predators and the regulation of insect herbivore populations. *Annual Review of Entomology* **43**: 421–447.

2001. Source–sink dynamics for a generalist insect predator in habitats with strong higher-order predation. *Ecological Monographs* **71**: 93–116.

Rosenheim, J. A., H. K. Kaya, L. E. Ehler, J. J. Marois, and B. A. Jaffee. 1995. Intraguild predation among biological-control agents: theory and evidence. *Biological Control* **5**: 303–335.

Sabelis, M. W. 1992. Arthropod predators. In M. J. Crawley (ed.) *Natural Enemies: The Population Biology of Predators, Parasites and Diseases*. Oxford, UK: Blackwell Scientific Publications, pp. 225–264.

Sabelis, M. W. and J. van der Meer. 1986. Local dynamics of the interaction between predatory mites and two-spotted spider mites. In J. A. J. Metz and O. Diekmann (eds.) *Dynamics of Physiologically Structured Populations*. Berlin, Germany: Springer-Verlag, pp. 322–344.

Sabelis, M. W. and P. C. J. van Rijn. 1997. Predation by insects and mites. In T. Lewis (ed.) *Thrips as Crop Pests*. Wallingford, UK: CAB International, pp. 259–354.

Snyder, W. E. and A. R. Ives. 2001. Generalist predators disrupt biological control by a specialist parasitoid. *Ecology* **82**: 705–716.

Stapel, J. O., A. M. Cortesero, C. M. DeMoraes, J. H. Tumlinson, and W. J. Lewis. 1997. Extrafloral nectar, honeydew and sucrose effects on searching behavior and efficiency of *Microplitis croceipes* (Hymenoptera: Braconidae) in cotton. *Environmental Entomology* **26**: 617–623.

Steffan-Dewenter, I. and T. Tscharntke. 2002. Insect communities and biotic interactions on fragmented calcareous grasslands: a mini review. *Biological Conservation* **104**: 275–284.

Stelzl, M. 1991. Investigations on food of Neuroptera adults (Neuropteroidea, Insecta) in Central Europe: with a short discussion of their role as natural enemies of insect pests. *Journal of Applied Entomology / Zeitschrift für angewandte Entomologie* **111**: 469–477.

Sullivan, D. J. and W. Volkl. 1999. Hyperparasitism: multitrophic ecology and behavior. *Annual Review of Entomology* **44**: 291–315.

Takimoto, G., T. Iwata, and M. Murakami. 2002. Seasonal subsidy stabilizes food web dynamics: balance in a heterogeneous landscape. *Ecological Research* **17**: 433–439.

Tilman, D. 1982. *Resource Competition and Community Structure*. Princeton, NJ: Princeton University Press.

Tylianakis, J. M., R. K. Didnam, and S. D. Wratten. 2004. Improved fitness of aphid parasitoids receiving resource subsidies. *Ecology* **85**: 658–666.

Ueno, T. 1999a. Host-feeding and acceptance by a parasitic wasp (Hymenoptera: Ichneumonidae) as influenced by egg load and experience in a patch. *Evolutionary Ecology* **13**: 33–44.

1999b. Multiparasitism and host feeding by solitary parasitoid wasps (Hymenoptera: Ichneumonidae) based on the pay-off from parasitized hosts. *Annals of the Entomological Society of America* **92**: 601–608.

Van Baalen, M. and M. W. Sabelis. 1995. The milker–killer dilemma in spatially structured predator–prey interactions. *Oikos* **74**: 391–400.

Van Baalen, M., V. Krivan, P. C. J. van Rijn, and M. W. Sabelis. 2001. Alternative food, switching predators and the persistence of predator–prey systems. *American Naturalist* **157**: 512–524.

Van den Bosch, F. and Diekmann, O. 1986. Interactions between egg-eating predator and prey: the effect of the functional response and of age structure. *IMA Journal of Mathematics Applied in Medicine and Biology* **3**: 53–69.

Van Nouhuys, S. and I. Hanski. 2002. Colonization rates and distances of a host butterfly and two specific parasitoids in a fragmented landscape. *Journal of Animal Ecology* **71**: 639–650.

Van Rijn, P. C. J. and L. K. Tanigoshi. 1999. Pollen as food for the predatory mites *Iphiseius degenerans* and *Neoseiulus cucumeris* (Acari: Phytoseiidae): dietary range and life history. *Experimental and Applied Acarology* **23**: 785–802.

Van Rijn, P. C. J., F. M. Bakker, W. A. D. van der Hoeven, and M. W. Sabelis. 2005. Is arthropod predation exclusively satiation-driven? *Oikos* **109**: 101–116.

Van Rijn, P. C. J., Y. M. van Houten, and M. W. Sabelis. 2002. How plants benefit from providing food to predators even when it is also edible to herbivores. *Ecology* **83**: 2664–2679.

Veech, J. A. 2001. The foraging behavior of granivorous rodents and short-term apparent competition among seeds. *Behavioral Ecology* **12**: 467–474.

Vet, L. E. M., A. G. de Jong, E. Franchi, and D. R. Papaj. 1998. The effect of complete versus incomplete information on odor discrimination in a parasitic wasp. *Animal Behaviour* **55**: 1271–1279.

Vos, M., B. W. Kooi, D. L. DeAngelis, and W. M. Mooij. 2004. Inducible defences and the paradox of enrichment. *Oikos* **105**: 471–480.

Wäckers, F. L. 2003. The effect of food supplements on parasitoid–host dynamics. Proc. Int. Symp. *Biological Control of Arthropods*, Honolulu, pp. 226–231.

Williams, T. and A. Polaszek. 1996. A re-examination of host relations in the Aphelinidae (Hymenoptera: Chalcidoidea). *Biological Journal of the Linnean Society* **57**: 35–45.

Xia, J. Y., R. Rabbinge, and W. van der Werf. 2003. Multistage functional responses in a ladybeetle–aphid system: scaling up from the laboratory to the field. *Environmental Entomology* **32**: 151–162.

9

Does floral nectar improve biological control by parasitoids?

GEORGE E. HEIMPEL AND MARK A. JERVIS

Introduction

The incorporation of plant diversity within agricultural systems has led to decreased insect pest densities in approximately 50% of studies in which monocultures and polycultures were directly compared (Risch *et al.* 1983; Andow 1991; Coll 1998; Gurr *et al.* 2000). One of the leading hypotheses explaining the observation of decreased pest densities under polycultures is that increased plant diversity can enhance the action of natural enemies of pests (the "enemies hypothesis" of Root 1973). Increased plant diversity can provide natural enemies with resources such as a favorable microclimate, alternative hosts or prey, or plant-based foods such as pollen, nectar, or honeydew (Landis *et al.* 2000). In this chapter, we focus on one of the more intuitively clear predictions encompassed within Root's enemies hypothesis – the idea that the presence of nectar-producing plants can improve biological control of pests by supplying parasitoids with sugar. Note that this idea includes two components: an outcome (improved biological control) and an underlying mechanism (nectar-feeding), both of which need to be demonstrated. We refer to the combined outcome and mechanism as the "parasitoid nectar provision hypothesis".

The hypothesis that plant diversification can decrease pest pressure by providing sugar to parasitoids that would otherwise be sugar-limited has its origins in anecdotal or semi-quantitative observations of increased parasitism rates and biological control in the vicinity of flowering plants. For instance, Wolcott (1941, 1942) noted that the mole cricket parasitoid *Larra analis* (Hymenoptera: Sphecidae) was dependent upon the nectar of two flowering weed species for establishment in Puerto Rico, and that *Tiphia* wasps (Hymenoptera: Tiphiidae)

Plant-Provided Food for Carnivorous Insects, ed. F. L. Wäckers, P. C. J. van Rijn, and J. Bruin. Published by Cambridge University Press.

were able to control their whitegrub hosts only in those parts of Haiti that supported flowering (as opposed to purely vegetative) populations of wild parsnip. Allen and Smith (1958) made similar observations in California, where *Cotesia medicaginis*, a braconid parasitoid of the alfalfa caterpillar, was seen feeding on inflorescences of various weeds within alfalfa fields and inflicted higher rates of parasitism where flowering weeds were abundant. A more experimental approach was taken in the 1950s in the former Soviet Union in which parasitoid action was found to be increased in the vicinity of deliberately planted areas of nectar-producing flowers in various orchard settings (Chumakova 1960; Powell 1986). In the 1960s, Leius used a number of elegant laboratory studies to demonstrate the role that nectar and other foodstuffs can play in increasing parasitoid longevity and fecundity (Leius 1961a,b, 1963), along with a field study showing an association between higher parasitism of codling moth and eastern tent caterpillars in abandoned apple orchards with a higher abundance and diversity of wildflowers in the understory (Leius 1967). Since these early observations and experiments, the potential role of nectar sources in biological control by parasitoids has been more formally recognized (e.g., Zandstra and Motooka 1978; Powell 1986; Van Emden 1990), and has received increased attention over the last decade (Gurr *et al.* 2000).

The parasitoid nectar provision hypothesis is particularly appealing for four main reasons. First, many monocultures are thought to be relatively devoid of sugar. Second, numerous laboratory studies have shown that parasitoids live far longer and can attack many more hosts when they are fed sugar than when they are sugar-starved. Typical lifespans range between 1 and 5 days for starved parasitoids and between 2 and 8 weeks for sugar-fed parasitoids (Jervis *et al.* 1996; Thompson 1999). Third, many parasitoid species utilize nectar under natural conditions (Wolcott 1942; Jervis *et al.* 1993; Gilbert and Jervis 1998; Jervis 1998). Lastly, the use of nectar by parasitoids does not occur at the expense of attacks on pests, as can occur with the use of alternative hosts in diversified habitats (Bugg *et al.* 1987; Corbett and Plant 1993). These considerations have combined to produce an expectation that biological control can be improved by the incorporation of flowering cover crops or other sources of sugar to parasitoids in the field (Froggatt 1902; Box 1927; Wolcott 1942; Allen and Smith 1958; Hagen 1986; Powell 1986; Van Emden 1990; Jervis *et al.* 1993; Landis *et al.* 2000). To what extent, however, is this expectation met? Are parasitoids really sugar-limited in the field? If they are, can sugar limitation be alleviated by incorporation of nectar sources in field settings? And further, does alleviating sugar limitation necessarily lead to improved biological control?

Nectar can be obtained from plants that are either deliberately sown to provide parasitoids with nectar (e.g., Chumakova 1960; Powell 1986; Gurr *et al.* 2000;

Berndt *et al.* 2002; Lee and Heimpel 2003) or that are present naturally as "weeds", and tolerated or encouraged either as a parasitoid food provision strategy or to increase plant diversity for other reasons (Zandstra and Motooka 1978; Altieri and Whitcomb 1979; Andow 1988; Bugg and Waddington 1994; Nentwig 1998; Bugg and Pickett 1998). In most cases, the sown plants are non-crop plants (or crop plants that are not harvested), but in some cases, they can be harvested as part of an intercropping system (e.g., Gallego *et al.* 1983; Letourneau 1987). In Florida and Hawaii, flowering *Crotolaria* (Fabaceae) have been used as "trap crops" around field crops and orchards with the dual purpose of retaining pest stink bugs and providing nectar for tachinids that attack these stink bugs (Leeper 1974).

We restrict our discussion to floral nectar for brevity, but many of the same principles will apply to plants with extrafloral nectar if it is not produced by the crop plant itself (Attsat and O'Dowd 1976; Hespenheide 1985; Rogers 1985). We also do not review work on parasitoid honeydew-feeding or the use of sugar sprays that mimic honeydew (Hagen 1986; Evans 1993) since these sugar sources are usually decoupled from plant diversity (but see Box 1927; Idoine and Ferro 1988).

We begin by outlining a series of behavioral and ecological requirements that are needed to validate the parasitoid nectar provision hypothesis, and then go on to evaluate the extent to which these requirements are met. We recognize five requirements for validation of the parasitoid nectar provision hypothesis. These requirements are similar to Wratten *et al.*'s (2003) hierarchical research outcomes needed to demonstrate success in nectar provisioning. Wratten *et al.*'s scheme emphasizes parasitoid aggregation to nectar sources, improved parasitoid fitness, increased parasitism, and reduced pest densities below an economic threshold. In our view, parasitoid aggregation is not needed per se, and we are more concerned with validating that the mechanisms leading to nectar-mediated improvement of biological control actually occur in the field than whether the magnitude of the effect is strong enough to drive pests below economic thresholds. Fundamentally, however, our approach is very similar to that of Wratten *et al.* (2003) and we reach similar conclusions.

Requirements for validation of the parasitoid nectar provision hypothesis

Sugar limitation in parasitoids

The parasitoid nectar provision hypothesis rests on the implicit assumption that sugar is a limiting resource under managed conditions such as agriculture, forestry, and urban landscape settings. Thus, parasitoids that can obtain sufficient sugar from, for example, floral or extrafloral nectar from the crop plant itself (e.g., Lingren and Lukefahr 1977; Treacy *et al.* 1987; Stapel *et al.* 1997), or homopteran honeydew (Box 1927; Zoebelein 1956; Idoine and Ferro 1988) may not stand to gain much from additional nectar supplied through interplantings. Also, if non-starvation mortality of adult parasitoids is so great in the field that the lifespan of parasitoids is routinely reduced to starvation or sub-starvation levels (Heimpel *et al.* 1997b; Rosenheim 1998), supplemental sugar will have less impact on lifespan in the field.

A small minority of parasitoid species appear to have little need for sugar as adults. Females of the bethylid *Goniozus nephantidis*, for instance, have a median laboratory age of approximately 30 days whether they are starved or honey-fed (Hardy *et al.* 1992). Species that are strictly pro-ovigenic (i.e., emerge with their lifetime egg complement mature: Jervis *et al.* 2001) have characteristically short lifespans and may have little need for sugar feeding in some field conditions. An important caveat here is that strict pro-ovigeny is rare among parasitoids (Jervis *et al.* 2001). The females of some host-feeding species can use host meals in lieu of sugar meals (Heimpel and Collier 1996), although in these cases, males still require sugar meals (e.g., Ode *et al.* 1996). Conversely, females of a number of host-feeding species can starve to death in 1–2 days despite imbibing substantial amounts of host fluids during this time (Heimpel *et al.* 1994, 1997a; Heimpel and Collier 1996).

Feeding on floral nectar in the field

The parasitoid nectar provision hypothesis obviously relies on parasitoids feeding on nectar in the field, but nectar-feeding can by no means be taken for granted. Although there are many reports of parasitoids feeding on nectar in the field (e.g., Jervis *et al.* 1993), whether or not adults of a given parasitoid species will use nectar from given flower species will depend upon factors such as flower morphology (Jervis *et al.* 1993; Idris and Grafius 1995; Orr and Pleasants 1996; Wäckers *et al.* 1996; Patt *et al.* 1997; Baggen *et al.* 1999), timing of nectar production (Bowie *et al.* 1995; Lee and Heimpel 2003), parasitoid mouthpart structure (Gilbert and Jervis 1998; Jervis 1998), the ability of parasitoids to locate inflorescences (Lewis and Takasu 1990; Jervis *et al.* 1993; Takasu and Lewis 1993, 1995, 1996; Wäckers 1994; Jervis and Kidd 1996; Patt *et al.* 1999; Jacob and Evans 2000), and the appropriate feeding response (Wäckers 1999; Beach *et al.* 2003)

Enhanced fecundity by female parasitoids that use nectar

For nectar provision to be effective, nectar consumption must lead to an increase in realized fecundity in female parasitoids. Realized fecundity of parasitoids is determined by the host encounter rate, pre-imaginal and imaginal egg maturation rates, oviposition behavior and lifespan (Driessen and Hemerik

1992; Godfray 1994; Heimpel *et al.* 1998; Jervis *et al.* 2001). Under most field conditions, increased lifespan will lead to higher fecundity because more time is available for host location and egg maturation. However, if parasitoids are able to realize maximum fecundity before starvation occurs (as may occur when the ratio of hosts to parasitoids is very high), lifespan may not limit fecundity.

Increased parasitism rates and decreased pest densities in the presence of nectar

Higher parasitism levels can be expected to lead to decreased pest densities under most conditions if the nectar source has no direct effect on pest numbers. It is not unlikely, however, for pest numbers to change as a direct result of the presence of nectar sources, and in this case the density-dependent response of the parasitoid needs to be taken into account. Lower pest densities in the presence of supplemental nectar are predicted by Root's (1973) "resource concentration hypothesis", which posits that specialist herbivores are attracted and/or retained by monocultures of their host plant to a greater extent than by diversified systems. This response has been documented in a number of pest insects (Kareiva 1983; Andow 1991). Another mechanism that could lead to lower pest numbers in the presence of nectar is increased predation on the pest in the presence of the nectar source. This is Root's (1973) enemies hypothesis applied to predators.

Density-dependent parasitism

What happens when parasitism is layered onto these determinants of herbivore density? Even before we consider nectar-feeding, it is useful to spell out the potential roles of different classes of density-dependent parasitism in the context of the herbivore response to plant diversity. Consider first the case in which mechanisms that are not related to parasitoid attack lead to lower herbivore numbers in the diverse (floral nectar-enhanced) habitat. In this case, density-independent parasitism will lead to increased (numerical) herbivore suppression in the diverse habitat, and inversely density-dependent parasitism will increase this suppression even further (Fig. 9.1A). Positive density dependence, however, could compromise the suppression occurring via resource concentration or predators by providing the herbivores in the diverse habitat with a refuge from parasitism. This latter effect could also occur if parasitoids were less able to locate hosts in more complex environments (Sheehan 1986; Andow and Risch 1987). Consider next the case where herbivore populations are higher in the presence of nectar sources (we describe some of these situations below). In this case, interactions with parasitoid density dependence are the opposite of those just described. Positive density dependence can diminish the extent to

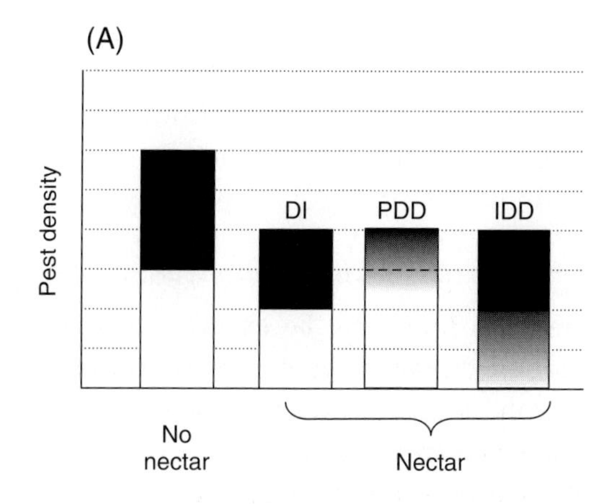

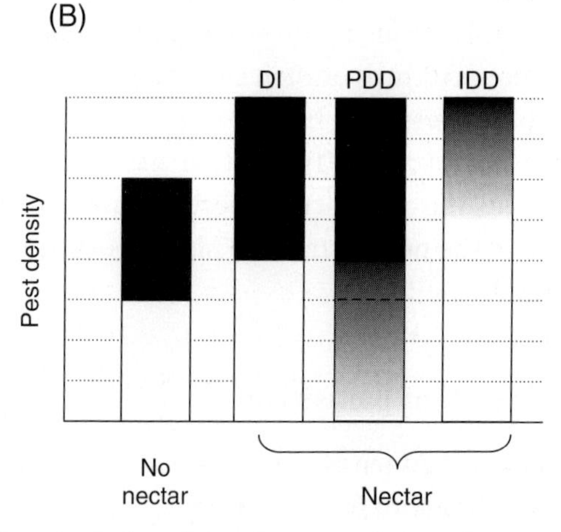

Figure 9.1 Hypothetical portions of pest populations that are parasitized and unparasitized when the pest response to nectar in the absence of parasitoids is negative (A) and positive (B), in which darkened areas represent parasitized pests. The left-hand columns in both panels represent pest densities in the absence of nectar sources with 50% of pests parasitized. The three columns to the right in both panels represent parasitized and unparasitized pest numbers in the presence of nectar sources with three classes of parasitoid response: density independent (DI), positive density dependence (PDD), and inverse density dependence (IDD). The dashed line in the PDD column of (A) represents the parasitism level below which pest densities are higher in the presence of nectar despite a negative response of pests to the nectar source. The dashed line in the PDD column of (B) represents the parasitism level above which pest densities are lower in the presence of nectar despite a positive response of pests to the nectar source.

which herbivore numbers are enhanced by the floral resource, and inverse density dependence can amplify this trend (Fig. 9.1B). Less predictable reactions would occur if the form or magnitude of density dependence changes with habitat or of the range of host densities present.

Nectar sources may have positive effects on pest populations (Zhao *et al.* 1992; Jervis *et al.* 1993, 1996; Chaney 1998; Baggen *et al.* 2000). This can happen, for instance, when herbivores benefit from the nectar themselves, either through direct feeding (Baggen and Gurr 1998; Baggen *et al.* 1999; Romeis and Wäckers 2000) or by nectar-associated reduction of a predator. As an illustration of the latter mechanism, Stephens *et al.* (1998) found that plantings of buckwheat in New Zealand peach and nectarine orchards increased the abundance of brown lacewing parasitoids, thus potentially reducing densities of a biological control agent. In a study conducted in Finland, Helenius (1990) found that the bird cherry-oat aphid, *Rhopalosiphum padi*, responded positively to interplanting oats with fava bean, with aphid densities roughly doubled in oat–bean mixtures with respect to oat monocrops. In this case, parasitism of *R. padi* was found to be inversely density dependent so that the pest-attraction effect of the polyculture was exacerbated by concomitant lower parasitism (as in the far-right column in Fig. 9.1B). Baggen and Gurr (1998) also found a positive response of an herbivore (the potato tuber moth, *Phthorimae operculella*) to a flower source. In this case though, parasitism by the encyrtid *Copidosoma koehleri* was also higher near the flowers. Baggen and Gurr were not able to distinguish between the hypotheses that *C. koehleri* was responding to the flowers, or that it was responding to hosts in a positively density-dependent manner.

Parasitoids themselves may also be at increased risk of predation by sit-and-wait predators. Instances of predation events on parasitoids visiting flowers have included an ichneumonid being attacked by a crab spider on an inflorescence of white dryad (*Dryas integrifolia*; Rosaceae) (Kevan 1973) and an ichneumonid being attacked by a damsel bug on the umbellifer *Angelica sylvestris* (Jervis 1990).

Parasitism rates can also interact with insect movement in ways that affect the parasitoid nectar provision hypothesis. These interactions can be varied and complex, and remain relatively unexplored both empirically and theoretically. Consider first the case in which nectar acts mainly by attracting parasitoids from nectar-poor (unmodified) to nectar-rich (modified) areas, and in which both of these areas lie within the pest-infested habitat being managed. Assume further that parasitoid lifespan is enhanced in the vicinity of the nectar. In this case, increased parasitism in the presence of nectar may come at the expense of decreased parasitism in nectarless areas within the managed habitat (Jervis *et al.* 1992a). Field-wide (or average) parasitism levels should still be higher within the managed habitat than in fields completely devoid of nectar in this scenario, unless mutual interference among parasitoids leads to sharp declines in efficiency near the nectar sources. In an alternative scenario (modeled by Corbett and Plant 1993), highly mobile parasitoids disperse throughout the field after feeding upon sugar. Here, the nectar sources enhance parasitism throughout the field. Such a benefit is not likely to be seen from parasitoids with low mobility, but neither is there a great risk of parasitoid depletion in nectar-poor areas or mutual interference in nectar-rich areas when parasitoids have low mobility. Lastly, nectar sources may attract parasitoids from non-agricultural settings. In this case, trade-offs between nectar-rich and nectar-poor areas within the agricultural field may not occur, but unanticipated interactions at the interface of agricultural and non-agricultural habitats may be produced.

In order to address these issues, more data on parasitoid dispersal patterns are needed. Particularly important are absolute distances traveled, and how these distances are affected by nutritional state. Topham and Beardsley (1975) estimated that the "commuting range" that the tachinid fly *Lixophaga sphenophori* moved between hosts and nectar sources was not much more than 60 m. If pests respond to the nectar source, the spatial scale of this response will be important as well.

In the next section, we will evaluate the empirical support for each of the five conditions outlined above.

Experimental support

Sugar limitation in parasitoids

We recognize the following classes of evidence that can be used to assess sugar limitation in parasitoids: (1) inspection of the habitat for sugar sources and observations of parasitoid behavior, (2) manipulation of the habitat coupled with specific outcomes, and (3) analyses of parasitoid gut contents.

It is sometimes apparent that a given habitat is devoid of sugar sources. For instance, large tracts of loblolly pine plantations in the southeastern USA apparently harbor neither flowering plants nor honeydew-producing Homoptera and it has been argued that this lack of sugar sources has resulted in a general paucity of parasitoids in these habitats (Stephen *et al.* 1996). Almond orchards in central California support annual flowering weeds in late winter and early spring, but these plants senesce before summer, which is when most pest problems occur (G. E. Heimpel, unpublished data). The almond trees themselves can be a source of sugar, through both floral and extrafloral nectar, but again, these sources only produce sugar in late winter and early spring (G. E. Heimpel, unpublished data). Finally, the only honeydew-producing Homoptera in the system, the European fruit lecanium, *Parthenolecanium corni*, is rare in most orchards so that honeydew is virtually non-existent (G. E. Heimpel, unpublished data). This habitat appeared completely devoid of naturally occurring sugar sources during most of the year, and no instances of sugar-feeding by *Aphytis* and *Encarsia* parasitoids were noted despite hundreds of hours of observation (Heimpel *et al.* 1998; G. E. Heimpel, unpublished data).

Our last example comes from the Spanish province of Andalucia, where farmers practice drastic control of non-crop vegetation in olive orchards. Disk-harrowing and/or herbicide treatment of inter-tree areas keeps herbaceous vegetation to a minimum. Usually, this is done once or twice per season so as to conserve soil nutrients and water for the olive trees, and to facilitate harvesting of the olive fruits (usually done with machinery). It is thought to seriously deprive some olive pest parasitoids of potential sugar sources, particularly because honeydew deposits are often highly localized (Jervis *et al.* 1992a; Jervis and Kidd 1993), and it has been postulated as a factor underlying the poor performance of classical biological control in European olives (<20% of all natural enemy introductions have resulted in any degree of control) (Jervis and Kidd 1993).

Several authors have argued that annual cropping systems can be inhospitable habitats for parasitoids, in part because of a lack of sugar sources (e.g., Van den Bosch and Telford 1964; Lewis *et al.* 1998; Landis and Marino 1999). For example, in early-season corn in Michigan (USA), the ichneumonid *Eriborus terebrans* appears to be excluded from the centers of corn fields, possibly in part because of a need to stay near field edges (particularly woodlots) that harbor sugar sources. Interestingly, *E. terebrans* colonized central portions of the field later in the season, perhaps in response to increases in aphid honeydew, floral nectar from weeds, and/or increased shade (Landis and Haas 1992; Dyer and Landis 1996, 1997). Edge effects have been discussed in more detail by Van Emden (1988) and Jervis and Kidd (1996); see also Wilkinson and Landis (Chapter 10).

While increased parasitism associated with nectar sources is consistent with sugar limitation in the absence of nectar, it may also reflect factors such as improved microclimate, increased water intake, increased overwintering sites, differences in pest levels coupled with density-dependent parasitism (Fig. 9.1), the presence of alternative hosts, or other community-level effects such as reduced pressure from predators or hyperparasitoids. Also, in one set of studies in which herbicides were used to generate a flower-free treatment for the tachinid *L. sphenophori*, it was determined that some (but not all) of the herbicides were toxic to the fly (Leeper 1974; Topham and Beardsley 1975). Lower parasitism levels or parasitoid abundances in the absence of nectar can be taken as indirect evidence of sugar limitation only if great care is taken to alter nothing but the nectar itself. This tactic was employed by Irvin *et al.* (2000), in which parasitism of the light brown apple moth and parasitoid abundance were compared in plots planted to flowering buckwheat and plots planted to flowering buckwheat from which the inflorescences had been clipped off (see also Gurr *et al.* 1999). In this experiment, abundance of the parasitoid *Dolichogenidea tasmanica* (Braconidae) was significantly higher in buckwheat + flowers than in clipped buckwheat or herbicide control plots, but there were no significant differences in parasitism rate between the plots planted to buckwheat with and without flowers. Parasitism levels of both classes of buckwheat treatments were significantly higher than in the herbicide control plots on one sampling date, however, mirroring previous results in which unclipped buckwheat was compared with herbicide controls (Stephens *et al.* 1998). Together, these results suggest that nectar was either attracting or retaining parasitoids, possibly with concomitant increases in parasitoid lifespan, and that fecundity may have been enhanced by the presence of the plant in general, but not necessarily the inflorescences.

Cappuccino *et al.* (1999) drew a similar conclusion in a study in which effects of sugar sprays were compared with effects of a flowering understory on parasitism of the spruce budworm, *Choristoneura fumiferana,* in a Canadian boreal forest. They found that parasitism rates were significantly increased by the application of sugar-water, but not by allowing flowering plants to grow as an understory (Cappuccino *et al.* 1999). These results suggest some level of sugar limitation by the parasitoids, but also that this sugar limitation cannot be alleviated by the nectar source(s).

Some studies have shown higher field parasitism rates of cotton pests on cotton varieties that have extrafloral nectar versus varieties that do not (e.g., Treacy *et al.* 1987). To the extent that there are no other important varietal differences between the nectaried and nectariless strains of cotton, these results provide strong circumstantial evidence that parasitoids are sugar-limited in fields of nectariless cotton.

Another experimental approach involves sugar sprays, as described above for the work of Cappuccino *et al.* (1999). The most likely explanation for higher parasitism and/or parasitoid abundance in the presence of sugar sprays is alleviation of sugar limitation. The most comprehensive work in this area is that of Evans and associates (Evans 1993; Evans and Swallow 1993; Evans and England 1996; Jacob and Evans 1998). In these studies, sugar-water was sprayed into alfalfa fields and this led to increased parasitism of the alfalfa weevil by the ichneumonid *Bathyplectes curculionis*, a parasitoid that is likely to be heavily dependent on feeding for ovigenesis (Dowell 1978; Jervis *et al.* 2001). The experiments were done on a rather small spatial scale so that it is likely that treatment differences were due to a combination of movement of parasitoids from control plots to sugar-spray plots and increased longevity of parasitoids that were fed upon the sugar-water.

In either case though, there is a strong implication that *B. curculionis* wasps do not have an abundant sugar source in the absence of sugar sprays.

Other cases of increased parasitism or parasitoid abundance in the field as a result of sugar sprays include *Trichogramma minutum* in a Canadian forest setting within large field cages (Smith *et al.* 1986) and parasitoids of the fall armyworm, *Spodoptera frugiperda*, attacking corn in Honduras (Cañas and O'Neil 1998). In the case of *Trichogramma brassicae* in cabbage, however, sugar sprays had no effect on parasitism of *Pieris rapae* eggs (Lundgren *et al.* 2002). Negative results of sugar-spray studies can be attributed to any of at least three causes: (1) the parasitoids were not sugar-limited in the absence of sugar sprays, (2) the parasitoids were not using the sprayed sugar, or (3) complex interactions in which a positive effect of the sugar sprays is canceled out by responses of other insects (such as the hosts or natural enemies of the parasitoid) to the sugar sprays. An especially creative method of delivering food to parasitoids foraging high in the canopy of pine plantations was developed by Stephen and Browne (2000), who incorporated a synthetic food solution (Eliminade™) into plastic "paint balls", which were shot into the canopy from the ground using paint-ball guns. In addition, Stephen and Browne incorporated food dye into the food spray, which was visible within parasitoid guts upon dissection, and were able to show that 76% of southern pine beetle parasitoids collected from trees sprayed using the paint-ball gun had fed upon the artificial food. Such a high rate of feeding is consistent with low sugar availability within the pine plantations.

Another way that parasitism data can be used to get circumstantial evidence of sugar limitation is to show that: (1) parasitism is lower in the absence of nectar, and (2) nectar is used by parasitoids when it is present in the field. Chumakova (1960) used this approach to argue that lifespan and fecundity of the San Jose scale parasitoid *Aphytis proclia* was limited by a lack of sugar in apple orchards in southern Russia. Chumakova showed that parasitism was 15 times higher in parts of on orchard undersown with *Phacelia tanacetifolia*, and she supplemented this finding with observations of *A. proclia* (and other parasitoids) feeding on *P. tanacetifolia* nectar in the field. She also used laboratory studies to show that *P. tanacetifolia* nectar could sustain *A. proclia* longevity and fecundity for 1–2 weeks whereas starved *A. proclia* (or females allowed to host-feed only) lived at most 5 days and produced very few eggs. Although this study remains one of the most complete of its kind, it suffers from a lack of replication and from the fact that host densities were noticeably higher in the portion of the orchard planted with *P. tanacetifolia* (Chumakova 1960). Thus, increased parasitism levels in the nectar-supplemented plot could have in principle been due, at least in part, to positively density-dependent parasitism (see Fig. 9.1B). Having said this, there is little evidence for positive density dependence in *Aphytis* spp.

(Smith and Maelzer 1986; Reeve and Murdoch 1985; Murdoch *et al.* 1984, 1995; but see Huffaker *et al.* 1986), including *A. proclia* (Pedata *et al.* 1995).

A more direct way of determining whether parasitoids are sugar-limited under field conditions is to use analyses of gut contents and ask whether sugar is present in the gut or not. An absence of gut sugars can be taken as evidence of sugar limitation for the individual that was sampled although there is no way of knowing whether that individual would have remained sugar-limited or, if so, whether it would have died of starvation. An important caveat here is that, as mentioned above, a minority of parasitoid species may not require sugar meals to reach maximum longevity or fecundity.

Methods of detecting gut sugars in parasitoids and other insects were reviewed by Jervis *et al.* (1992b) and include cold anthrone tests and various forms of chromatography. An important feature of all of these tests is that they can distinguish between fructose and other sugars. Fructose is not present in insect hemolymph and can therefore be used as a marker for gut sugars in whole-insect preparations (Van Handel 1984; Olson *et al.* 2000). Alternatively, insect guts can be dissected out and analyzed directly (Lewis and Domoney 1966). Both anthrone and chromatography tests have been used on field-caught individuals of a number of species of biting flies (e.g., Lewis and Domoney 1966; Van Handel 1972, Yuval and Schlein 1986, Burkett *et al.* 1999, Hunter and Ossowski 1999; Heimpel *et al.* 2004), but the application of these methodologies to parasitoids is in its infancy. The first parasitoid subjected to anthrone tests was the braconid wasp *Macrocentrus grandii*; in this species, the cold anthrone test was used to determine whether an individual had fed upon fructose-containing sugars (Olson *et al.* 2000; Fadamiro and Heimpel 2001). It is also possible to determine whether individual parasitoids are close to starvation or relatively well fed by quantifying levels of hemolymph sugars (presumably mainly trehalose) and glycogen with hot anthrone tests (Olson *et al.* 2000; Fadamiro and Heimpel 2001). Chromatographic methodologies are just now being developed to detect gut and other sugar levels in parasitoids (Wäckers and Steppuhn 2003; Heimpel *et al.* 2004).

Analyses of gut sugars in field-caught parasitoids using anthrone and chromatography methods are beginning to reveal great differences in the proportion of sugar-starved parasitoids in the field. While a small fraction of *M. grandii*, *Aphytis aonidiae*, and *Trichogramma ostriniae* collected from corn fields interplanted with flowering buckwheat tested positive for gut sugars (not more than 20%), much higher levels of gut sugars were found for a number of parasitoids attacking cabbage pests, both in the vicinity of flower plantings and in areas that had no obvious sugar sources (Lee and Heimpel 2003; Wäckers and Steppuhn 2003; Heimpel *et al.* 2004). The results of gut sugar analyses from these and other field-caught parasitoids are summarized in Table 9.1.

Table 9.1 *Results of studies in which field-caught parasitoids were subjected to analyses for gut sugars*

Parasitoid species	Habitat	Fraction of individuals testing positive for gut sugars	Method of analysis[a]	Reference
Macrocentrus grandii	Corn w/ and w/o flowering buckwheat (*Fagopyrum esculentum* (Polygonaceae))	0–0.2	Cold anthrone tests (quantitative)	G. E. Heimpel, unpublished data
Cotesia glomerata	Cabbage w/ and w/o flowering buckwheat	0.7–0.8	Cold anthrone tests (quantitative)	J. C. Lee and G. E. Heimpel, unpublished data
C. rubecula	Cabbage w/ and w/o flowering buckwheat	0.7–1	Cold anthrone tests (quantitative), HPLC	J. C. Lee and G. E. Heimpel, unpublished data; Wäckers and Steppuhn 2003
Diadegma insulare	Cabbage w/ and w/o flowering buckwheat	0.5–0.9	Cold anthrone tests (quantitative), HPLC	J. C. Lee and G. E. Heimpel, unpublished data; Heimpel *et al.* 2004
Pteromalus puparum	Cabbage w/ and w/o flowering buckwheat	0.3–0.6	Cold anthrone tests (quantitative)	J. C. Lee and G. E. Heimpel, unpublished data
Aphytis aonidiae	Almonds w/ and w/o flowering dill	0.04	Cold anthrone tests (qualitative)	G. E. Heimpel and J. A. Rosenheim, unpublished data
Aphelinus albipodus	Soybean field	0.18	Cold anthrone tests (qualitative)	Heimpel *et al.* 2004
Trichogramma ostriniae	Corn w/ and w/o buckwheat	0.14	Cold anthrone tests (qualitative)	G. E. Heimpel, unpublished data
Anagrus erythroneurae	Vineyard	0.30–0.85	Cold anthrone tests (qualitative)	J. A. Rosenheim and J. Hodgen, unpublished data

[a] In qualitative cold anthrone tests, sugar levels are determined using spectrophotometry (Olson *et al.* 2000), and in qualitative cold anthrone tests, anthrone-mediated color changes associated with sugar feeding are scored visually as presence/absence (Van Handel 1972). HPLC, high-performance liquid chromatography.

Feeding on floral nectar in the field

Many of the approaches outlined above can simultaneously tell us whether parasitoids are likely to be sugar-limited in the field and whether they are able to utilize nectar from flowers that are sown for the purposes of feeding them. To a first approximation, a demonstration of increased parasitism and/or parasitoid abundance in the presence of a nectar source is consistent both with sugar limitation (in the absence of nectar) and that parasitoids utilize the nectar. However, as was discussed above, caveats abound, and unambiguously establishing nectar-feeding in the field can be a formidable task.

What is the evidence for nectar-feeding in the field? We have already mentioned cold anthrone tests, but these cannot distinguish between sugars from nectar (floral or extrafloral), honeydew, or other rarer sources such as plant exudates, fruit juices, or sugar sprays. While chromatography methods (GLC, gas–liquid chromatography and HPLC, high-performance liquid chromatography) can tell us much more about the specific kinds of sugar present in the gut, nectars rarely have "signature" sugars (Table 9.2). Instead, they tend to have varying levels of sucrose, fructose, and glucose (Percival 1961; Van Handel *et al.* 1972; Harborne 1988), sugars that are also present in the other sources just mentioned. In specific cases where the habitat from which a parasitoid is captured is well defined so that it is known that there are no honeydew-producing Homoptera or other sugar sources present, it may be possible to rule out all sugar sources but nectar. However, even if honeydew is present, it typically contains signature sugars (most notably erlose and melezitose: Zoebelein 1956, Burkett *et al.* 1999), so it may be possible to infer nectar-feeding by their absence (Table 9.2).

The most common way to ascertain nectar-feeding in the field has been direct observation of parasitoids (Jervis and Kidd 1996). Jervis *et al.* (1993) documented over 1000 cases of sugar feeding from 250 parasitoid species in a diverse array of habitats including arable fields and forests, mainly in Wales (UK). In this study, care was taken to confirm nectar-feeding as opposed to pollen-feeding or resting. Similarly, Hassan (1967) documented feeding on inflorescences (nectar and/or pollen) from 148 parasitoid species from various natural habitats in Germany. Bugg (1987) observed various species of unidentified parasitoids feeding on inflorescences of the soapbark tree, *Quillaja saponaria*, in California (USA) and noted that nectar was fed upon to a greater extent than pollen. Other observations of nectar-feeding in particular systems have been published (e.g., Allen and Smith 1958; Leeper 1974). A number of studies report flower visitation by parasitoids (Györfi 1951; Bugg and Wilson 1989; Maingay *et al.* 1991; Al-Doghairi and Cranshaw 1999; Tooker and Hanks 2000). In these reports, true nectar-feeding is not separated from pollen-feeding or resting, but it is likely that some if not most of these parasitoids were nectar-feeding. In observations of 42 individuals of 18 parasitoid species visiting flowers in the Canadian arctic, Kevan (1973) reported that 17 were nectar-feeding, 12 were resting, and 13 were performing other activities (either "unknown" or "running at base of corolla"). No pollen-feeding was noted. There are also numerous studies in which parasitoid abundance and/or diversity is highest in the presence of flowering plants (e.g., Van Emden 1962; Helenius 1990; Stephens *et al.* 1998; Gurr *et al.* 1999; Baggen *et al.* 2000; Irvin *et al.* 2000; Braman *et al.* 2002; but see Bowie *et al.* 1999; Berndt *et al.* 2002). These sorts of results are consistent with, but not indicative of, nectar-feeding.

A particularly elegant way to demonstrate feeding from particular plants by parasitoids involves experimentally marking the plants and subsequently identifying the marker substance in adult parasitoids. To our knowledge, this has only been attempted once. Freeman-Long *et al.* (1998) used rubidium (Rb), a trace element, to mark various flowering plants near agricultural fields at three sites in central California (USA). They then captured insects within these fields and analyzed them for the mark. Among the marked insects were individuals of three parasitoid species (*Hyposoter* sp., *Trichogramma* sp., and *Macrocentrus* sp.), and they were captured at distances of up to 80 m from the marked nectar source. Although this study is consistent with feeding in association with the marked plants, nectar can still not be singled out as the only possible food. Presumably Rb would mark pollen as well as nectar in this kind of study, and is also passed on to herbivores feeding on the Rb-laced plants as well as parasitoids feeding on these marked herbivores (Payne and Wood 1984; Jackson *et al.* 1988; Hopper 1991; Hopper and Woolson 1991; Jackson 1991; Corbett and Rosenheim 1996). The mark can therefore be obtained by pollen-feeding or host-feeding. Nectar appears to be a more universal food for parasitoids than either pollen or hosts, however (Gilbert and Jervis 1998; Jervis 1998) so that the most likely source of the Rb mark in Freeman-Long *et al.*'s (1998) study was nectar.

Enhanced fecundity in the presence of nectar

Increased parasitoid fecundity in the presence of nectar is listed as a requirement for validation of the parasitoid nectar provision hypothesis. Care needs to be taken though, in considering how differences in fecundity may interact with parasitoid movement. If parasitoids are attracted to flower-rich areas from flower-poor areas, and aggregation results, then increased rates of parasitism can be observed in the vicinity of flowers even without an effect of nectar on fecundity. If there is no effect on fecundity, biological control will only be improved if parasitoids are attracted from outside the managed habitat (all else being equal). In this case, therefore, the parasitoid nectar provision hypothesis can be valid even without an increase in parasitoid fecundity in the presence of nectar.

Table 9.2 *Sugars found in honeydew and nectar*

Sugar[a]	Honeydew	Extrafloral nectar	Floral nectar
Monosaccharides			
Fructose	+	+	+
Galactose	+	+	+
Glucose	+	+	+
Mannose	-	-	+ (traces)
Rhamnose	-	+	-
Xylose	+	-	-
Disaccharides			
Laminaribiosose	+ (coccids)		
Maltose	+		+
Melibiose	-	-	+
Sucrose	+		+
Trehalose	+		-
Trehalulose	+		-
Turanose	+		-
Trisaccharides			
Bemisiose	+ (aleyrodids)		-
Erlose	+	-	-
Fructomaltose	+ (coccids)		
Melezitose	+	+ (rare)	+ (rare)
Raffinose	+	-	+ (rare)

Tetrasaccharides			
Bemisiotetrose	+ (aleyrodids)		-
Eriose	+ (coccids)		
Maltosucrose	+ (coccids)		
Stachyose	+	-	+ (Orchidaceae only)
Pentasaccharides			
Cryptose	+ (coccids)		
Diglucomelezitose	+ (aleyrodids)		-
Maltotriosucrose	+ (coccids)		
Hexasaccharides			
Lasiose	+ (coccids)		
Polysaccharides			
Dextrins	+ (Flatidae)		

[a] Erlose, melezitose, trehalulose, and (in some situations) xylose can be used as signature sugars for honeydew, although caution should be exercised in the case of melezitose, as it occurs in a few floral nectars (see Baker and Baker 1983). Stachyose is apparently restricted to Orchidaceae (Baker and Baker 1983), so can be used in situations where such plants are known to be absent from the habitat. Bemisiose, bemisiotetrose, and diglucomelezitose are apparently taxonomically restricted honeydew sugars that can also be used as signatures where the hosts (whiteflies) are known to be present in the habitat. Turanose and trehalose are probably not useful honeydew indicators: the former has been found in the guts of insects fed on pure sucrose (Janzen and Hunter 1998), while the latter is a major constituent of insect hemolymph (i.e., it is a likely contaminant in the gut contents of dissected insects). Rhamnose can be used as a signature sugar for extrafloral nectar. Melibiose can be used as a signature sugar of floral nectar.

Sources: Honeydews: Gray and Fraenkel 1954; Wolf and Ewart 1955; Ewart and Metcalf 1956; Bacon and Dickinson 1957; Auclair 1964; Maurizio 1975; Bellardio *et al.* 1978; Liebig 1979; Lombard *et al.* 1979 1987; Manino *et al.* 1985; Moller and Tilley 1989; Byrne and Miller 1990; Nemec and Stary 1990; Koptur 1992; Hendrix and Wei 1992, 1994; Davidson *et al.* 1994; Wei *et al.* 1996, 1997; Wilkinson *et al.* 1997; Costa *et al.* 1999; Fiori *et al.* 2000; Henneberry *et al.* 2000; Wäckers and Steppuhn 2003; Heimpel *et al.* 2004; nectars: Wykes 1952; Percival 1961; Van Handel *et al.* 1972; Baker and Baker 1983.

Laboratory demonstrations of increased fecundity in the presence of nectar are numerous (reviewed by Jervis *et al.* 1996) and form the basis of the expectation that incorporation of nectar sources in managed habitats can increase parasitism of pests. Estimates of parasitoid fecundity in the field are rare, however, under any conditions (Driessen and Hemerik 1992; Heimpel *et al.* 1998; Ellers *et al.* 1998; Casas *et al.* 2000) and as far as we know, there have been no comparisons of field fecundity of parasitoids in habitats with and without flowers. This is clearly an area worthy of further study.

Higher parasitism associated with nectar sources is consistent with increased parasitoid fecundity in the presence of nectar. If nectar-feeding itself can be demonstrated, and there are no complications due to direct effects on pest numbers (Fig. 9.1) or parasitoid aggregation, the circumstantial evidence for increased fecundity under the parasitoid nectar provision hypothesis is strong if parasitism is enhanced. Even if the differences in these parameters between habitats with and without nectar are due in part to attraction or retention near nectar, it is likely that increases in fecundity play a role in observed differences. Separating the relative roles of parasitoid dispersal and fecundity in contributing to effects of nectar (and other components of plant diversity) on parasitism levels remains an important and unexplored area of parasitoid ecology.

Increased parasitism levels in the presence of nectar

We have noted throughout this chapter that increased levels of parasitism of pest insects have been documented in a number of systems, and that this may or may not be directly attributable to nectar-feeding by parasitoids. Indeed, although we suspect that nectar-feeding is the mechanism driving these results in many, if not most cases, we know of no single study that has definitively shown that nectar-feeding by parasitoids has led to increased levels of pest parasitism in managed systems. In Table 9.3, we list cases in which parasitism levels were compared in nectar-enhanced and control settings, along with a description of the information needed to confirm the role of nectar-feeding in affecting parasitism levels. Increases in parasitism were reported in 7 of 20 cases, but a concomitant decrease in pest density was only documented in one of these cases (Gallego *et al.* 1983). In this case, however, no causal link of the nectar-feeding per se has been made to the population-level patterns. In some as-yet unpublished studies, sugar feeding is documented, but does not translate into increased parasitism or decreased pest densities (Table 9.3).

Decreased pest levels in the presence of nectar

Increased parasitism is expected to lead to decreased pest densities in the current or subsequent generation (depending upon the pest life stage from

Table 9.3 *Parasitoids for which field parasitism rates with and without floral nectar sources are available*

Parasitoid	Host	Flower	Habitat	Increased parasitism?	Decreased pest density?	Parasitoid sugar-limited in the field?	Parasitoid feeds on nectar in the field?	Increased parasitoid fecundity in the field?	Reference
Lixophaga sphenophori (Diptera: Tachinidae)	New Guinea sugarcane weevil (*Rhabdoscelus obscurus* (Coleoptera: Curculionidae))	*Euphorbia* spp. (Euphorbiaceae)	Sugar cane fields	Yes	?	?	Yes	?	Topham and Beardsley 1975
Vorria ruralis (Diptera: Tachinidae)	Cabbage looper (*Rhabdoscelus obscurus* (Coleoptera: Curculionidae))	Buckwheat (*Fagopyrum esculentum* (Polygonaceae))	Cabbage/ soybean field	Variable	No	?	?	?	J. C. Lee and G. E. Heimpel, unpublished data
Diadegma insulare (Hymenoptera: Ichneumonidae)	Diamond-back moth (*Plutella xylostella* (Lepidoptera: Plutellidae))	Buckwheat	Cabbage/ soybean field	Variable	No	No	Probable	?	Zhao *et al.* 1992; J. C. Lee and G. E. Heimpel, unpublished data

Table 9.3 *(cont.)*

Parasitoid	Host	Flower	Habitat	Increased parasitism?	Decreased pest density?	Parasitoid sugar-limited in the field?	Parasitoid feeds on nectar in the field?	Increased parasitoid fecundity in the field?	Reference
Cotesia glomerata and *C. rubecula* (Hymenoptera: Braconidae)	Cabbage white (*Pieris rapae* (Lepidoptera: Pieridae))	Buckwheat	Cabbage/ soybean field	No	No	No	Probable	?	Zhao *et al.* 1992; J. C. Lee and G. E. Heimpel, unpublished data
Dolichogenidea tasmanica (Hymenoptera: Braconidae)	Light brown apple moth (*Epiphyas postvittana* (Lepidoptera: Tortricidae))	Buckwheat	Various orchards	Variable	?	Yes	?	?	Stephens *et al.* 1998; Irvin *et al.* 2000
D. tasmanica, *Glyptapanteles demeter*, *Meteorus pulchicoris* (Hymenoptera: Braconidae)	Unidentified leafminers (Lepidoptera: Tortricidae))	Buckwheat	Vineyard	No	?	?	?	?	Berndt *et al.* 2002

Macrocentrus grandii (Hymenoptera: Braconidae)	European corn borer (*Ostrinia nubilalis* (Lepidoptera: Crambidae))	Buckwheat	Corn field	No	No	Yes	No	?	G. E. Heimpel, unpublished data
Pteromalus puparum (Hymenoptera: Pteromalidae)	Cabbage white	Buckwheat	Cabbage/ soybean field	No	No	No	Probable	?	J. C. Lee and G. E. Heimpel, unpublished data
Chrysocharis prodice, *Cirrospilus vittatus* (Hymenoptera: Eulophidae)	Banded apple pygmy moth (*Stigmella malella* (Lepidoptera: Nepticulidae))	Various (mainly Apiaceae)	Apple orchards	No	No	?	?	?	Gruys 1982
Colpoclypeus florus (Hymenoptera: Eulophidae)	Leafroller complex (Lepidoptera: Tortricidae)	Various (mainly Apiaceae)	Apple orchards	No	No	?	?	?	Gruys 1982

Table 9.3 *(cont.)*

Parasitoid	Host	Flower	Habitat	Increased parasitism?	Decreased pest density?	Parasitoid sugar-limited in the field?	Parasitoid feeds on nectar in the field?	Increased parasitoid fecundity in the field?	Reference
Copidosoma koehleri (Hymenoptera: Encyrtidae)	Potato tuber moth (*Phthorimae operculella* (Lepidoptera: Gelechiidae))	Borage (*Borago officinalis* (Boraginaceae))	Potato plots	Yes	No	?	?	?	Baggen and Gurr 1998; Gurr *et al.* 1999
Aphelinus mali (Hymenoptera: Aphelinidae)	Wooly apple aphid (*Eriosoma lanigerum* (Homoptera: Aphididae))	*Phacelia* (Hydrophyllaceae), *Eryngium* spp. (Apiaceae)	Apple orchard	Yes	?	?	?	?	Telenga 1958, cited by Powell 1986
Aphytis aonidiae (Hymenoptera: Aphelinidae)	San Jose scale (*Quadraspitiotus perniciosus* (Homoptera: Diaspididae))	Buckwheat, dill	Almond orchard	No	No	Yes	No	Presume no	G. E. Heimpel, unpublished data
A. proclia	San Jose scale	*Phacelia tanacitifolia* (Hydrophyllaceae)	Apple orchard	Yes	?	Probable	Probable	?	Chumakova 1960

Trichogramma ostriniae (Hymenoptera: Trichogrammatidae)	European corn borer	Buckwheat	Corn field	No	?	Yes	No	Presume no	G. E. Heimpel, unpublished data
Anogrus spp. (Hymenoptera: Mymaridae)	Grape leafhoppers (*Erythronaera* spp.) (Homopetra: Cicadellidae)	Various	Vineyards	Variable	?	?	?	?	Nicholls *et al.* 2000; English-Loeb *et al.* 2003
Unidentified aphid parasitoids	Bird cherry-oat aphid (*Rhopalosiphum padi* (Homoptera: Aphididae))	Faba bean (*Vicia faba* (Fabaceae))	Wheat field	Yes	No	?	?	?	Helenius 1990
Various	Eastern tent caterpillar (*Malacosoma americanum* (Lepidoptera: Lasciocampidae), codling moth (*Cydia pomonella* (Lepidoptera: Tortricidae))	Various	Apple orchards	Yes	?	?	?	?	Leius 1967

Table 9.3 *(cont.)*

Parasitoid	Host	Flower	Habitat	Increased parasitism?	Decreased pest density?	Parasitoid sugar-limited in the field?	Parasitoid feeds on nectar in the field?	Increased parasitoid fecundity in the field?	Reference
Various	Spruce bud worm (*Choristoneura fumiferana* (Lepidoptera: Tortricidae))	Various	Forest	No	?	Yes	?	?	Cappuccino *et al.* 1999
Various	Spruce bud worm (*Choristoneura fumiferana* (Lepidoptera: Tortricidae))	Various	Forest	No	?	Yes	?	?	Cappuccino *et al.* 1999
Various	Coconut leafminer (*Promecotheca cumingii* (Coleoptera: Hispidae))	*Callopogonium* sp., *Centrosema* sp. (Fabaceae)	Palm plantations	Yes	Yes	?	?	?	Gallego *et al.* 1983

Studies of intercropping or cover cropping that were not specifically aimed at evaluating floral nectar sources are not included. Studies that report only parasitoid abundance with and without nectar sources are not included either, because it is often not clear whether these parasitoids are attacking pests in the managed area. Where available, information on pest densities and parasitoid sugar use and fecundity are provided. Not all entries in the table reflect the results of properly replicated or statistically analyzed data.

which the parasitoid emerges). However, at least three circumstances can lead to the unexpected scenario in which increased rates of parasitism do not lead to lower pest densities in habitats with augmented nectar sources. One of these has been mentioned above: if herbivore densities are higher in association with nectar sources for reasons not related to parasitoid attack, then increased parasitism levels near nectar may not bring herbivore densities below the levels found in the absence of nectar. Only strong positive density-dependent parasitism can lead to increased pest suppression in this case (Fig. 9.1A). Second, increased levels of parasitism that are inflicted on a pest developmental stage preceding one that is subjected to strong density-dependent mortality may be completely decoupled from herbivore suppression (May *et al.* 1981). For example, Van Hamburg and Hassell (1984) showed that the density-dependent mortality acting after egg mortality can lead to no improvement in control or even the raising of pest larval density above the level achieved under conditions under which egg mortality is increased. Thus, it may be that observing increased suppression of a pest's feeding stage may be more difficult with parasitoids attacking and emerging from eggs than parasitoids attacking larvae. The third reason complements the second reason by stressing the effect of mortality late in the host life cycle. If the parasitoid allows normal or increased levels of host growth post-parasitism (koinobiosis: Godfray 1994), or if hosts are attacked in a post-feeding stage, increased parasitism in association with floral nectar may have no effect on short-term suppression of pest damage. This scenario is especially relevant for annual crops with univoltine pests in which the pest colonizes fields from a stable, external habitat every season. For these reasons, lower pest densities cannot be taken for granted when parasitism rates are higher, although this is certainly expected.

Even when increased parasitism in nectar-rich areas clearly leads to lower pest densities, parasitoid movement from nectar-poor to nectar-rich areas can amplify differences between the two habitats, possibly leading to spurious results. This is a potentially serious problem given our relatively poor understanding of parasitoid dispersal (Corbett and Plant 1993; Hirose 1998; Hastings 2000; Roland 2000) and the logistical limitations on performing field experiments correctly in the face of highly mobile parasitoids (Van Emden 1990; Corbett 1998; Hirose 1998). Ideally, experimental plots should be distant enough from one another so that dispersal between plots of naturally occurring parasitoids is minimal (this is needed not only to keep parasitoids from moving from control plots to nectar-rich plots, but to prevent the opposite movement as well). In a practical setting, attraction of parasitoids to nectar-rich habitats may very well be desirable since these parasitoids may be attracted from non-crop areas, or even if they are initially attracted from other cropping areas, they may return to those areas.

However, parasitoid movement can interfere with our ability to interpret correctly the outcome of field experiments.

Conclusions

We have come a long way since the landmark reviews of Powell (1986) and Van Emden (1990). Innovations have included increased understanding of the roles of flower structure, and various aspects of parasitoid biology in mediating interactions between floral nectar and parasitoids. Advances have been made in the field as well, and recent studies in vineyards and orchards of Australia and New Zealand have allowed new insights (Baggen and Gurr 1998; Gurr *et al.* 1999; Irvin *et al.* 2000; Berndt *et al.* 2002; Wratten *et al.* 2003). Annual field buckwheat, *Fagopyrum esculentum*, is emerging as a sort of model plant in some of these and other studies (see Table 9.3), and although this will facilitate some cross-study comparison, other flowers need to be assessed (Baggen and Gurr 1998; Baggen *et al.* 1999). Another positive trend is the use of biochemical methodologies previously used only in studies on biting flies to analyze the dynamics of carbohydrate allocation in laboratory studies (Olson *et al.* 2000; Fadamiro and Heimpel 2001) (Table 9.2) and, more importantly, to characterize parasitoid sugar feeding in the field (Table 9.1).

Future directions in applied studies of parasitoid nectar-feeding will have to be aimed at improving our ability to unambiguously evaluate the parasitoid nectar provision hypothesis. Critical issues include interactions between dispersal and nectar-feeding, density dependent parasitism in the context of nectar-enhanced and nectar-poor habitats, responses of pests to nectar sources in the field, parasitoid fecundity in the field, and practical considerations such as logistic and economic feasibility of incorporating flowering cover crops into agricultural or other systems (see also Gurr *et al.* 2000; Wratten *et al.* 2003; Gurr *et al.*, Chapter 11). We have tried to stress in this review that the nectar provision hypothesis contains a great deal of complexity despite its intuitive appeal. It is surely this complexity that has impeded a clear demonstration of the hypothesis up to this point.

Acknowledgments

We thank Jillian Hodgen, Jana Lee, Jay Rosenheim, Felix Wäckers, and Livy Williams for sharing unpublished data, Geoff Gurr, Steve Wratten, Felix Wäckers, and Livy Williams for sending reprints, Al Venables and Richard Dickinson for advice on sugar biochemistry, and Carol Boggs, Henry Fadamiro, Jana Lee, Roy van Driesche, Paul van Rijn, and Felix Wäckers for comments on the manuscript.

References

Al-Doghairi, M. A. and W. S. Cranshaw. 1999. Surveys on visitation of flowering landscape plants by common biological control agents in Colorado. *Journal of the Kansas Entomological Society* **72**: 190–196.

Allen, W. W. and R. F. Smith. 1958. Some factors influencing the efficiency of *Apanteles medicaginis* Muesebeck (Hymenoptera: Braconidae) as a parasite of the alfalfa caterpiller, *Colias philodice eurytheme* Boisduval. *Hilgardia* **28**: 1–41.

Altieri, M. A. and W. H. Whitcomb. 1979. The potential use of weeds in the manipulation of beneficial insects. *HortScience* **14**: 12–18.

Andow, D. A. 1988. Management of weeds for insect manipulation in agroecosystems. In M. A. Altieri and M. Liebman (eds.) *Weed Management in Agroecosystems: Ecological Approaches.* Baton Rouge, FL: CRC Press, pp. 265–302.

1991. Vegetational diversity and arthropod population response. *Annual Review of Entomology* **36**: 561–586.

Andow, D. A. and S. J. Risch. 1987. Parasitism in diversified agroecosystms: phenology of *Trichogramma minutum* (Hymenoptera: Trichogrammatidae). *Entomophaga* **32**: 255–260.

Atsatt, P. R. and D. J. O'Dowd. 1976. Plant defense guilds. *Science* **193**: 24–29.

Auclair, J. L. 1964. Aphid feeding and nutrition. *Annual Review of Entomology* **8**: 439–490.

Bacon, J. S. D. and B. Dickinson. 1957. The origin of melezitose: a biochemical relationship between the lime tree (*Tilia* spp.) and an aphid (*Eucallipterus tiliae* L.). *Biochemical Journal* **66**: 289–297.

Baggen, L. R. and G. M. Gurr. 1998. The influence of food on *Copidosoma koehleri* (Hymenoptera: Encyrtidae), and the use of flowering plants as a habitat management tool to enhance biological control of potato moth, *Phthorimaea operculella* (Lepidoptera: Gelechiidae). *Biological Control* **11**: 9–17.

Baggen, L. R., G. M. Gurr, and A. Meats. 1999. Flowers in tri-trophic systems: mechanisms allowing selective exploitation by insect natural enemies for conservation biological control. *Entomologia Experimentalis et Applicata* **91**: 155–161.

2000. Field observations on selective food plants in habitat manipulation for biological control of potato moth by *Copidosoma koehleri* Blanchard (Hymenoptera: Encyrtidae). In A. D. Austin and M. Dowton (eds.) *Hymenoptera: Evolution, Biodiversity and Biological Control.* Collingwood, Australia: CSIRO, pp. 388–395.

Baker, H. G. and I. Baker. 1983. Floral nectar sugar constituents in relation to pollinator type. In C. E. Jones and R. J. Little (eds.). *Handbook of Experimental Pollination Biology.* New York: Van Nostrand Reinhold, pp. 117–141.

Beach, J. P., L. Williams III, D. L. Hendrix, and L. D. Price. 2003. Different food sources affect the gustatory response of *Anaphes iole*, an egg parasitoid of *Lygus* spp. *Journal of Chemical Ecology* **29**: 1203–1222.

Bellardio, F., A. Lombard, A. Patetta, and C. Vidano. 1978 Indagini sui carboidrati presenti in foglie di *Tilia cordata*, melata di *Eucallipterus tiliae* e relativo miele. *Apicoltura moderna* **69**: 5–12.

Berndt, L. A., S. D. Wratten, and P. G. Hassan. 2002. Effects of buckwheat flower on leafroller (Lepidoptera: Tortricidae) parasitoids in a New Zealand vineyard. *Agricultural and Forest Entomology* **4**: 39–45.

Bowie, M. H., S. D. Wratten, and A. J. White. 1995. Agronomy and phenology of 'companions plants' of potential for enhancement of insect biological control. *New Zealand Journal of Crop and Horticultural Science* **23**: 423–427.

Bowie, M. H., G. M. Gurr, Z. Hossain, L. R. Baggen, and C. M. Frampton. 1999. Effects of distance from field edge on aphidophagous insects in a wheat crop and observations on trap design and placement. *International Journal of Pest Management* **45**: 69–73.

Box, H. E. 1927. The introduction of braconid parasites of *Diatraea saccharalis* Fabr., into certain of the West Indian islands. *Bulletin of Entomological Research* **18**: 365–371.

Braman, S. K., A. F. Pendley, and W. Corley. 2002. Influence of commercially available wildflower mixes on beneficial arthropod abundance and predation in turfgrass. *Environmental Entomology* **31**: 564–572.

Bugg, R. L. 1987. Observations on insects associated with a nectar-bearing Chilean tree, *Quillaja saponaria* Molina (Rosaceae). *Pan-Pacific Entomologist* **63**: 60–64.

Bugg, R. L. and C. H. Pickett. 1998. Introduction: enhancing biological control – habitat management to promote natural enemies of agricultural pests. In C. H. Pickett and R. L. Bugg (eds.) *Enhancing Biological Control*. Berkeley, CA: University of California Press, pp. 1–23.

Bugg, R. L. and C. Waddington. 1994. Using cover crops to manage arthropod pests of orchards: a review. *Agriculture, Ecosystems and Environment* **50**: 11–28.

Bugg, R. L. and L. T. Wilson. 1989. *Ammi visnaga* (L.) Lamarck (Apiaceae): associated beneficial insects and implications for biological control, with emphasis on the bell-pepper agroecosystem. *Biological Agriculture and Horticulture* **6**: 241–268.

Bugg, R. L., L. E. Ehler, and L. T. Wilson. 1987. Effect of common knotweed (*Polygonum aviculare*) on abundance and efficiency of insect predators of crop pests. *Hilgardia* **55**: 1–53.

Burkett, D. A., D. L. Kline, and D. A. Carlson. 1999. Sugar meal composition of five north central Florida mosquito species (Diptera: Culicidae) as determined by gas chromatagraphy. *Journal of Medical Entomology* **36**: 462–467.

Byrne, D. N. and W. B. Miller. 1990. Carbohyrate and amino acid composition of phloem sap and honeydew produced by *Bemisia tabaci*. *Journal of Insect Physiology* **36**: 433–439.

Cañas, L. A. and R. J. O'Neil. 1998. Applications of sugar solutions to maize, and the impact of natural enemies on fall armyworm. *International Journal of Pest Management* **44**: 59–64.

Cappuccino, N., M. J. Houle, and J. Stein. 1999. The influence of understory nectar sources on parasitism of the spruce budworm *Choristoneura fumiferana* in the field. *Agricultural and Forest Entomology* **1**: 33–36.

Casas, J., R. M. Nisbet, S. Swarbrick, and W. W. Murdoch. 2000. Eggload dynamics and oviposition rate in a wild population of a parasitic wasp. *Journal of Animal Ecology* **69**: 185–193.

Casas, J., C. Driessen, N. Mandon, *et al.* 2003. Energy dynamics of a parasitoid foraging in the wild. *Journal of Animal Ecology* **72**: 691–697.

Chaney, W. E. 1998. Biological control of aphids in lettuce using in-field insectaries. In C. H. Pickett and R. L. Bugg (eds.) *Enhancing Biological Control*. Berkeley, CA: University of California Press, pp. 73–84.

Chumakova, B. M. 1960. Additional nourishment as a factor increasing activity of insect-pests parasites. *Proceedings of All Union Scientific Research Institute of Plant Protection* **15**: 57–70. (Translated from Russian.)

Coll, M. 1998. Parasitoid activity and plant species composition in intercropped systems. In C. H. Pickett and R. L. Bugg (eds.) *Enhancing Biological Control*. Berkeley, CA: University of California Press, pp. 85–120.

Corbett, A. 1998. The importance of movement in the response of natural enemies to habitat manipulation. In C. H. Pickett and R. L. Bugg (eds.) *Enhancing Biological Control*. Berkeley, CA: University of California Press, pp. 25–48.

Corbett, A. and R. E. Plant. 1993. Role of movement in the response of natural enemies to agroecosystem diversification: a theoretical evaluation. *Environmental Entomology* **22**: 519–531.

Corbett, A. and J. A. Rosenheim. 1996. Impact of a natural enemy overwintering refuge and its interactions with the surrounding landscape. *Ecological Entomology* **21**: 155–164.

Costa, H. S., N. C. Toscano, D. L. Hendrix, and T. J. Henneberry. 1999. Patterns of honeydew droplet production by nymphal stages of *Bemisia argentifolii* (Homoptera: Aleyrodidae) and relative composition of honeydew sugars. *Journal of Entomological Science* **34**: 305–313.

Davidson, E. W., B. J. Segura, T. Steele, and D. L. Hendrix. 1994. Microorganisms influence the composition of honeydew produced by the silverleaf whitefly, *Bemisia argentifolii*. *Journal of Insect Physiology* **40**: 1069–1076.

Dowell, R. V. 1978. Ovary structure and reproductive biologies of larval parasitoids of the alfalfa weevil (Coleoptera: Cucurulionidae). *Canadian Entomologist* **109**: 641–648.

Driessen, G. and L. Hemerik. 1992. The time and egg budget of *Leptopilina clavipes*, a parasitoid of larval *Drosophila*. *Ecological Entomology* **17**: 17–27.

Dyer, L. E. and D. A. Landis. 1996. Effects of habitat, temperature, and sugar availability on longevity of *Eriborus terebrans* (Hymenoptera: Ichneumonidae). *Environmental Entomology* **25**: 192–201.

1997. Influence of noncrop habitats on the distribution of *Eriborus terebrans* (Hymenoptera: Ichneumonidae) in cornfields. *Environmental Entomology* **26**: 924–932.

Ellers, J., J. J. M. van Alphen, and J. G. Sevenster. 1998. A field study of size–fitness relationships in the parasitoid *Asobara tabida*. *Journal of Animal Ecology* **67**: 318–324.

English-Loeb, C., M. Rhainds, T. Martinson, and T. Ugine. 2003. Influence of flowering cover crops on *Anagrus* parasitoids (Hymenopetera: Mymaridae) and *Erythronevra* leafhoppers (Homopetra: Cicadellidae) in New York vineyards. *Agricultural and Forest Entomology* **5**: 173–181.

Evans, E. W. 1993. Indirect interactions among phytophagous insects: aphids, honeydew and natural enemies. In A. D. Watt, S. Leather, N. J. Mills, and K. F. A. Walters (eds.) *Individuals, Populations and Patterns in Ecology*. Andover, UK: Intercept Press, pp. 287–298.

Evans, E. W. and S. England. 1996. Indirect interactions in biological control of insects: pests and natural enemies in alfalfa. *Ecological Applications* **6**: 920–930.

Evans, E. W. and J. G. Swallow. 1993. Numerical responses of natural enemies to artificial honeydew in Utah alfalfa. *Environmental Entomology* **22**: 1392–1401.

Ewart, W. H. and R. L. Metcalf. 1956. Preliminary studies of sugars and amino acids in the honeydews of five species of coccids feeding on citrus in California. *Annals of the Entomological Society of America* **9**: 441–447.

Fadamiro, H. Y. and G. E. Heimpel. 2001. Effects of partial sugar deprivation on lifespan and carbohydrate mobilization in the parasitoid *Macrocentrus grandii* (Hymenoptera: Braconidae). *Annals of the Entomological Society* **94**: 909–916.

Fiori, J., G. Serra, P. Sabatini, *et al.* 2000. Analisi con HPLC di destrine in mieli di melata di *Metcalfa pruinosa*. *Industrie Alimentari* **34**: 463–466.

Freeman-Long, R., A. Corbett, C. Lamb, *et al.* 1998. Beneficial insects move from flowering plants to nearby crops. *California Agriculture* September–October: 23–26.

Froggatt, W. W. 1902. A natural enemy of the sugar-cane beetle in Queensland. *New South Wales Agricultural Gazette* **13**: 63.

Gallego, V. C., C. R. Baltazar, E. P. Cadapan, and R. G. Abad. 1983. Some ecological studies on the coconut leafminer, *Promecotheca cumingii* Baly (Coleoptera: Hispidae) and its hymenopterous parasitoids in the Philippines. *Philippine Entomologist* **6**: 471–494.

Gilbert, F. and M. A. Jervis. 1998. Functional, evolutionary and ecological aspects of feeding-related mouthpart specializations in parasitoid flies. *Biological Journal of the Linnean Society* **63**: 495–535.

Godfray, H. C. J. 1994. *Parasitoids: Behavioral and Evolutionary Ecology*. Princeton, NJ: Princeton University Press.

Gray, H. E. and G. Fraenkel. 1954. The carbohydrate components of honeydew. *Physiological Zoology* **27**: 56–65.

Gruys, P. 1982. Hits and misses: the ecological approach to pest control in orchards. *Entomologia Experimentalis et Applicata* **31**: 70–87.

Gurr, G. M., S. D. Wratten, and P. Barbosa. 2000. Success in conservation biological control of arthropods. In G. Gurr and S. D. Wratten (eds.) *Biological Control:*

Measures of Success. Dordrecht, the Netherlands: Kluwer Academic Publishers, pp. 105–132.

Gurr, G. M., S. D. Wratten, N. A. Irvin, *et al.* 1999. Habitat manipulation in Australasia: recent biological control progress and prospects for adoption. Proc. 6th Australasian Applied Entomological Research Conference, vol. 2, pp. 7225–7234.

Györfi, J. 1951. Die Schlupfwespen und der Unterwuchs des Waldes. *Zeitschrift für angewandte Entomologie* **33**: 32–47.

Hagan, K. S. 1986. Ecosystem analysis: plant cultivars (HPR), entomophagous species and food supplements. In D. J. Boethel and R. D. Eikenbary (eds.) *Interactions of Plant Resistance and Parasitoids and Predators of Insects*. Chichester, UK: Ellis Horwood, pp. 151–197.

Harborne, J. B. 1988. *Introduction to Ecological Biochemistry*. London: Academic Press.

Hardy, I. C. W., N. T. Griffiths and H. C. J. Godfray. 1992. Clutch size in a parasitoid wasp: a manipulation experiment. *Journal of Animal Ecology* **61**: 121–129.

Hassan, V. E. 1967. Untersuchung über die Bedeutung der Kraut- und Strauchshicht als Nahrungsquelle für Imagines entomophager Hymenoptera. *Zeitschrift für angewandte Entomologie* **60**: 238–265.

Hastings, A. 2000. Parasitoid spread: lessons for and from invasion biology. In M. A. Hochberg and A. R. Ives (eds.) *Parasitoid Population Biology*. Princeton, NJ: Princeton University Press, pp. 70–82.

Helenius, J. 1990. Incidence of specialist natural enemies of *Rhopalosiphum padi* (L.) (Hom., Aphididae) on oats in monocrops and mixed intercrops with faba bean. *Journal of Applied Entomology* **109**: 136–143.

Heimpel, G. E. and T. R. Collier. 1996. The evolution of host-feeding behaviour in insect parasitoids. *Biological Reviews* **71**: 373–400.

Heimpel, G. E., J. C. Lee, Z. Wu, *et al.* 2004. Gut sugar analysis in field-caught parasitoids: adapting methods used on biting flies. *International Journal of Pest Management* **50**: 193–198.

Heimpel, G. E., M. Mangel, and J. A. Rosenheim. 1998. Effects of time limitation and egg limitation on lifetime reproductive success of a parasitoid in the field. *American Naturalist* **152**: 273–289.

Heimpel, G. E., J. A. Rosenheim, and J. M. Adams. 1994. Behavioral ecology of host feeding in *Aphytis* parasitoids. *Norwegian Journal of Agricultural Sciences* (Suppl.) **16**: 101–115.

Heimpel, G. E., J. A. Rosenheim, and D. Kattari. 1997a. Adult feeding and lifetime reproductive success in the parasitoid *Aphytis melinus*. *Entomologia Experimentalis et Applicata* **83**: 305–315.

Heimpel, G. E., J. A. Rosenheim, and M. Mangel. 1997b. Predation on adult *Aphytis* parasitoids in the field. *Oecologia* **110**: 346–352.

Hendrix, D. L. and Y.-A. Wei. 1992. Homopteran honeydew sugar composition is determined by both the insect and plant species. *Comparative Biochemistry and Physiology B* **101**: 23–27.

1994. Bemisiose: an unusual trisaccharide in *Bemisia* honeydew. *Carbohydrate Research* **253**: 329–334.

Henneberry, T. J., L. Forlow Jech, T. de la Torre, and D. L. Hendrix. 2000. Cotton aphid (Homoptera: Aphididae) biology, honeydew production, sugar quality and quantity, and relationships to sticky cotton. *Southwestern Entomologist* **25**: 161–174.

Hespenheide, H. A. 1985. Insect visitors to extrafloral nectaries of *Byttneria aculeata* (Sterculiaceae): relative importance and roles. *Ecological Entomology* **10**: 191–204.

Hirose, Y. 1998. Conservation biological control of mobile pests: problems and tactics. In P. Barbosa (ed.) *Conservation Biological Control*. San Diego, CA: Academic Press.

Hopper, K. R. 1991. Ecological applications of elemental labeling: analysis of dispersal, density, mortality and feeding. *Southwestern Entomologist* (Suppl.) **14**: 71–83.

Hopper, K. R. and E. A. Woolson. 1991. Labeling a parasitic wasp, *Microplitis croceipes* (Hymenoptera: Braconidae) with trace elements for mark–recapture studies. *Annals of the Entomological Society of America* **84**: 255–262.

Huffaker, C. B., C. E. Kennet, and R. L. Tassan. 1986. Comparisons of parasitism and densities of *Parlatoria oleae* (1952–1982) in relation to ecological theory. *American Naturalist* **128**: 379–393.

Hunter, F. F. and A. M. Ossowski. 1999. Honeydew sugars in wild-caught female horse flies (Diptera: Tabanidae). *Journal of Medical Entomology* **36**: 896–899.

Idoine, K. and D. N. Ferro. 1988. Aphid honyedew as a carbohydrate source of *Edovum puttleri* (Hymenoptera: Eulophidae). *Environmental Entomology* **17**: 941–944.

Idris, A. B. and E. Grafius. 1995. Wildflowers as nectar sources for *Diadegma insulare* (Hymenoptera: Ichneumonidae), a parasitoid of diamondback moth (Lepidoptera: Yponomeutidae). *Environmental Entomology* **24**: 1726–1735.

Irvin, N. A., S. D. Wratten, and C. M. Frampton. 2000. Understorey management for the enhancement of the leafroller parasitoid *Dolichogenidea tasmanica* (Cameron) in orchards at Canterbury, New Zealand. In A. D. Austin and M. Dowton (eds.) *Hymenoptera: Evolution, Biodiversity and Biological Control*. Collingwood, Australia: CSIRO, pp. 396–402.

Jackson, C. G. 1991. Elemental markers for entomophagous insects. *Southwestern Entomologist* (Suppl.) **14**: 65–70.

Jackson, C. G., A. C. Cohen, and C. L. Verdugo. 1988. Labeling *Anaphes ovijentatus* (Hymenoptera: Mymaridae) an egg parasite of *Lygus* (Hemiptera: Miridae), with rubidium. *Annals of the Entomological Society of America* **81**: 919–922.

Jacob, H. S. and E. W. Evans. 1998. Effects of sugar spray and aphid honeydew on field populations of the parasitoid *Bathyplectes curculionis* (Thomson) (Hymenoptera: Ichneumonidae). *Environmental Entomology* **27**: 1563–1568.

2000. Influence of experience on the response of *Bathyplectes curculionis* (Hymenoptera: Ichneumonidae), a nonaphidophagous parasitoid, to aphid odor. *Biological Control* **19**: 237–244.

Janzen, T. A. and F. F. Hunter. 1998. Honeydew sugars in wild-caught female deer flies (Diptera: Tabanidae). *Journal of Medical Entomology* **35**: 685–689.

Jervis, M. A. 1990. Predation of *Lissonota coracinus* (Gmelin) (Hymenoptera: Ichneumonidae) by *Dolichonabis limbatus* (Dahlbom) (Hemiptera: Nabidae). *Entomologist's Gazette* **41**: 231–233.

1998. Functional and evolutionary aspects of mouthpart structure in parasitoid wasps. *Biological Journal of the Linnean Society* **63**: 461–493.

Jervis, M. A. and N. A. C. Kidd. 1993. Integrated pest management in European olives: new developments. *Antenna* **17**: 108–114.

1996. Phytophagy. In M. A. Jervis and N. A. C. Kidd (eds.) *Insect Natural Enemies*. London: Chapman and Hall, pp. 375–394.

Jervis, M. A., G. E. Heimpel, P. N. Ferns, J. A. Harvey, and N. A. C. Kidd. 2001. Life-history strategies in parasitoid wasps: a comparative analysis of "ovigeny". *Journal of Animal Ecology* **70**: 442–458.

Jervis, M. A., N. A. C. Kidd, M. G. Fitton, T. Huddleston, and H. A. Dawah. 1993. Flower-visiting by hymenopteran parasitoids. *Journal of Natural History* **27**: 67–105.

Jervis, M. A., N. A. C. Kidd, and G. E. Heimpel. 1996. Parasitoid adult feeding and biological control: a review. *Biocontrol News and Information* **17**: 1N–22N.

Jervis, M. A., N. A. C. Kidd, P. McEwen, M. Campos, and C. Lozano. 1992a. Biological control strategies in olive pest management. In P. T. Haskell (ed.) *Research Collaboration in European IPM Systems*. Farnham, UK: British Crop Protection Council, pp. 31–39.

Jervis, M. A., N. A. C. Kidd, and M. Walton. 1992b. A review of methods for determining dietary range in adult parasitoids. *Entomophaga* **37**: 565–574.

Kareiva, P. 1983. Influence of vegetation texture on herbivore populations: resource concentration and herbivore movement. In R. F. Denno and M. S. McClure (eds.) *Variable Plants and Herbivores in Natural and Managed Systems*. New York: Academic Press, pp. 259–289.

Kevan, P. G. 1973. Parasitoid wasps as flower visitors in the Canadian high arctic. *Anzeiger Schädlungskunde* **46**: 3–7.

Koptur, S. 1992. Extrafloral nectary-mediated interactions between insects and plants. In E. Bernays (ed.) *Insect–Plant Interactions*. Boca Raton, FL: CRC Press, pp. 81–129.

Landis, D. A. and M. Haas. 1992. Influence of landscape structure on abundance and within-field distribution of *Ostrinia nubilalis* Hübner (Lepidoptera: Pyralidae) larval parasitoids in Michigan. *Environmental Entomology* **21**: 409–416.

Landis, D. A. and P. C. Marino. 1999. Landscape structure and extra-field processes: impact on management of pests and beneficials. In J. R. Ruberson (ed.) *Handbook of Pest Management*. New York: Marcel Dekker, pp. 79–104.

Landis, D. A., S. D. Wratten, and G. M. Gurr. 2000. Habitat management to conserve natural enemies of arthropod pests in agriculture. *Annual Review of Entomology* **45**: 175–201.

Lee, J. C. and G. E. Heimpel. 2003. Nectar availability and parasitoid sugar feeding. Proc. 1st Int. Symp. *Biological Control of Arthropods*, Honolulu, pp. 220–225.

Leeper, J. R. 1974. Adult feeding behavior of *Lixophaga sphenophori*, a tachinid parasite of the New Guinea sugarcane weevil. *Proceedings of the Hawaiian Entomological Society* **21**: 403–412.

Leius, K. 1961a. Influence of food on fecundity and longevity of adults of *Itoplectis conquisitor* (Say) (Hymenoptera: Ichneumonidae). *Canadian Entomologist* **93**: 771–780.

1961b. Influence of various foods on fecundity and longevity of adults of *Scambus buolianae* (Htg.) (Hymenoptera: Ichneumonidae). *Canadian Entomologist* **93**: 1079–1084.

1963. Effects of pollens on fecundity and longevity of adult *Scambus buolianae* (Htg.) (Hymenoptera: Ichneumonidae). *Canadian Entomologist* **95**: 202–207.

1967. Influence of wild flowers on parasitism of tent caterpillar and codling moth. *Canadian Entomologist* **99**: 444–446.

Letourneau, D. K. 1987. The enemies hypothesis: tritrophic interactions and vegetational diversity in tropical agroecosystems. *Ecology* **68**: 1616–1622.

Lewis, D. J. and C. R. Domoney. 1966. Sugar meals in Phlebotominae and Simuliidae (Diptera). *Proceedings of the Royal Entomological Society (London) A* **41**: 175–179.

Lewis, W. J. and K. Takasu. 1990. Use of learned odours by a parasitic wasp in accordance with host and food needs. *Nature* **348**: 635–636.

Lewis, W. J., J. O. Stapel, A. M. Cortesero, and K. Takasu. 1998. Understanding how parasitoids balance food and host needs: importance to biological control. *Biological Control* **11**: 175–183.

Liebig, G. 1979. Gaschromatographie und enzymatische Untersuchungen des Zuckerspektrums des Honigtaus von *Buchnera pectinatae* (Nordl.). *Apidologie* **10**: 213–225.

Lingren, P. D. and M. J. Lukefahr. 1977. Effects of nectariless cotton on caged populations of *Campoletis sonorensis*. *Environmental Entomology* **6**: 586–588.

Lombard, A., M. Buffa, F. Belliardo, A. Patetta, and F. Marletto. 1979. I carboidrati della melata di *Tuberolachnus salignus* Gmel. *Apicoltura Moderna* **70**: 1–6.

Lombard, A., M. Buffa, A. Patetta, A. Manino, and F. Marletto. 1987. Some aspects of the carbohydrate composition of callaphidid honeydew. *Journal of Apicultural Research* **26**: 233–237.

Lundgren, J. G., G. E. Heimpel, and S. A. Bomgren. 2002. Comparison of *Trichogramma brassicae* (Hymenoptera: Trichogrammatidae) augmentation with organic and synthetic pesticides for control of cruciferous Lepidoptera. *Environmental Entomology* **31**: 1231–1239.

Maingay, H. M., R. L. Bugg, R. W. Carlson, and N. A. Davidson. 1991. Predatory and parasitic wasps (Hymenoptera) feeding at flowers of sweet fennel (*Foeniculum vulgare* Miller var. *dulce* Battandier and Tabut, Apiaceae) and spearmint (*Mentha spicata* L., Lamiaceae) in Massachusetts. *Biological Agriculture and Horticulture* **7**: 363–383.

Manino, A., A. Patetta, F. Marletto, A. Lombard, and M. Buffa. 1985. Sequential carbohydrate variations from larch phloem sap to honeydew and to honeydew honey. *Apicoltura* **1**: 93–103.

Maurizio, A. 1975. How bees make honey. In E. Crane (ed.) *Honey*. London: Heinemann, pp. 77–105.

May, R. M., M. P. Hassell, R. M. Anderson, and D. W. Tonkyn. 1981. Density dependence in host-parasitoid models. *Journal of Animal Ecology* **50**: 855–865.

Moller, H. and J. A. V. Tilley. 1989. Beech honeydew: seasonal variation and use by wasps, honey bees, and other insects. *New Zealand Journal of Zoology* **16**: 289–302.

Murdoch, W. W., J. D. Reeve, C. B. Huffaker, and C. E. Kennet. 1984. Biological control of olive scale and its relevance to ecological theory. *American Naturalist* **123**: 371–392.

Murdoch, W. W., R. F. Luck, S. L. Swarbrick, *et al.* 1995. Regulation of an insect population under biological control. *Ecology* **76**: 206–217.

Nemec V. and P. Stary. 1990. Sugars in honeydew. *Biológia (Bratislava)* **45**: 259–264.

Nentwig, W. 1998. Weedy plant species and their beneficial arthropods: potential for manipulation in field crops. In C. H. Pickett and R. L. Bugg (eds.) *Enhancing Biological Control*. Berkeley, CA: University of California Press, pp. 49–72.

Ode, P. J., M. F. Antolin, and M. R. Strand. 1996. Sex allocation and sexual asymmetries in intra-brood competition in the parasitic wasp *Bracon hebetor*. *Journal of Animal Ecology* **65**: 690–700.

Olson, D. M., H. Fadamiro, J. G. Lundgren, and G. E. Heimpel. 2000. Effects of sugar feeding on carbohydrate and lipid metabolism in a parasitoid wasp. *Physiological Entomology* **25**: 17–26.

Orr, D. B. and J. M. Pleasants. 1996. The potential of native prairie plant species to enhance the effectiveness of the *Ostrinia nubilalis* parasitoid *Macrocentrus grandii*. *Journal of the Kansas Entomological Society* **69**: 133–143.

Patt, J. M., G. C. Hamilton, and J. H. Lashomb. 1997. Foraging success of parasitoid wasps on flowers: interplay of floral architecture and searching behavior. *Entomologia Experimentalis et Applicata* **83**: 21–30.

1999. Response of two parasitoid wasps to nectar odors as a function of experience. *Entomologia Experimentalis et Applicata* **90**: 1–8.

Payne, J. A. and B. W. Wood. 1984. Rubidium as a marking agent for the hickory shuckworm (Lepidoptera: Tortricidae). *Environmental Entomology* **13**: 1519–1521.

Pedata, P. A., M. S. Hunter, H. C. J. Godfray, and G. Viggiani. 1995. The population dynamics of the white peach scale and its parasitoids in a mulberry orchard in Campania, Italy. *Bulletin of Entomological Research* **85**: 531–539.

Percival, M. S. 1961. Types of nectar in angiosperms. *New Phytologist* **60**: 235–281.

Powell, W. 1986. Enhancing parasitoid activity in crops. In J. Waage and D. Greathead (eds.) *Insect Parasitoids*. London: Academic Press, pp. 319–340.

Reeve, J. D. and W. W. Murdoch. 1985. Aggregation by parasitoids in the successful control of California red scale: a test of theory. *Journal of Animal Ecology* **54**: 797–816.

Risch, S. J., D. A. Andow, and M. Altieri. 1983. Agroecosystems diversity and pest control: data, tentative conclusions, and new research directions. *Environmental Entomology* **12**: 625–629.

Rogers, C. A. 1985. Extrafloral nectar: entomological implications. *Bulletin of the Entomological Society of America* **31**: 15–20.

Roland, J. 2000. Landscape ecology of parasitism. In M. A. Hochberg and A. R. Ives (eds.) *Parasitoid Population Biology*. Princeton, NJ: Princeton University Press, pp. 83–100.

Romeis, J. and F. L. Wäckers. 2000. Feeding responses by female *Pieris brassicae* butterflies to carbohydrates and amino acids. *Physiological Entomology* **25**: 247–253.

Root, R. B. 1973. Organization of a plant–arthropod association in simple and diverse habitats: the fauna of collards *(Brassica oleracea)*. *Ecological Monographs* **43**: 95–124.

Rosenheim, J. A. 1998. Higher order predators and the regulation of insect herbivore populations. *Annual Review of Entomology* **43**: 421–448.

Sheehan, W. 1986. Response by specialist and generalist natural enemies to agroecosystem diversification: a selective review. *Environmental Entomology* **15**: 456–461.

Smith, A. D. M. and D. A. Maelzer. 1986. Aggregation of parasitoids and density-independence of parasitism in field populations of the wasp *Aphytis melinus* and its host, the red scale, *Aonidiella aurantii*. *Ecological Entomology* **11**: 425–434.

Smith, S. M., M. Hubbes, and J. R. Carrow. 1986. Factors affecting inundative release of *Trichogramma minutum* Ril. against the spruce budworm. *Journal of Applied Entomology* **101**: 29–39.

Stapel, J. O., A. M. Cortesero, C. M. de Moraes, J. H. Tumlinson, and W. J. Lewis. 1997. Extrafloral nectar, honeydew, and sucrose effects on searching behavior and efficiency of *Microplitis croceipes* (Hymenoptera: Braconidae) in cotton. *Environmental Entomology* **26**: 617–623.

Stephen, F. M. and L. E. Browne. 2000. Application of Eliminade™ parasitoid food to boles and crowns of pines (Pinaceae) infested with *Dendroctonus frontalis* (Coleoptera: Scolytidae). *Canadian Entomologist* **132**: 983–985.

Stephen, F. M., M. P. Lih, and L. E. Browne. 1996. Biological control of southern pine beetle through enhanced nutrition of its adult parasitoids. In Proc. North American Forest Insect Work Conference, Publication no.160, Texas Forest Service, Lufkin, TX, pp. 34–35.

Stephens, M. J., C. M. France, S. D. Wratten, and C. Frampton. 1998. Enhancing biological control of leafrollers (Lepidoptera: Tortricidae) by sowing buckweat *(Fagopyrum esculentum)* in an orchard. *Biocontrol Science and Technology* **8**: 547–558.

Takasu, K. and W. J. Lewis. 1993. Host- and food-foraging of the parasitoid *Microplitis croceipes*: learning and physiological state effects. *Biological Control* **3**: 70–74.

1995. Importance of adult food sources to host searching of the larval parasitoid *Microplitis croceipes*. *Biological Control* **5**: 25–30.

1996. The role of learning in adult food location by the larval parasitoid *Microplitis croceipes* (Hymenoptera: Braconidae). *Journal of Insect Behavior* **9**: 265–281.

Thompson, S. N. 1999. Nutrition and culture of entomophagous insects. *Annual Review of Entomology* **44**: 561–592.

Tooker, J. F. and L. M. Hanks. 2000. Flowering plant hosts of adult hymenopteran parasitoids of central Illinois. *Annals of the Entomological Society of America* **93**: 580–588.

Topham, M. and J. W. Beardsley Jr. 1975. Influence of nectar source plants on the New Guinea sugarcane weevil parasite, *Lixophaga sphenophori* (Villenueve). *Proceedings of the Hawaiian Entomological Society* **22**: 145–154.

Treacy, M. F., J. H. Benedict, M. H. Walmsley, J. D. Lopez, and R. K. Morrison. 1987. Parasitism of bollworm (Lepidoptera: Noctuidae) eggs on nectaried and nectariless cotton. *Environmental Entomology* **16**: 420–423.

Van den Bosch, R. and A. D. Telford. 1964. Environmental modification and biological control. In P. DeBach (ed.) *Biological Control of Insect Pests and Weeds*. New York: John Wiley, pp. 459–488.

Van Emden, H. F. 1962. Observations on the effect of flowers on the activity of parasitic Hymenoptera. *Entomologists' Monthly Magazine* **98**: 265–270.

1988. The potential for managing indigenous natural enemies of aphids on field crops. *Philosophical Transactions of the Royal Society of London Series* B **318**: 183–201.

1990. Plant diversity and natural enemy efficiency in agroecosystems. In M. Mackauer, L. E. Ehler, and J. Roland (eds.) *Critical Issues in Biological Control*. Andover, UK: Intercept Press, pp. 63–80.

Van Hamburg, H. and M. P. Hassell. 1984 Density dependence and the augmentative release of egg parasitoids against graminaceous stem borers. *Ecological Entomology* **9**: 101–108.

Van Handel, E. 1972. The detection of nectar in mosquitoes. *Mosquito News* **32**: 458.

1984. Metabolism of nutrients in the adult mosquito. *Mosquito News* **44**: 573–579.

Van Handel, E., J. S. Haeger, and C. W. Hansen. 1972. The sugars of some Florida nectars. *American Journal of Botany* **59**: 1030–1032.

Wäckers, F. L. 1994. The effect of food deprivation on the innate visual and olfactory preferences in the parasitoid *Cotesia rubecula*. *Journal of Insect Physiology* **40**: 641–649.

1999. Gustatory response by the hymenopteran parasitoid *Cotesia glomerata* to a range of nectar- and honeydew-sugars. *Journal of Chemical Ecology* **12**: 2863–2877.

Wäckers, F. L. and A. Steppuhn. 2003. Characterizing nutritional state and food source use of parasitoids collected in fields with high and low nectar availability. *IOBC/WPRS Bulletin* **26**: 209–214.

Wäckers, F. L., A. Bjornsten, and S. Dorn. 1996. A comparison of flowering herbs with respect to their nectar accessibility for the parasitoid *Pimpla turionellae*. *Proceedings of Experimental and Applied Entomology* **7**: 177–182.

Wei, Y. A., D. L. Hendrix, and R. Nieman. 1996. Isolation of a novel tetrasaccharide, bemisiotetrose, and glycine betaine from silverleaf whitefly honeydew. *Journal of Agricultural and Food Chemistry* **44**: 3214–3218.

1997. Diglucomelezitose, a novel pentasaccharide in silverleaf whitefly honeydew. *Journal of Agricultural and Food Chemistry* **45**: 3481–3486.

Wilkinson, T. L., D. A. Ashford, J. Pritchard, and A. E. Douglas. 1997. Honeydew sugars and osmoregulation in the pea aphid *Acyrthosiphon pisum*. *Journal of Experimental Biology* **200**: 2137–2143.

Wolcott, G. N. 1941. The establishment in Puerto Rico of *Larra americana* Saussere. *Journal of Economic Entomology* **34**: 53–56.

1942. The requirements of parasites for more than hosts. *Science* **96**: 317–318.
Wolf, J. P. and W. H. Ewart. 1955. Carbohydrate composition of the honeydew of *Coccus hesperidum* L.: evidence for the existence of two new oligosaccharides. *Archives of Biochemistry and Biophysics* **58**: 365–372.
Wratten, S. D., L. Berndt, G. Gurr, *et al.* 2003. Adding floral diversity to enhance parasitoid fitness and efficacy. Proc. 1st Int. Symp. *Biological Control of Arthropods*, Honolulu, pp. 211–214.
Wykes, G. R. 1952. An investigation of the sugars present in the nectar of flowers of various species. *New Phytologist* **51**: 210–215.
Yuval, B. and Y. Schlein. 1986. Leishmaniasis in the Jordan Valley. III. Nocturnal activity of *Phlebotomus papatasi* (Diptera: Psychodidae) in relation to nutrition and ovarian development. *Journal of Medical Entomology* **23**: 411–415.
Zandstra, B. H. and P. S. Motooka. 1978. Beneficial effects of weeds in pest mangement. *Pest Articles and News Summary* **24**: 333–338.
Zhao, J. Z., G. S. Ayers, E. J. Grafius, and F. W. Stehr. 1992. Effects of neighboring nectar-producing plants on populations of pest Lepidoptera and their parasitoids in broccoli plantings. *Great Lakes Entomologist* **25**: 253–258.
Zoebelein, G. 1956. Der Honigtau als Nahrung der Insekten. I. *Zeitschrift für angewandte Entomologie* **38**: 369–416.

10

Habitat diversification in biological control: the role of plant resources

T. K. WILKINSON AND D. A. LANDIS

Introduction

Modern agricultural production practices frequently lead to the simplification of agricultural landscape structure. When landscapes are developed for agricultural uses, it is common for the prior ecosystems to be highly fragmented or completely replaced by relatively simple habitats dominated by a few plant species (Merriam 1988; Pogue and Schnell 2001). This process has generally intensified in the last half century due to increased use of mechanization and chemical inputs. As a consequence of both landscape simplification and intensive use of pesticides, predator and parasitoid populations that may otherwise suppress herbivore populations can be rendered inefficient. Recently, there has been significant interest in diversifying agricultural landscapes to benefit natural enemy communities, reduce reliance on pesticides, and increase agricultural sustainability (Altieri and Letourneau 1982; Pickett and Bugg 1998; Gurr *et al.* 2000; Landis *et al.* 2000). Typically these practices focus on providing specific resources to natural enemies through selective addition of habitats to the crop and surrounding environment. Such habitats are designed to provide natural enemies with appropriate food and shelter, and are typically accomplished by manipulation of plant species, populations, or communities, and the resources that they provide. At the same time, these habitats need to deny similar benefits to herbivores.

Habitat diversification benefits natural enemies in a variety of ways. Resources may be provided directly (e.g., floral nectar, pollen), indirectly (e.g., increased host or prey availability), or as emergent properties of habitat diversification (e.g., moderated microclimates). Furthermore, these resources can occur in agricultural ecosystems at several scales; i.e., at the field, farm, or

Plant-Provided Food for Carnivorous Insects, ed. F. L. Wäckers, P. C. J. van Rijn, and J. Bruin.
Published by Cambridge University Press.

landscape level. Although it can be difficult to isolate the impacts of a particular habitat diversification practice, increasing habitat diversity is generally viewed as an important factor in stabilizing pest and natural enemy interactions (Altieri 1994). Because individual natural enemy species may require specific resources at different times and spatial scales, not all attempts to manipulate habitat diversity are equally effective. In this chapter we examine habitat management as a tool for favoring natural enemy communities and for increasing the effectiveness of biological control. Many of the concepts discussed apply across both annual and perennial cropping systems. We use the terms "crop" or "crop area" to define individual units of production such as a field, orchard, or vineyard, regardless of whether they are annual or perennial in nature.

Habitat management

The need for habitat management

Annual cropping systems are the clearest example of how agricultural production systems can result in habitat simplification and reduced effectiveness of natural enemies. Annual crop fields undergo seasonal tilling of soil to prepare for crop planting. Herbicides and cultivation are then commonly used to eliminate weeds. After planting, weed, insect, and disease control measures are applied to remove or reduce unwanted competitors. Finally, at harvest, a large export of biomass occurs and may be followed by tillage and crop residue burial to reduce opportunities for pest overwintering and prepare for the following growing season. In ecological terms, each of these practices constitutes a significant disturbance (Pickett and White 1985). The extreme frequency and uniformity of these disturbances results in highly simplified ecosystems where natural enemy population density is often too low to successfully control pest populations (Helenius 1998; Landis and Menalled 1998). Though perennial-cropping systems may undergo fewer disturbances and can frequently support a more complex arthropod community (Szentkiralyi and Kozar 1991) they also benefit from habitat diversification (Smith and Papacek 1991; Liang and Huang 1994).

Benefits of habitat management

Habitat management has been defined as a subset of conservation biological control practices that aim to manipulate habitats to improve availability of the resources required by natural enemies for optimal performance as biological control agents (Landis *et al.* 2000). Typically, habitat management involves manipulating plants in or around the crop ecosystem to provide alternate prey or hosts, non-host food resources including pollen and nectar, shelter from adverse conditions such as tillage, pesticides, etc., and favorable microclimates for in-season and overwintering survival (Pickett and Bugg 1998). Habitat management may occur at the within-crop, within-farm, or landscape levels, as needed to provide resources at the optimal time or place for natural enemies.

Diversification of agricultural landscapes to include more natural areas can have other positive effects as well. Daily *et al.* (1997) reviewed the benefits that natural ecosystems provide to human societies. Services that directly influence agriculture include: purification of air and water, mitigation of droughts and floods, generation and preservation of soils and renewal of soil fertility, detoxification and decomposition of wastes, pollination of crops and natural vegetation, dispersal of seeds, cycling and movement of nutrients, and control of most potential agricultural pests. The benefits of ecosystem services also accrue to society in general in terms of maintenance of biodiversity, protection of coastal shores from erosion, protection from sun's ultraviolet rays, moderation of weather extremes, and provision of esthetic beauty (Daily *et al.* 1997). Recently work has examined the value of the ecosystem services provided to society by natural habitats and estimates suggest that to replace these functions would cost society US$33 trillion annually (Costanza *et al.* 1997).

Habitat diversification and natural enemies

The presence of diverse habitats in or near crops can be important in sustaining natural enemy populations until conditions are suitable or until prey species become established in the crop area. On the other hand, habitat diversification may disrupt some natural enemies by making their prey more difficult to locate through increased spatial or olfactory complexity of the habitat (Sheehan 1986). The dispersal ability of a natural enemy may influence its response to habitat diversity. Highly mobile natural enemies or those that require a wide range of resources are likely to utilize habitats over a broader spatial scale than less mobile natural enemies or those with more restricted resource needs (Russell 1989). In relation to the agroecosystems in which they exist, natural enemies are very small and must locate hosts on a single plant out of many using a combination of visual and olfactory cues (Vet *et al.* 1991; Vet and Dicke 1992; Wäckers and Lewis 1994). The surface complexity of the crop plant and surrounding vegetation may also affect a natural enemy's ability to locate hosts or prey (Casas and Djemai 2002). As leaf surface complexity (area and number of structures) increases, a natural enemy must search more to find the same number of prey. For example, leaf surface area per unit soil area may be greater in complex habitats than in simple systems (Sheehan 1986). Andow and Risch (1987) suggest that *Trichogramma minutum* search efficiency was higher in monocultures than in a bean, red leaf clover, and squash polyculture as a result of decreased plant surface area. The combined olfactory, structural, and geometric complexity of diverse vegetation may thus combine to limit natural enemy effectiveness.

Plant-provided food for natural enemies

Plant resources such as pollen, floral and extrafloral nectar, and plant sap are directly utilized by parasitoids and predators for energy, reproduction, and survival during periods of prey scarcity. Limburg and Rosenheim (2001) found that green lacewing larvae *Chrysoperla plorabunda* feed on extrafloral nectar on almonds and cotton leaves when aphid prey are scarce. Pemberton and Vandenberg (1993) observed 41 species of coccinellid adults feeding on extrafloral nectaries of various plants. The predatory mite *Iphiseius degenerans* uses castor bean as a year-round host plant feeding on pollen to sustain reproduction and extrafloral nectar to increase longevity when prey are limited (Van Rijn and Tanigoshi 1999). Predators may also use plant sap as a water or nutrient source in order to sustain life during times of prey scarcity. Females of the predatory bug *Podisus maculiventrus* provided with plant material survived four times longer than starved females (Legaspi and O'Neil 1993). Ruberson *et al.* (1986) found that both nymphs and adult *P. maculiventrus* had enhanced survival and development rates and reproductive success when provided with potato leaves with prey and water versus prey or water alone.

For parasitoids to maximize reproductive success they must minimize foraging time (Lewis *et al.* 1998). Parasitoids that require floral resources for survival must choose between searching for food or hosts, particularly when these resources are spatially separated (see also Olson *et al.*, Chapter 5). Habitat manipulation may increase proximity of floral resources and host/prey species and should maximize a natural enemy's ability to control pest populations within an agroecosystem. For example, *Costesia rubecula*, a parasitoid of the cabbage white butterfly, *Pieris rapae*, requires a sugar meal approximately once per day to prevent starvation (Siekmann *et al.* 2001). Maintaining access to such resources in *Brassica* cropping systems is likely to increase the impact of this parasitoid. Similarly, Costamagna and Landis (2004) found that honey-fed *Glyptapanteles militaris* and *Meteorus communis* adult females lived significantly longer than water-fed females and *M. communis* parasitized significantly more hosts because of their increased longevity.

Provision of alternative food within the agroecosystem is of particular economic importance when growers are paying to release natural enemies. Trichogrammatid egg parasitoids are used in inundative release programs worldwide for the control of lepidopteran pests (Li 1994). Because of their small size and weak flight capabilities they are unable to travel long distances in search of pollen and nectar (Gurr and Nicol 2000). For example, the parasitoid *Trichogramma carverae* is typically released twice in vineyards costing $85 (Australian) per hectare. Since these parasitoids only live 1–3 days without food, releases must be closely synchronized with the presence of host eggs (Gurr and Wratten 1999). If floral resources are available, such parasitoids may persist for longer periods, increasing the success of inundative releases.

Alternate prey or hosts

Certain prey species may only be available at particular times during the year requiring predators to locate alternative prey for year-round survival. For example, predatory mites and their alternative prey are often found on herbaceous forbs and woody plants bordering orchards (Coli *et al.* 1994). Similarly, the aphid parasitoid *Diaeretiella rapae* can successfully switch between 13 host species that utilize a variety of plant families. The presence of alternate hosts bordering the crop increases the stability of the parasitoid population when the aphid pest is scarce (Pike *et al.* 1999). Parasitoids may have to leave the crop area to find a host that overwinters in a suitable lifestage. *Colpoclypeus florus* is a eulophid parasitoid of the Pandemis leafroller, *Pandemis pyrusana*, and the obliquebanded leafroller, *Choristoneura rosaceana*, both pests of apples. Although *C. florus* is capable of high rates of parasitism it cannot overwinter on *P. pyrusana* or *C. rosaceana* and must leave the orchard ecosystem to find an overwintering host. In the state of Washington (USA), Pfannenstiel *et al.* (2000) found *C. florus* overwintering on the strawberry leafroller, *Ancylis comptana*, which itself overwinters on a wild rose, *Rosa woodsii*. In their study area *R. woodsii* only occurs in riparian areas. Orchards close to riparian areas are colonized by *C. florus* earlier in the season and achieve higher overall parasitism than orchards far from such habitats. By planting *R. woodsii* close to these distant orchards, they were able to increase early season parasitism by *C. florus*.

Overwintering habitats

Non-crop areas often provide suitable overwintering habitat for natural enemies. Pfiffner and Luka (2000) found the upper 5 cm of soil in arable fields to be routinely frozen, whereas the soil of adjacent wildflower margins never froze. Carabids, staphylinids, and spiders were only found in wildflower margins during the overwintering period. Bordering vegetation may also influence natural enemy dispersal from overwintering sites. Varchola and Dunn (2001) studied movement of carabids in corn surrounded by hedgerows and grassy field borders. Overwintering success of carabid beetles was increased by thick vegetative cover compared to bare or open ground. As corn emerged, carabid activity-density and species richness increased more in corn adjacent to complex hedges than to grassy borders. However, after the corn canopy closed the opposite was true. They hypothesized that closure of the corn canopy likely made conditions for moving between habitats more favorable by affording the carabids greater protection from weather and other factors. Varchola and Dunn (1999) also found that roadside grass habitats provided carabids with refuges. They found that though complex roadside grass habitats had greater species richness, the abundance of carabids was not different from simple roadside grass habitats until after corn canopy closure, when corn became the preferred habitat.

Favorable microclimates

Natural enemies may also require access to moderated microclimates in summer months. Dyer and Landis (1996) demonstrated that *Eriborus terebrans*, an ichneumonid parasitoid of the European corn borer, *Ostrinia nubilalis*, required both a source of sugar and a moderated microclimate for maximum survival in corn agroecosystems. Whereas herbaceous field edges likely provided adequate sources of floral nectar and aphid honeydew to sustain *E. terebrans*, the temperature in herbaceous edges was frequently too high for wasp survival. Only wooded field edges provided the right combination of food resources and moderated temperatures to maximize *E. terebrans* survival (Fig. 10.1).

Habitat diversification and herbivores

Poorly designed habitat management practices have the potential to favor secondary pests or diseases, and to cause crop competition or nutrient effects that could be detrimental to the crop (Bugg and Waddington 1994; Prokopy 1994). Plant resource manipulations must be carefully designed and assessed to avoid an increase in pest damage (Baggen and Gurr 1998). For example, habitat management can in some cases increase the natural enemies that attack beneficial insects. When buckwheat was planted in orchards the oviposition activity of the leafroller parasitoid *Dolichogenidea tasmanica* was increased but it also attracted large numbers of *Anacharis* sp., which parasitize beneficial brown lacewings (Stephens *et al.* 1998). In some cases, plant resources may favor pests more than the natural enemies. Zhao *et al.* (1992) found that eggs and larvae of *P. rapae* and larvae of the diamondback moth, *Plutella xylostella*, were more abundant in broccoli interplanted with various nectar-bearing plants than in broccoli monocultures.

Alternatively, increasing habitat complexity in the field may reduce the incidence of insect pests by decreasing the ease in which herbivores can locate

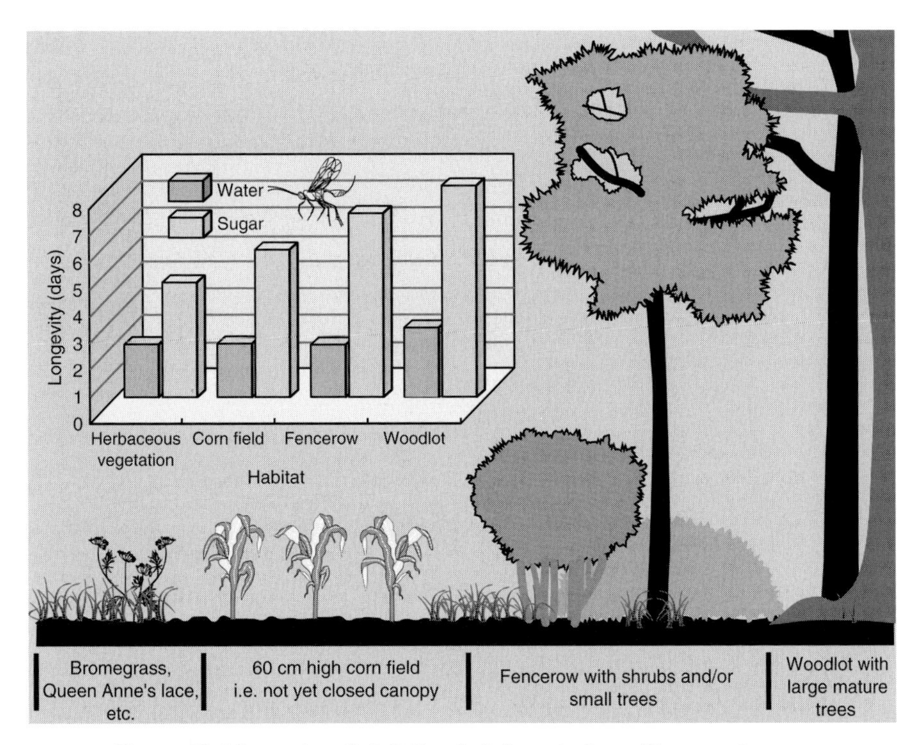

Figure 10.1 Longevity of adult female *Eriborus terebrans* (Hymenoptera: Ichneumonidae) in various habitats within an agroecosystem. Without sugar resources wasps live a maximum of 2 days. Survival increases with provision of sugar and is maximized when both food and shelter are provided in the moderated microclimate of fencerow and woodlot habitats (Dyer and Landis 1996).

crop plants. Horn (1981) found that fewer green peach aphids, *Myzus persicae*, colonized weedy collard (*Brassica oleracea*) patches than cultivated patches, whereas aphidophagous predators would colonize weedy areas even at low aphid densities. Trap cropping can also be used to divert herbivore populations away from valued crops (All 1999). However, trap cropping can also divert specialist natural enemies and increase damage to adjacent crops if timing of trap crop spraying or destruction kills large numbers of natural enemies (Hokkanen 1991).

The effects of scale

Plant resources for natural enemies may be located within the crop area itself, at the crop margin, or external to the production area (Fig. 10.2). The optimum location depends on the frequency of resource use and the

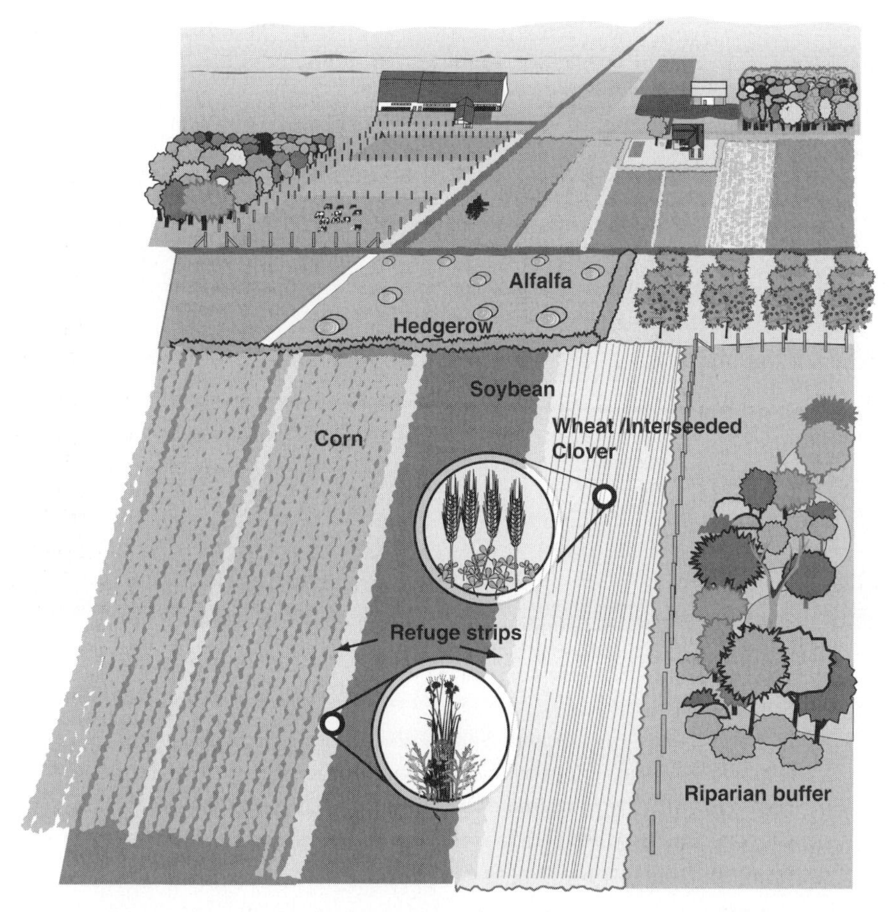

Figure 10.2 Agricultural landscape with natural enemy resource habitats integrated at the field, farm, and landscape level. Interseeded cover crops and refuge strips provide resources within fields. Hedgerows and riparian buffers provide corridors and stable habitats at the farm scale. A diversity of crop and animal production systems interspersed with natural areas provides natural enemy resources at the landscape scale.

dispersal capabilities of the natural enemy. For example, when the natural enemy has excellent dispersal abilities, overwintering sites may be located many kilometers from the target prey/host (Corbett *et al.* 1996). On the other hand, food resources that are required on a daily basis may need to be provided in close proximity to the target prey/host regardless of the natural enemy's dispersal ability. Because natural enemies may operate on different spatial scales than their hosts or prey, the size of the target crop and vegetation habitats in relation to the natural enemies' ability to disperse is critical. Doak (2000) found that larger patches of the early successional shrub *Dryas drummondii* (Rosaceae) experienced higher parasitism of the geometrid caterpillar *Itame andersoni* by ichneumonid wasps. In contrast, greater parasitism by tachinid flies occurred in medium-sized patches and smaller patches saw greater parasitism by braconid wasps. It was suggested that these patterns might have been influenced by the relative dispersal ability of these natural enemies. Kruess and Tscharntke (2000) found that habitat fragmentation influenced parasitism of the weevil *Oxystoma ochropus*. Parasitism increased with increasing habitat area from 35% parasitism in small meadows to 70% in larger meadows. They also found that species richness of parasitoids declined more rapidly than herbivore species richness as fragmentation increased. With *et al.* (2002) found that coccinellid species differed in their ability to effectively track aphid prey in fragmented experimental landscapes. Patch shape as well as size may influence the dispersal capability of natural enemies. Grez and Prado (2000) compared I-shaped patches to square patches and found that surrounding vegetation affected immigration of coccinellids. Immigration increased when the surrounding vegetation did not provide the beetles with any resources, and emigration was reduced in square patches where beetles were less likely to reach patch edges.

Plant resources in the cultivated area

Perhaps the simplest means of increasing plant resources in cultivated areas is to manage for increased plant diversity in all or a portion of the crop area. In the United Kingdom, restricted pesticide applications in "conservation headlands" allowed increased plant diversity and significantly increased carabid abundance in crop fields (Hassall *et al.* 1992). In the European Union, farmers may take advantage of subsidy programs that encourage practices to enhance biodiversity of flora and fauna in crops. In the Netherlands such programs resulted in increased species richness of bees and hoverflies, although plant and bird populations were not enhanced (Kleijn *et al.* 2001).

Polycultures

Establishing polycultures is another technique to increase diversity of resources within the crop area. In contrast to monocultures, polycultures can result in increased natural enemy populations (52.7% of cases reviewed) (Andow 1991) and higher pest mortality (60% of cases reviewed) (Russell 1989). Coll and Bottrell (1996) collected more *Pediobius foveolatus*, parasitoids of the Mexican bean beetle, *Epilachna varivestis*, in a bean (*Phaseolus vulgaris*) and maize (*Zea mays*) polyculture than in bean monocultures. Both tenure time and parasitism by *P. foveolatus* were higher in the more shaded polycultures.

Cover crops

Cover crops can be used to suppress weeds, decrease erosion, reduce soil compaction, increase water infiltration and moisture retention, and improve soil quality. They are also used increasingly to manage natural enemy populations (Bugg and Waddington 1994; Hooks *et al.* 1998). Cover crops may be mowed or allowed to senesce naturally to encourage movement of natural enemies into the crop. Predation of cantaloupe herbivores by the big-eyed bug, *Geocoris punctipes*, increased as the cool-season cover crops in the field became less suitable habitats for predators (Bugg *et al.* 1991). Mowing of faba bean covers planted between hop strings caused dispersal of aphid predators into the surrounding hops, increasing the coccinellid-to-aphid ratio (Goller *et al.* 1997). Cover crops are also valuable in supporting natural enemies during parts of the season when crop plants do not. Hairy vetch (*Vicia villosa*) in pecan orchards is capable of supporting two generations of the convergent lady beetle, *Hippodamia convergens*, feeding on pea aphids, *Acyrthosiphon pisum* (Tedders 1983). In cotton, relay intercropping with wheat, canola, and sorghum allowed predators to remain in the cotton field throughout the entire season. Predators could move between each of the crops as the other senesced, allowing predator numbers to increase earlier in relay cotton and suppress aphid populations (Parajulee *et al.* 1997; Parajulee and Slosser 1999).

Understory and ground covers

Perennial orchard and vineyard systems frequently employ understory or ground-cover vegetation to achieve many of the same benefits as cover crops. Alternative prey sources found in ground covers are often important in supporting predator populations when pest levels in the orchard or vineyard are low. For example, vetch planted in the understory of pecan orchards increased *H. convergens* populations in the canopy resulting in control of pecan aphid (*Melanocallis caryaefoliae*) populations (Tedders 1983). Ground covers grown in the understory of apple, pecan, and citrus orchards have successfully improved biological control of some fruit and nut pests (Bugg and Waddington 1994). Many parasitoids and predators benefit from flowers produced by ground covers. A mixture of plants may increase available resources for natural enemies by producing variability in flowering time, plant height, and phenology. Natural enemies can be further manipulated through altering ground-cover mowing. Bugg and Waddington (1994) described a system where mixtures of plants with varying phenologies planted between rows of apple trees supported diverse populations of natural enemies, reducing spider mite infestations. Mowing the ground cover in strips over time moved natural enemies into the canopy and promoted re-establishment of the cover crop.

Refuge habitats

Providing specific refuge habitats in crop areas can increase natural enemy density and effectiveness. Spider populations are often enhanced when provided with refuge habitats and can be particularly effective as early-season predators (Sunderland and Samu 2000). Spider and carabid densities have been increased by providing portable refuges and leaving weed residues in the crop. Halaj *et al.* (2000) used modular refuges constructed from wire baskets stuffed with straw that produced a significant increase in spider density and species richness, resulting in 33% less arthropod damage to adjacent soybean plants. Weed residues (piles, mulches, or strips) in upland rice have also been shown to increase spider and carabid densities and activity (Afun *et al.* 1999). Similarly, strips of flowering plants in winter wheat fields enhanced spider populations and decreased aphid populations by 4% per day (Jmhasly and Nentwig 1995). Additional crop plants can also be interspersed in the cultivated area, within rows, or as strips within a field or orchard. Yu-hua *et al.* (1997) planted strips of alfalfa and rape within rows of apple trees and found that the increased ground cover created a refuge for predators key in controlling secondary pests. As mite pests reached economically injurious levels the alfalfa strips were mowed, driving mite predators into the apple trees. This effectively increased the predator/prey ratio, which in turn increased predation and lowered mite populations below the economic injury level. This system allowed insecticide and miticide applications to be reduced by 50% and 70% respectively (Yu-hua *et al.* 1997). Similarly, planting alfalfa strips in cotton fields resulted in a greater number of predators on adjacent cotton and encouraged natural enemies to establish in the crop prior to arrival of the pests, *Helicoverpa* spp. (Mensah 1999).

Refuge strips have also been used to increase the density of predaceous ground beetles in corn and soybean fields (Carmona and Landis 1999; Menalled *et al.* 2001). Lee *et al.* (2001) studied carabid beetle activity-density in the presence or absence of refuges, in corn fields with or without soil insecticide treatments. They found that the refuge strips consistently supported higher carabid activity-density than the control strips and that insecticide use significantly decreased carabid activity-density in treated crop areas. The presence of refuges did not increase the activity-density of carabids in untreated plots; however, refuges did allow carabid populations in treated areas to recover from insecticide disturbance. In the United Kingdom, "beetle banks" to enhance populations of predaceous ground beetles have been researched for many years (Thomas *et al.* 1991). Collins *et al.* (2001) found that 4-year-old beetle banks within a winter wheat crop increased generalist predators and reduced cereal aphid populations but did not prevent an outbreak when aphids arrived late in the season.

Plant resources at the crop edge

Plant resources for natural enemies are sometimes planted along the border of the crop to avoid direct competition or loss of crop area. White *et al.* (1995) planted strips of *Phacelia tanacetifolia* along the edges of cabbage fields to attract hoverflies (Syrphidae), important predators of aphid pests. Adult hoverflies require nectar for energy and pollen for reproductive development. Cabbage fields with a *P. tanacetifolia* boundary strip had an increase in the number of hoverfly adults and eggs and a decrease in aphid populations. "Insectary hedgerows" composed of a diversity of flowering shrubs and herbaceous plants have been studied for their ability to provide continuous sources of pollen, nectar, and shelter for natural enemies (Long *et al.* 1998). Marking studies showed that natural enemies utilize these habitats and subsequently disperse into adjacent crops. Finally, the well-known role of hedgerows in providing resources to natural enemies has been previously reviewed (Burel and Baudry 1990; Dix *et al.* 1995).

Flowering plants must be carefully chosen to provide resources for natural enemies but not for pests. Natural enemies may discriminate between plants based on floral structure, color, and types of sugars produced. Hoverfly species are particularly attracted to white or yellow and usually prefer flowers in the family Apiaceae (Umbelliferae) (Colley and Luna 2000). Unlike many bees that have mouthparts adapted to reach deeply hidden floral resources, most natural enemies must utilize plant species with easily accessible pollen and nectar. Plants in the Apiaceae have open floral structures with exposed nectaries and anthers easily accessible to a wide variety of parasitoids (Tooker and Hanks 2000) and can support diverse parasitoid assemblages (Bugg and Wilson 1989; Maingay *et al.* 1991). Plant sugars used by natural enemies are often different from those used by pest species, therefore, it may be possible to select plant resources that only appeal to natural enemies. The parasitoid *Cotesia glomerata* can use sugar sources that have no nutritional value to its host *Pieris brassicae* (Wäckers 2001). Baggen *et al.* (1999) found that phacelia (*Phacelia tanacetifolia*), nasturtium (*Tropaeoleum majus*), and borage (*Borago officinalis*) are attractive to the parasitoid *Copidosoma koehleri* but not to its host the potato moth, *Phthorimaea opercula*.

Plant resources in the agricultural landscape

Because natural enemy populations in crops are subject to periodic population declines, habitats surrounding the crop may be important sources for recolonization following disturbances such as insecticide applications (Landis and Menalled 1998). In this way, populations of natural enemies in crop and nearby non-crop habitats may function as genetically interconnected local populations or metapopulations (Hanski and Gilpin 1997). The value of these habitats as refuges for natural enemies is often unappreciated.

Crop diversity

Manipulation of crop diversity in the larger agroecosystem can affect natural enemy populations and their dispersal. Specifically, choice of crops and crop rotations can enhance natural enemy populations in the agroecosystem. Nault and Kennedy (2000) found that populations of the predaceous twelve-spotted ladybird beetle, *Coleomegilla maculata*, moved from corn into adjacent potato fields and that corn-potato crop rotations decreased Colorado potato beetle, *Leptinotarsa decemlineata*, populations.

Corridors

Certain habitats in the agricultural landscape may act as corridors to enhance population exchanges between cultivated and non-cultivated areas. Connectivity allows dispersal rates of natural enemies to keep up with changing pest populations and the ability of natural enemies to quickly recolonize an agricultural crop can reduce the probability of pest outbreaks (Tscharntke 2000). Mite predators such as *Amblyseius fallacis* disperse aerially into orchards from bordering woodlots and hedgerows, contributing to colonization into the orchard and to subsequent biological control of pest mite species (Coli *et al.* 1994). Nicholls *et al.* (2001) established corridors consisting of a wide variety of flowering plants from adjacent forest habitats into vineyards. These corridors provided natural enemies with alternative food sources throughout the season and aided in the continuous dispersal of generalist predators and parasitoids, thereby decreasing the numbers of leafhoppers and thrips in the vineyard.

Landscape diversity

Complexity, age, and composition of surrounding habitats can affect natural enemy populations entering the agroecosystem. Dyer and Landis (1997) found that *Eriborus terebrans* was more abundant at edges of corn fields bordered by woodlots than at edges bordered by herbaceous vegetation. Szentkiralyi and Kozar (1991) found that orchards surrounded by complex vegetation such as woodlots contained more natural enemy species than orchards surrounded by less complex vegetation. The successional stage of the vegetation bordering agroecosystems is often an important factor in the abundance of natural enemies that are able to inhabit an area. Hassall *et al.* (1992) found that "uncropped headlands" that were 10 years of age had 80% greater spider diversity than 4-year-old sites, which themselves had 35% greater spider diversity than the 1-year-old sites. Pfiffner and Luka (2000) contrasted arthropod populations in arable fields and adjacent semi-natural field margins. They found that undisturbed 2-year-old wildflower strips contained most arthropod species. They attributed this to the greater structural diversity of the annual, biennial, and perennial plant species that these strips contained.

Landscape spatial complexity plays an important role in increasing natural enemy abundance and reducing the frequency and severity of insect outbreaks (Roland and Taylor 1997; Landis and Marino 1999; Kruess and Tscharntke 2000). Thies and Tscharntke (1999) studied parasitoids of the rape pollen beetle, *Meligethes aeneus*, in Germany. They found parasitism was reduced when non-crop habitat fell below 20% of the total landscape and that the amount of non-crop habitat in a 1.5 km radius of the target field was most important (Thies *et al.* 2003). In Michigan (USA), studies indicate that cornfields in complex agricultural landscapes experience greater parasitism of armyworm *Pseudaletia unipuncta* in years when *Meteorus communis* dominates the parasitoid community (Marino and Landis 1996; Menalled *et al.* 1999). However, when *Glypapanteles militaris* is the predominant species no landscape effect on parasitism is observed (Costamagna 2002; Menalled *et al.* 2003). Östman *et al.* (2001) studied bird cherry-oat aphid, *Rhopalosiphum padi*, and predator populations in Sweden. They found that ground-dwelling predators reduced aphid populations in both organic and conventional spring barley fields. Remarkably, landscape structure was even more important than production practices in determining the abundance of these predators. Fields in landscapes with high non-crop habitat complexity experienced better biological control than those in simple landscapes comprised primarily of crops. Similarly, Elliott *et al.* (2002) found that the amount of grasslands, woods, and other non-crop habitats was the most important factor in determining predator abundance in alfalfa fields in South Dakota (USA).

Conclusions

Management of plant resources and habitat diversity offers significant potential to conserve natural enemies and improve biological control in agroecosystems (Landis *et al.* 2000). However, to do it successfully requires detailed knowledge of the types and timing of plant resources that natural enemies require as well as information on the scale and spatial arrangement to most reliably favor natural enemies over pests. Providing plant resources directly in crop areas may be key for those natural enemies where dispersal is limited or for resources that are needed on a frequent basis such as food. Because natural enemy populations outside of cultivated areas are often key in recolonizing the crop after a disturbance, manipulating plant resources outside of crop areas may be required as well. In both cases, careful selection of plant materials and management strategies are necessary to assure that they are compatible with the cropping systems and consistently favor the natural enemy, and not the pest. Though habitat manipulation has some negative impacts, many of these are small compared with the impact insecticides have on insect pest populations, natural enemies, and the environment. Habitat diversification elicits complex ecological interactions that can be difficult to study and even harder to predict. However, this should not preclude their use. Rather, to increase our ability to reliably utilize habitat diversification, more effort needs to be placed on implementing these practices in a variety of agricultural situations coupled with careful evaluation of the long-term effects on pest populations and crop production.

Acknowledgments

We wish to thank Matthew O'Neal, Donald Sebolt, Tibor Bukovinszky, Geoff Gurr, and the editors for helpful reviews of the manuscript, Steve Deming for the artwork, and Molly Sklapsky for technical assistance. Support for this work has been provided by Michigan State University GREEEN Project #GR99-062, through a US Department of Agriculture Special Fruit Grant to Michigan State University and by the Michigan Agricultural Experiment Station.

References

Afun, J. V. K., D. E. Johnson, and A. Russell-Smith. 1999. The effects of weed residue management on pests, pest damage, predators and crop yield in upland rice in Côte d'lvoire. *Biological Agriculture and Horticulture* **17**: 47–58.

All, J. N. 1999. Cultural approaches to managing pests. In J. Ruberson (ed.) *Handbook of Pest Management*. New York: Marcel Dekker, pp. 395–415.

Altieri, M. A. 1994. *Biodiversity and Pest Management in Agroecosystems*. New York: Food Products Press.

Altieri, M. A. and D. K. Letourneau. 1982. Vegetation management and biological control in agroecosystems. *Crop Protection* **1**: 405–430.

Andow, D. A. 1991. Vegetational diversity and arthropod population response. *Annual Review of Entomology* **36**: 561–586.

Andow, D. A. and S. J. Risch. 1987. Parasitism in diversified agroecosystems: phenology of *Trichogramma minutum* (Hymenoptera: Trichogrammatidae). *Entomophaga* **32**: 255–260.

Baggen, L. R. and G. M. Gurr. 1998. The influence of food on *Copidosoma koehleri*, and the use of flowering plants as a habitat management tool to enhance biological control of potato moth, *Phthorimaea operculella*. *Biological Control* **11**: 9–17.

Baggen, L. R., G. M. Gurr, and A. Meats. 1999. Flowers in tri-trophic systems: mechanisms allowing selective exploitation by insect natural enemies. *Entomologia Experimentalis et Applicata* **91**: 155–161.

Bugg, R. L. and C. Waddington. 1994. Using cover crops to manage arthropod pests of orchards: a review. *Agriculture, Ecosystems and Environment* **50**: 11–28.

Bugg, R. L. and L. T. Wilson. 1989. *Ammi visnaga* (L.) Lamarck (Apiaceae): associated beneficial insects and implications for biological control, with emphasis on the bell-pepper ecosystem. *Biological Agriculture and Horticulture* **6**: 241–268.

Bugg, R. L., F. L. Wäckers, K. E. Brunson, J. D. Dutcher, and S. C. Phatak. 1991. Cool-season cover crops relay intercropped with cantaloupe: influence on a generalist predator, *Geocoris punctipes* (Hemiptera: Lygaeidae). *Journal of Economic Entomology* **84**: 408–416.

Burel, F. and J. Baudry. 1990. Hedgerow networks as habitats for colonization of abandoned agricultural land. In R. H. G. Bunce and D. C. Howard (eds.) *Species Dispersal in Agricultural Landscapes*. London: Belhaven Press, pp. 238–255.

Carmona, D. M. and D. A. Landis. 1999. Influence of refuge habitats and cover crops on seasonal activity density of ground beetles (Coleoptera: Carabidae) in field crops. *Environmental Entomology* **28**: 1145–1153.

Casas, J. and I. Djemai. 2002. Canopy architecture and multitrophic interactions. In T. Tscharntke and B. A. Hawkins (eds.) *Multitrophic Level Interactions*. Cambridge, UK: Cambridge University Press, pp. 174–196.

Coli, W. M., R. A. Ciurlino, and T. Hosmer. 1994. Effect of understory and border vegetation composition on phytophagous and predatory mites in Massachusetts commercial apple orchards. *Agriculture, Ecosystems and Environment* **50**: 49–60.

Coll, M. and D. G. Bottrell. 1996. Movement of an insect parasitoid in simple and diverse plant assemblages. *Ecological Entomology* **21**: 141–149.

Colley, M. R. and J. M. Luna. 2000. Relative attractiveness of potential beneficial insectary plants to aphidophagous hover flies (Diptera: Syrphidae). *Environmental Entomology* **29**: 1054–1059.

Collins, K. L., N. D. Boatman, A. W. Wilcox, J. M. Holland, and K. Chaney. 2001. Influence of beetle banks on cereal aphid predation in winter wheat. *Agriculture, Ecosystems and Environment* **93**: 337–350.

Corbett, A., B. C. Murphy, J. A. Rosenheim, and P. Bruins. 1996. Labeling an egg parasitoid, *Anagrus epos* (Hymenoptera: Mymaridae), with Rubidium within an overwintering refuge. *Environmental Entomology* **25**: 29–38.

Costamagna, A. C. 2002. Agricultural landscape complexity has mixed effects on patterns of parasitoid abundance and diversity. M.Sc. thesis, Michigan State University, East Lansing, MI.

Costamagna, A. C. and D. A. Landis. 2004. Effect of food resources on adult *Glyptapanteles militaris* and *Meteorus communis* (Hymenoptera: Braconidae), parasitoids of *Pseudaletia unipuncta* (Lepidoptera: Noctuidae). *Environmental Entomology* **33**: 128–137.

Costanza R., R. dArge, R. deGroot, *et al.* 1997. The value of the world's ecosystem services and natural capital. *Nature* **387**: 253–260.

Daily, G. C., S. Alexander, P. R. Ehrlich, *et al.* 1997. *Ecosystem Services: Benefits Supplied to Human Societies by Natural Ecosystems.* Washington, DC: Ecological Society of America.

Dix, M. E., R. J. Johnson, M. O. Harrell, 1995. Influence of trees on abundance of natural enemies of insect pests: a review. *Agroforestry Systems* **29**: 03–11.

Doak, P. 2000. The effects of plant dispersion and prey density on parasitism rates in a naturally patchy habitat. *Oecologia* **122**: 556–567.

Dyer, L. E. and D. A. Landis. 1996. Effects of habitat, temperature, and sugar availability on longevity of *Eriborus terebrans* (Hymenoptera: Ichneumonidae). *Environmental Entomology* **25**: 1192–1201.

1997. Influence of non-crop habitats on the distribution of *Eriborus terebrans* (Hymenoptera: Ichneumonidae) in cornfields. *Environmental Entomology* **26**: 924–932.

Elliott, N. C., R. W. Kieckhefer, G. J. Michels, and K. L. Giles. 2002. Predator abundance in alfalfa fields in relation to aphids, within-field vegetation, and landscape matrix. *Environmental Entomology* **31**: 253–260.

Goller, E., L. Nunnenmacher, and H. E. Goldbach. 1997. Faba beans as a cover crop in organically grown hops: influence on aphids and aphid antagonists. *Biological Agriculture and Horticulture* **15**: 279–284.

Grez, A. A. and E. Prado. 2000. Effect of plant patch shape and surrounding vegetation on the dynamics of predatory coccinellids and their prey *Brevicoryne brassicae* (Hempitera: Aphididae). *Environmental Entomology* **29**: 1244–1250.

Gurr, G. M. and H. I. Nicol. 2000. Effect of food on longevity of adults of *Trichogramma carverae* Oatman and Pinto and *Trichogramma* nr *brassicae* Bezdenko (Hymenoptera: Trichogrammatidae). *Australian Journal of Entomology* **39**: 185–187.

Gurr, G. M. and S. D. Wratten. 1999. Integrated biological control: a proposal for enhancing success in biological control. *International Journal of Pest Management* **45**: 81–84.

Gurr, G. M., S. D Wratten, and P. Barbosa. 2000. Success in conservation biological control of arthropods. In G. Gurr and S. D. Wratten (eds). *Biological Control: Measures of Success.* Dordrecht, the Netherlands: Kluwer Academic Publishers, pp. 105–132.

Halaj, J., A. B. Cady, and G. W. Uetz. 2000. Modular habitat refugia enhance generalist predators and lower plant damage in soybeans. *Environmental Entomology* **29**: 383–393.

Hanski, I. A. and M. E. Gilpin. 1997. *Metapopulation Biology: Ecology, Genetics, and Evolution.* San Diego, CA: Academic Press.

Hassall, M., A. Hawthorne, M. Maudsley, P. White, and C. Cardwell. 1992. Effects of headland management on invertebrate communities in cereal fields. *Agriculture, Ecosystems and Environment* **40**: 155–178.

Helenius, J. 1998. Enhancement of predation through within-field diversification. In C. H. Pickett and R. L. Bugg (eds.) *Enhancing Biological Control: Habitat Management to Promote Natural Enemies of Agriculture Pests.* Berkeley, CA: University of California Press, pp. 121–160.

Hokkanen, H. M. T. 1991. Trap cropping in pest management. *Annual Review of Entomology* **36**: 119–138.

Hooks, C. R. R., H. R. Valenzuela, and J. Defrank. 1998. Incidence of pests and arthropod natural enemies in zucchini grown with living mulches. *Agriculture, Ecosystems and Environment* **69**: 217–231.

Horn, D. J. 1981. Effect of weedy backgrounds on colonization of collards by green peach aphid, *Myzus persicae*, and its major predators. *Environmental Entomology* **10**: 285–289.

Jmhasly, P. and W. Nentwig. 1995. Habitat management in winter wheat and evaluation of subsequent spider predation on insect pests. *Acta Oecologica* **16**: 389–403.

Kleijn, D., F. Berendse, R. Smit, and N. Gilissen. 2001. Agri-environment schemes do not effectively protect biodiversity in Dutch agricultural landscapes. *Nature* **413**: 723–725.

Kruess, A. and T. Tscharntke. 2000. Species richness and parasitism in a fragmented landscape: experiments and field studies with insects on *Vicia sepium*. *Oecologia* **122**: 129–137.

Landis, D. A. and P. Marino. 1999. Landscape structure and extra-field processes: impact on management of pests and beneficials. In J. Ruberson (ed.) *Handbook of Pest Management*. New York: Marcel Dekker, pp. 79–104.

Landis, D. A. and F. D. Menalled. 1998. Ecological considerations in the conservation of effective parasitoid communities in agriculture systems. In P. Barbosa (ed.) *Conservation Biological Control*. San Diego, CA: Academic Press, pp. 101–121.

Landis, D. A., S. D. Wratten, and G. M. Gurr. 2000. Habitat management to conserve natural enemies of arthropod pests in agriculture. *Annual Review of Entomology* **45**: 175–201.

Lee, J. C., F. D. Menalled, and D. A. Landis. 2001. Refuge habitats modify impact of insecticide disturbance on carabid beetle communities. *Journal of Applied Ecology* **38**: 472–483.

Legaspi, J. C. and R. J. O'Neil. 1993. Life history of *Podisus maculiventris* given low numbers of *Epilachna varivestis* as prey. *Environmental Entomology* **22**: 1192–1200.

1994. Developmental response of nymphs of *Podisus maculiventris* (Heteroptera: Pentatomidae) reared with low numbers of prey. *Environmental Entomology* **23**: 374–380.

Lewis, W. J., J. O. Stapel, A. M. Cortesero, and K. Takasu. 1998. Understanding how parasitoids balance food and host needs: importance in biological control. *Biological Control* **11**: 175–183.

Li, L.-Y. 1994. Worldwide use of *Trichogramma* for biological control on different crops: a survey. In E. Wajnberg and S. A. Hassan (eds.) *Biological Control with Egg Parasitoids*. Wallingford, UK: CAB International, pp. 37–53.

Liang, W. and M. Huang. 1994. Influence of citrus orchard ground cover plants on arthropod communities in China: a review. *Agriculture, Ecosystems and Environment* **50**: 29–37.

Limburg, D. D. and J. A. Rosenheim. 2001. Extrafloral nectar consumption and its influence on survival and development of an omnivorous predator, larval *Chrysoperla plorabunda* (Neuroptera: Chrysopidae). *Environmental Entomology* **30**: 595–604.

Long, R. F., A. Corbett, C. Lamb, *et al.* 1998. Beneficial insects move from flowering plants to nearby crops. *California Agriculture* **52**: 23–26.

Maingay, H., R. L. Bugg, R. W. Carlson, and N. A. Davidson. 1991. Predatory and parasitic wasps (Hymenoptera) feeding at flowers of sweet fennel (*Foeniculum vulgare* Miller var. *dulce* Battandier and Trabut, Apiaceae) and spearmint (*Mentha spicata* L., Lamiaceae) in Massachusetts. *Biological Agriculture and Horticulture* **7**: 363–383.

Marino, P. C. and D. A. Landis. 1996. Effect of landscape structure on parasitoid diversity and parasitism in agroecosystems. *Ecological Applications* **6**: 276–284.

Menalled, F. D., A. C. Costamagna, P. C. Marino, and D. A. Landis. 2003. Temporal variation in the response of parasitoids to agricultural landscape structure. *Agricultural Ecosystems and Environment* **96**: 29–35.

Menalled, F. D., J. C. Lee, and D. A. Landis. 2001. Herbaceous filter strips in agroecosystems: implications for carabid beetle (Coleoptera: Carabidae) conservation and invertebrate weed seed predation. *Great Lakes Entomologist* **34**: 77–91.

Menalled, F. D., P. C. Marino, S. H. Gage, and D. A. Landis. 1999. Does agricultural landscape structure affect parasitism and parasitoid diversity? *Ecological Applications* **9**: 634–641.

Mensah, R. K. 1999. Habitat diversity: implications for the conservation and use of predatory insects of *Helicoverpa* spp. in cotton systems in Australia. *International Journal of Pest Management* **45**: 91–100.

Merriam, G. 1988. Landscape dynamics in farmland. *Trends in Ecology and Evolution* **3**: 16–20.

Nault, B. A. and G. G. Kennedy. 2000. Seasonal changes in habitat preference by *Coleomegilla maculata*: implications for Colorado potato beetle management in potato. *Biological Control* **17**: 164–173.

Nicholls, C. I., M. Parrella, and M. A. Altieri. 2001. The effects of a vegetational corridor on the abundance and dispersal of insect biodiversity within a northern California organic vineyard. *Landscape Ecology* **16**: 133–146.

Östman, O., B. Ekbom, and J. Bengtsson. 2001. Landscape heterogeneity and farming practice influence biological control. *Basic and Applied Ecology* **2**: 365–371.

Parajulee, M. N. and J. E. Slosser. 1999. Evaluation of potential relay strip crops for predator enhancement in Texas cotton. *International Journal of Pest Management* **45**: 275–286.

Parajulee, M. N., R. Montandon, and J. E. Slosser. 1997. Relay intercropping to enhance abundance of insect predators of cotton aphid (*Aphis gossypii* Glover) in Texas cotton. *International Journal of Pest Management* **43**: 227–232.

Pemberton, R. W. and N. J. Vandenberg. 1993. Extrafloral nectar feeding by ladybird beetles (Coleoptera: Coccinellidae). *Proceedings of the Entomological Society of Washington* **95**: 139–151.

Pfannenstiel, R. S., T. R. Unruh, and J. F. Brunner. 2000. Biological control of leafrollers: prospects using habitat manipulation. *Washington Horticulture Association* **95**: 144–149.

Pfiffner, L. and H. Luka. 2000. Overwintering of arthropods in soils of arable fields and adjacent semi-natural habitats. *Agriculture, Ecosystems and Environment* **78**: 215–222.

Pickett, C. H. and R. L. Bugg. 1998. *Enhancing Biological Control: Habitat Management to Promote Natural Enemies of Agriculture Pests*. Berkeley, CA: University of California Press.

Pickett, S. T. A. and P. S. White. 1985. *The Ecology of Natural Disturbance and Patch Dynamics*. San Diego, CA: Academic Press.

Pike, K. S., P. Stary, T. Miller, *et al.* 1999. Host range and habitats of the aphid parasitoid *Diaeretiella rapae* (Hymenoptera: Aphididae) in Washington State. *Environmental Entomology* **28**: 61–71.

Pogue, D. W. and G. D. Schnell. 2001. Effects of agriculture on habitat complexity in a prairie–forest ecotone in the southern great plains of North America. *Agriculture, Ecosystems and Environment* **87**: 287–298.

Prokopy, R. J. 1994. Integration in orchard pest and habitat management: a review. *Agriculture, Ecosystems and Environment* **50**: 1–10.

Roland, J. and P. D. Taylor. 1997. Insect parasitoid species respond to forest structure at different spatial scales. *Nature* **386**: 710–713.

Ruberson, J. R., M. J. Tauber, and C. A. Tauber. 1986. Plant feeding by *Podisus maculiventris* (Heteroptera: Pentatomidae): effect on survival, development, and preoviposition period. *Environmental Entomology* **15**: 894–897.

Russell, E. P. 1989. Enemies hypothesis: a review of the effect of vegetational diversity on predatory insects and parasitoids. *Environmental Entomology* **18**: 590–599.

Sheehan, W. 1986. Response by specialist and generalist natural enemies to agroecosystem diversification: a selective review. *Environmental Entomology* **15**: 456–461.

Siekmann, G., B. Tenhumberg, and M. A. Keller. 2001. Feeding and survival in parasitic wasps: sugar concentration and timing matter. *Oikos* **95**: 425–430.

Smith, D. and D. F. Papacek. 1991. Studies of the predatory mite *Amblyseius victoriensis* (Acarina: Phytoseiidae) in citrus orchards in south-east Queensland: control of *Tegolophus australis* and *Phyllocoptruta oleivora* (Acarina: Eriophyidae), effect of pesticides, alternative host plants and augmentative release. *Experimental and Applied Acarology* **12**: 195–217.

Stephens, M. J., C. M. France, S. D. Wratten, and C. Frampton. 1998. Enhancing biological control of leafrollers (Lepidoptera: Tortricidae) by sowing buckwheat (*Fagopyrum esculentum*) in an orchard. *Biocontrol Science and Technology* **8**: 547–558.

Sunderland, K. and F. Samu. 2000. Effects of agricultural diversification on the abundance, distribution, and pest control potential of spiders: a review. *Entomologia Experimentalis et Applicata* **95**: 1–13.

Szentkiralyi, F. and F. Kozar. 1991. How many species are there in apple insect communities? Testing the resource diversity and intermediate disturbance hypotheses. *Ecological Entomology* **16**: 491–503.

Tedders, W. L. 1983. Insect management in deciduous orchard ecosystems: habitat manipulation. *Environmental Management* **7**: 29–34.

Thies, C. and T. Tscharntke. 1999. Landscape structure and biological control in agroecosystems. *Science* **285**: 893–895.

Thies, C., I. Steffan-Dewenter, and T. Tscharntke. 2003. Effects of landscape context on herbivory and parasitism at different spatial scales. *Oikos* **101**: 18–25.

Thomas, M. B., S. D. Wratten, and N. W. Sotherton. 1991. Creation of "island" habitats in farmland to manipulate populations of beneficial arthropods: predator densities and emigration. *Journal of Applied Ecology* **28**: 906–917.

Tooker, J. F. and L. M. Hanks. 2000. Flowering plant hosts of adult hymenopteran parasitoids of central Illinois. *Annals of the Entomological Society of America* **93**: 580–588.

Tscharntke, T. 2000. Parasitoid populations in the agricultural landscape. In M. E. Hochberg and A. R. Ives (eds.) *Parasitoid Population Biology*. Princeton, NJ: Princeton University Press, pp. 235–253.

Van Rijn, P. C. J. and L. K. Tanigoshi. 1999. The contribution of extrafloral nectar to survival and reproduction of the predatory mite *Iphiseius degenerans* on *Ricinus communis*. *Experimental and Applied Acarology* **23**: 281–296.

Varchola, J. M. and J. P. Dunn. 1999. Changes in ground beetle (Coleoptera: Carabidae) assemblages in farming systems bordered by complex or simple roadside vegetation. *Agriculture, Ecosystems and Environment* **73**: 41–49.

Varchola, J. M. and J. P. Dunn. 2001. Influence of hedgerow and grassy field borders on ground beetle (Coleoptera: Carabidae) activity in fields of corn. *Agriculture, Ecosystems and Environment* **83**: 153–163.

Vet, L. E. M. and M. Dicke. 1992. Ecology of infochemical use by natural enemies in a tritrophic context. *Annual Review of Entomology* **37**: 141–172.

Vet, L. E. M., F. Wäckers, and M. Dicke. 1991. How to hunt for hiding hosts: the reliability–detectability problem in foraging parasitoids. *Netherlands Journal of Zoology* **41**: 202–213.

Wäckers, F. L. 2001. A comparison of nectar- and honeydew sugars with respect to their utilization by the hymenopteran parasitoid *Cotesia glomerata*. *Journal of Insect Physiology* **47**: 1077–1084.

Wäckers, F. L. and W. J. Lewis. 1994. Olfactory and visual learning and their combined influence on host site location by the parasitoid *Microplitis croceipes* (Cresson). *Biological Control* **4**: 105–112.

With, K. A., D. M. Pavuk, J. L. Worchuck, R. K. Oates, and J. L. Fisher. 2002. Threshold effects of landscape structure on biological control in agroecosystems. *Ecological Applications* **12**: 52–65.

White, A. J., S. D. Wratten, N. A. Berry, and U. Weigmann. 1995. Habitat management to enhance biological control of *Brassica* pests by hoverflies (Diptera: Syrphidae). *Journal of Economic Entomology* **88**: 1171–1176.

Yu-hua, Y., Y. Yi, D. Xiang-ge, and Z. Bai-ge. 1997. Conservation and augmentation of natural enemies in pest management of Chinese apple orchards. *Agriculture, Ecosystems and Environment* **62**: 253–260.

Zhao, J. Z., G. S. Ayers, E. J. Grafius, and F. W. Stehr. 1992. Effects of neighboring nectar-producing plants on populations of pest Lepidoptera and their parasitoids in broccoli plantings. *Great Lakes Entomologist* **25**: 253–258.

11

Providing plant foods for natural enemies in farming systems: balancing practicalities and theory

G. M. GURR, S. D. WRATTEN, J. TYLIANAKIS, J. KEAN, AND M. KELLER

Introduction

The need for a theoretical foundation to biological control has often been emphasized (Waage 1990; DeBach and Rosen 1991; Ehler 1994; Sheehan 1994; Barbosa 1998; Gurr *et al.* 1998; Wratten *et al.* 1998; Berryman 1999; Hawkins and Cornell 1999; Landis *et al.* 2000). Enhancement of natural enemy efficacy with floral resources provides a perfect opportunity for this kind of input; however, a divide exists between ecological principles and the needs of practical agriculture. In order for advances in theory to be utilized, protocols based on both ecology and agricultural realism are needed. Partial information, based on anecdote, may lead to the accidental introduction of noxious weeds, and the enhancement of pest populations (Baggen and Gurr 1998) or higher-order predators/hyperparasitoids (Stephens *et al.* 1998). Thus the introduction of non-crop plants as nectar and pollen sources has the potential to cause harm as well as to provide benefits (see also Wilkinson and Landis, Chapter 10). Practical guidelines for employing plant foods in farming systems must be based on sound theoretical and empirical foundations, yet be easily integrated into agricultural and horticultural practice.

We review the ways in which researchers, as well as agronomists and farmers, have attempted to provide plant foods to natural enemies of pests. We discuss the various approaches to the use of flowering plants, and draw a distinction between "shotgun" and "directed" approaches. The shotgun approach is exemplified by a floristically diverse planting (e.g., vineyard ground cover) that is used in the expectation that there will be some net benefit to pest management

Plant-Provided Food for Carnivorous Insects, ed. F. L. Wäckers, P. C. J. van Rijn, and J. Bruin.

Table 11.1 *Contrasting possible steps in, and assumptions of, "directed" and "shotgun" approaches to conservation biological control using plant foods*

Directed approach	Shotgun approach
Field surveys to determine which natural enemies of the key pest are present	Introduce botanically diverse feature (e.g., mixed-species ground cover)
Literature review for available information on ecology of natural enemies and pests	Ignore/accept the risk of one or more plant species having a negative effect
Modeling to predict benefits and avoid risks	Assume/hope net effect of habitat manipulation for pest management is beneficial
Consultation with farmers to determine agronomic acceptability of possible resource plants and avoid risks (e.g., weed potential, product contamination, and toxicity to livestock)	Assume/hope net effect of habitat manipulation is, more broadly, beneficial
Laboratory assays to measure the effect of candidate plant species on important natural enemy species (e.g., longevity, fecundity, and flight propensity).	
Laboratory assays to measure the effect of candidate plant resources on target pest (e.g., to avoid nectar feeding by adult Lepidoptera or foliar feeding by larvae)	
Field experiments to check for unpredicted effects including enhancement of secondary pests or agonists of important natural enemy species	

(see examples in Bugg and Waddington 1994). This approach is informed only by a broad extrapolation from other studies and belief that "diversity helps", rather than an alternative view that "appropriate diversity helps". Reasons why this remains a common practice will be explored and contrasted with directed approaches that use empirical information, ecological (especially community-level) theory or modeling to guide practice. A typical directed approach might involve the steps summarized in Table 11.1. Here, issues like multitrophic level interactions (Janssen *et al.* 1998), intraguild competition (Janssen *et al.* 1998; Rosenheim 1998), resource subsidies (Polis and Strong 1996; Tylianakis *et al.* 2004), life-history omnivory (Polis 1994; Polis and Strong 1996), and ecological engineering (Gurr *et al.* 2004) are relevant. The final and least exploited of the directed approaches is use of modeling. Exploration of population models offers scope to better predict how, for example, enhancing longevity and fecundity of natural enemies with plant foods can affect pest densities.

Approaches to providing plant foods

The requirement of natural enemies for plant food may be divided into two broad categories. The first of these is as an alternative diet when prey are scarce. Although pollen may serve only as a relatively short-term stop-gap diet in some species, it may allow development and reproduction even in the absence of prey in the case of some phytoseiid mites (Van Rijn and Tanigoshi 1999b) and coccinellid beetles (Cottrell and Yeargan 1998). Nectar, also, can be utilized by predatory mites (Van Rijn and Tanigoshi 1999a). A practical application of this general phenomenon is in the citrus orchards in southeast Queensland (Australia) where rows of Rhodes grass ground cover are left to flower (Smith and Papacek 1991). Mowing of alternate rows ensures continuity of pollen availability for predatory mites. This, in turn, allows predator populations to persist within the orchards even when pest eriophyid mites are rare, so preventing future pest outbreaks. Reduced mowing frequency has also been shown to lead to increased abundance of predators and parasitoids in pear orchards (Horton *et al.* 2003). A contrasting study with the coccinellid *Coleomegilla maculata* demonstrated the potential importance of aggregation and enhanced tenure time in addition to dietary effects (Harmon *et al.* 2000). In that study, *C. maculata* aggregated to lucerne interspersed with dandelion. Low densities of aphids were associated with these patches and subsequent cage studies demonstrated that the pollen-producing flowers led to this effect primarily by an increase in tenure time of the omnivorous predator. Pollen-feeding may, however, be counterproductive to biological control if an abundant supply diverts the predator from carnivory as was suggested by work with *C. maculata* in sweet corn (Cottrell and Yeargan 1998).

The second broad category of food use by natural enemies is associated with ontogenetic changes in diet or life-history omnivory (Polis and Strong 1996; see below) in which the adult and immature life stage of species utilize different diets. The most familiar examples of this phenomenon come from parasitic Hymenoptera groups in which larvae are parasitic and adults feed on nectar. In the case of *Copidosoma koehleri*, an egg–larval parasitoid of the potato moth (*Phthorimaea operculella*) both longevity and fecundity are enhanced by nectar (Baggen and Gurr 1998) and enhanced longevity and fecundity have even been shown to translate into increased rates of rose-grain aphid (*Metopolophium dirhodum*) parasitism by *Aphidius rhopalosiphi* (Tylianakis *et al.* 2004). For synovigenic parasitoids, those able to develop more eggs than the complement with which they eclose, access to nectar is of obvious value. However, even strongly pro-ovigenic parasitoids such as *Trichogramma* spp. benefit from adult food via increased longevity and a higher level of activity minimizing the risk of females dying before depositing all eggs (Ashley and Gonzalez 1974; Begum *et al.* 2004a).

Hoverflies (Syrphidae) are also life-history omnivores. The larvae of many species are important predators of aphids but the adults feed on nectar and pollen, the latter being especially important for egg maturation. This has led to the planting of flowers in proximity to cereal crops to provide these resources (Hickman and Wratten 1996).

Although the focus of this chapter is the provision of plant foods to natural enemies in farming systems, it is critical to recognize that farmers have the principal aim of making a profit. This objective may be tempered by secondary aims such as seeking to maximize farm sustainability and alternative income, as well as reducing risk, workload, and environmental impact. Increasing natural enemy density or impact, and even reducing the densities and impact of pests, will be significant only within this far broader frame of reference. Accordingly, the title of this chapter is apt, for even the most ecologically elegant and well-proven technology for enhancing natural enemies will fail to be adopted if it is incompatible with the central aims of agriculture.

Fortunately, there is a growing body of evidence to suggest that many of the approaches used to enhance natural enemies via plant-provided foods are compatible with these aims and have benefits in the form of "multiple-function agricultural biodiversity" (Gurr *et al.* 2002). The types of benefits that may accrue when botanical diversity of agricultural systems is enhanced include: fixation of nitrogen by legume ground covers (e.g., Theunissen *et al.* 1995), weed suppression, pollination enhancement, secondary sources of income from introduced crops that may include trees (e.g., Peng *et al.* 1993) or from the enhanced faunal diversity such as game birds (Rands and Sotherton 1987; Sotherton and Rands 1987), wildlife conservation (Gorman and Reynolds 1993), enhanced esthetics (Kuiper 1997), and carbon sequestration (Dale *et al.* 1994).

An unfortunate consequence of the multiple benefits that *potentially* accrue from enhanced agricultural biodiversity is that it may be used as an excuse for adopting a simplistic "diversity is good" philosophy. This has been evident in some early studies of conservation biological control and is reflected in the availability of "insectary" seed mixes in the USA, Europe, and Australia (see examples in Bugg and Waddington 1994). The addition of any new plant species to farming systems – even if that species is native or naturalized – is associated with risks, including its use as an alternative host for a disease or pest of the crop. The work on *C. koehleri* discussed above showed that use of some plant species to provide nectar led to elevated pest damage because nectar was used by female potato moths increasing their longevity and lifetime fecundity (Baggen and Gurr 1998). Slugs and pest vertebrates also may be favored by introduced landscape features (e.g., weed strips: Balmelli *et al.* 1999). Another risk is that the resources provided may enhance the fourth trophic level (i.e., an agonist of a natural enemy). An instance of this was recently reported from New Zealand when buckwheat, *Fagopyrum esculentum* (Polygonaceae), was sown in orchards to enhance natural enemies of leafrollers (Tortricidae). A parasitoid *Anacharis* sp. (Hymenoptera: Figitidae) was enhanced, leading to lower densities of the beneficial lacewing (*Micromus tasmaniae*) (Stephens *et al.* 1998). Even without the risk of such direct adverse effects, establishing optimal food plant species (e.g., Wäckers *et al.* 1996; Colley and Luna 2000; Hulshof and Jurchenko 2000) and planting patterns – both temporal (e.g., Bowie *et al.* 1995) and spatial (Hossain *et al.* 2002) – is important to ensure maximum benefit with minimum cost and disruption to normal farming practice.

Questions such as which plant species to sow, and when and where to sow it, are important within the context of a given crop system but, even before this point is reached, there are several issues that are rarely explicitly addressed in conservation biological control research. These include the type of agricultural system in which conservation biological control is most likely to succeed, the pest taxa to which this approach is best suited, and which guilds of natural enemies are most amenable to enhancement. Such questions are prompted by Hawkins *et al.*'s (1999) attempt to address the broader question of whether biological control of insects is a natural phenomenon. Their analysis of the insect life-table literature suggested that when top–down control of herbivores occurred in agricultural systems it tended to be due to single parasitoid species for exotic pests on exotic plants. Such a scenario lends itself well to a directed approach (Table 11.1) because the intensive research required to identify the "right kind of diversity" is tractable. In contrast, native herbivores on native plant hosts were most frequently controlled by a suite of generalist predators. This scenario makes more difficult the use of a directed approach, because the resources required to identify the key requirements of more than a modest number of natural enemy species may not be available. This highlights the need to make more efficient the methods by which directed habitat-manipulation strategies are developed and the importance of developing robust theoretical generalities that may reduce the need for case-by-case research projects.

Advances in the discipline of plant-derived foods for natural enemies may make possible the management of all significant pests within a system by the combined use of complementary habitat-manipulation strategies to enhance a suite of different natural enemies. This is a challenging objective for biological control researchers but, in the interim, integrated pest management is increasingly accepted and allows approaches other than biological control (such as host plant resistance, cultural methods, and narrow-spectrum insecticides) to be used to manage pests for which natural enemy impact remains inadequate.

A shotgun approach to providing plant foods to natural enemies, in which multispecies communities of plants are sown or allowed to regenerate or persist, may be acceptable when practiced by farmers. It is indicative of good intentions and – although there is a risk of adverse effects – it is likely to have a net benefit in many cases. However, those involved in researching conservation biological control should adopt more directed approaches to understanding and using plant-derived food use by natural enemies. These directed approaches are explored below in three broad classes, namely based on empirical information, ecological theory, and mathematical modeling.

The importance of empirical information

Empirical information flows from rigorously conducted laboratory and field experiments examining the ecological requirements (especially limiting factors) of predators and parasitoids of significance in specific agricultural and horticultural systems. Evaluations of the effects of differing means of delivering appropriate plant foods are often made (e.g., Mensah 2002), and such an approach can contribute to, as well as draw from, ecological theory.

Several studies have demonstrated that non-crop resources enhance one or more components of natural enemy fitness (e.g., Jervis *et al.* 1993; Hickman *et al.* 1995; Dyer and Landis 1996; Irvin *et al.* 2000; Tylianakis *et al.* 2004), and alter the spatial distribution of natural enemies (Lövei *et al.* 1992; Thomas *et al.* 1992; Liang and Huang 1994). However, few studies both explored the effects of floral resources on natural enemies in the field and tested the mechanisms underlying these effects (e.g., Berndt *et al.* 2002; Tylianakis *et al.* 2004; Begum *et al.* 2004b).

Before a plant species can be used for the enhancement of natural enemy populations, it is imperative to assess the ability of the control agent to utilize its resources (Table 11.1). Studies have been conducted to determine the preferences of a species of natural enemy for different flower species (e.g., Colley and Luna 2000), thereby establishing the optimal species for a given control agent. Although this may also be possible through examination of feeding morphology and floral structure (e.g., Gilbert 1985; Patt *et al.* 1997a,b; Baggen *et al.* 1999), the importance of certainty that the floral resource can be used by the target species makes an empirical approach necessary.

Laboratory work needs to be complemented by field studies to check for possible negative and positive effects. An example of the former is that honeybees (*Apis mellifera*) may outcompete natural enemies for nectar or pollen, negating any benefit for pest management. Conversely, plants that would seem of little use in laboratory tests may actually be useful sources of floral resources. For example, bees may benefit natural enemies by their foraging "tripping" flowers and making nectaries accessible to less powerful species that may otherwise be unable to feed from flowers of plants such as lucerne (*Medicago sativa*). Other Hymenoptera (e.g., *Diadegma insulare*: Idris and Grafius 1997) may bite through the corolla to access nectaries, making available to other natural enemies nectar they otherwise could not reach. Many parasitoids of pests are tiny (e.g., *Trichogramma* spp.) so, though this is yet to be studied, even very small amounts of residual nectar in nectaries, or spilled by earlier foragers, could have a major beneficial effect on their fitness. Because such interspecific interactions can be difficult to predict, work in the field is essential to identify the pest management implications of candidate plant species. Any field use does, however, need to be preceded by an assessment of factors such as hazards (Table 11.1).

Understanding the spatial scale at which the benefits of floral resources may operate remains a major obstacle for the application of conservation biological control. Although numerous studies have demonstrated enhanced suppression of pest populations (Patt *et al.* 1997b; Irvin *et al.* 2000) or higher natural enemy density proximate to floral resources (Van Emden 1963; Smith 1969; Horn 1981; Bugg *et al.* 1991; Costello and Altieri 1995; Theunissen *et al.* 1995; Hickman and Wratten 1996; Lehmhus *et al.* 1996, 1999; Goller *et al.* 1997; Vidal 1997; Stephens *et al.* 1998), fewer have measured the maximum distance over which these effects can be observed (e.g., Chaney 1998; Bowie *et al.* 1999; Hossain *et al.* 1999, 2002). This type of information is difficult to predict, even with extensive knowledge of natural enemy searching and dispersal behavior, and it illustrates the value of field-based empirical studies, including marking and tracking of insect natural enemies (Lavandero *et al.* 2004).

Although empirical studies can provide extensive information to guide the selection and spatial arrangement of food plants, as well as the potential benefits to natural enemies, there are limits to the expectations that can be made of this approach. First, empirical studies generally yield species-specific information, and the time and expense associated with these studies make its use for all combinations of control agent and candidate plant species impossible. Second, though useful patterns may emerge from empirical studies, a directed approach to natural enemy enhancement requires an understanding of the mechanisms in operation and the fundamental principles that guide them. Ecological theory may provide this. We therefore consider next the relevant aspects of theory that may explain the enhancement of natural enemy efficacy with floral resource subsidies. We also examine the potential for ecological information to a second key area of conservation biological control, risk assessment.

Use of ecological theory to improve conservation biological control

Biological control is, by definition, an ecological phenomenon in which some trophic interactions or even entire food webs are manipulated for human benefit. However, the complexity inherent in any ecological system requires detailed understanding through theoretical and empirical analyses before interactions can be manipulated in a predictive manner. It is not enough merely to observe direct, "beneficial" predator–prey or parasitoid–host interactions and to attempt to recreate them in an agricultural setting. The mechanisms driving these interactions, and their indirect effects on other organisms, should be elucidated to confirm that a biological control program is both effective and environmentally responsible (Waage 1990; Ehler 1994; Sheehan 1994; Kareiva 1996; Simberloff and Stiling 1996; Berryman 1999; Gurr and Wratten 1999; Landis *et al.* 2000; Hopper 2001; Strong and Pemberton 2001).

Two main ecological principles are at the heart of natural enemy enhancement by floral foods. The first is the concept of life-history omnivory, whereby a species feeds at different trophic levels during different life-history stages (Polis and Strong 1996). This ecological phenomenon undermines the concept of discrete trophic levels. Many natural enemies, for example certain parasitoids, lacewings, and hoverflies, are carnivorous during their larval stage and become herbivorous as adults. Ecological theory has successfully demolished the discrete trophic-level paradigm and replaced it with complex multispecies food webs and interaction webs (e.g., Hawkins 1992; Polis 1994; Polis and Strong 1996; Janssen *et al.* 1998) (Fig. 11.1). It is the seemingly minor interactions with non-host or non-prey species that have been largely overlooked by classical biological control. Understanding these interactions not only reduces the probability of unforeseen environmental harm (e.g., Strong and Pemberton 2001), but also provides the theoretical tools necessary for successful biological control (Janssen *et al.* 1998; Lewis *et al.* 1998; Berryman 1999; Gurr and Wratten 1999, 2000; Landis *et al.* 2000).

The second component of ecological theory that is integral to conservation biological control is that of resource subsidies. Concomitant with the breakdown of the trophic-level paradigm was the understanding that many species obtain resources from outside their target habitat. These "spatial subsidies" (Polis and Strong 1996) allow an increase in consumer abundance, beyond that which can be sustained by the resources present within the local habitat alone (Polis 1994; Polis and Strong 1996). Although this concept was originally used to describe the input of detritus into stream ecosystems (Polis and Strong 1996), an analogous process occurs when natural enemies feed on non-crop plants. Floral

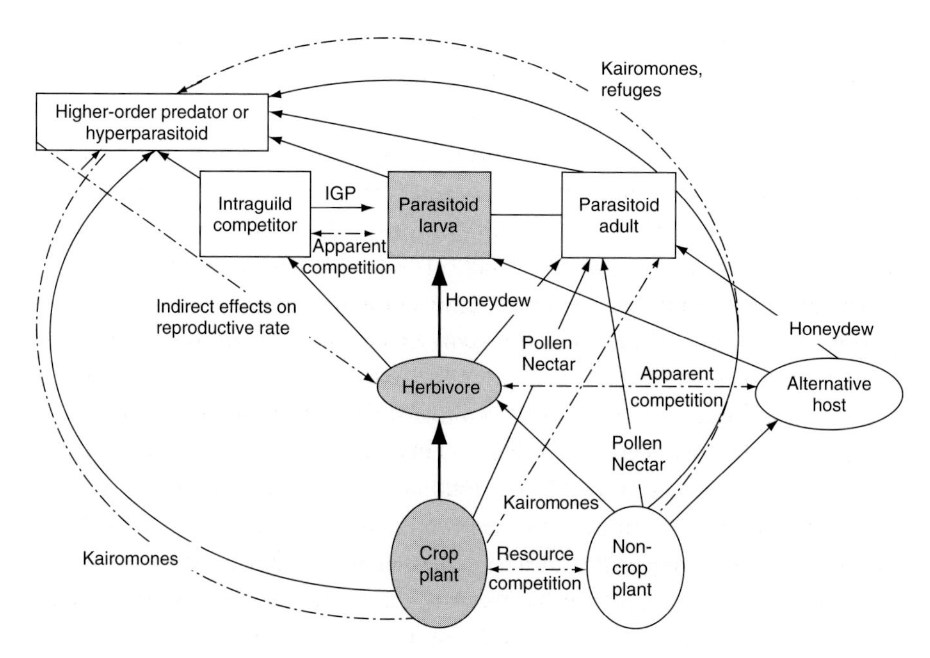

Figure 11.1 Simplified food web, depicting the potential interactions involving a hypothetical parasitoid (larva and adult) that receives resource subsidies (non-crop plant), its herbivorous host, and the crop plant on which the host feeds. Shaded areas and bold arrows represent a simplified food-chain concept. Arrows indicate the direction of energy flow in trophic interactions (solid lines) or the direction of some kind of benefit from non-trophic interactions (dotted lines). IGP, intraguild predation.

subsidies can occur within the crop habitat and should not necessarily be described as "spatial"; however, as they do not form part of the crop, they must be viewed as a distinct resource. The term "resource subsidies" is therefore adopted to describe the provision of any non-crop plant or resource, contained inside or outside of the crop environment, from which a natural enemy may derive benefits. This is particularly important in agricultural systems, where expansive monocultures are typical. The resources present in a simplified crop system alone may not be capable of sustaining a buoyant biological control agent population capable of exerting top–down (i.e., recipient) control. The impoverished nature of conventional agricultural ecosystems is highlighted by the significant positive impact resource subsidies can have on natural enemies (e.g., Hickman and Wratten 1996; Patt *et al.* 1997b; Baggen *et al.* 2000; Irvin *et al.* 2000; Tylianakis *et al.* 2004).

Resource subsidies allow the practical application of ecological theory to biological control. Several studies have shown that floral resources allow parasitoids to maximize their reproductive success via increases in longevity and egg load (Arthur 1944; Jervis *et al.* 1993, 2004; Dyer and Landis 1996; Wheeler 1996;

Heimpel *et al.* 1997; Jacob and Evans 2000; Johanowicz and Mitchell 2000; Sagarra *et al.* 2001), and that this may lead to reduced populations of arthropod pests in the field (Patt *et al.* 1997b; Irvin *et al.* 2000). The bottom–up (donor controlled) subsidy of parasitoids therefore improves top–down control of herbivores proximal to floral patches (Polis 1994). This seemingly paradoxical statement illustrates the concept that systems that appear to be structured by recipient control may often be partially or wholly donor controlled (Hawkins 1992; Polis 1994; Polis and Strong 1996; Rosenheim 1998). Therefore, food-web engineering by provision of floral resource subsidies (though often supported by bottom–up effects, e.g., the resource-concentration effects of Root (1972)) is not only effective in the enhancement of natural enemies, but its use is based on a sound foundation of ecological theory. This theoretical and mechanism-based approach to biological control is imperative if success rates are to improve.

Finally, ecological theory can inform the spatiotemporal positioning of floral resources so that they provide maximum benefits. Several theoretical models predict the minimum habitat requirements for the conservation of a species (e.g., Doncaster *et al.* 1996), and these can be related to the conservation of a natural enemy. Additionally, theory related to habitat fragmentation and species loss has been extended and experimentally tested to determine the effects of fragmentation and isolation on biological control. For example, the trophic-level hypothesis of island biogeography was supported by work in vetch meadows of varying size and level of isolation conducted in conjunction with potted vetch plants placed at various distances from meadows (Kruess and Tscharntke 2002). Parasitoids suffered more severely from habitat loss than herbivores, and their minimum area requirements were calculated to be higher than for herbivores. This was reflected in a decline in parasitism rate with increasing isolation. A broadly similar finding came from an earlier study with red clover. Herbivores within isolated patches experienced 19–60% of the parasitism measured for populations in non-isolated (connected) red clover patches (Kruess and Tscharntke 1994). Addressing such spatial issues will be an important challenge in future attempts to optimally provide floral resources to natural enemies. The types of practical decisions that could be informed by theory include the optimal spacing of floral strips within crops, determining the maximum size of a crop that can be "protected" by margin plantings rather than within-crop plantings, and layout of differing crop species in alley farming.

Temporal effects also may be important. Clearly, floral resource subsidies should coincide with the seasonality of a given predator or parasitoid and the phenology of crop plants, if the full potential for this type of enhancement is to be realized (Landis *et al.* 2000).

Use of ecological theory to assess risks

Risk analyses of biological control programs often center on host-range evaluation as a measure of the potential for non-target effects (e.g., Lynch and Ives 1999). Indeed natural enemy host-range expansion comprises a major environmental risk of classical biological control introductions (Secord and Kareiva 1996; Howarth 2000; Hopper 2001), and ecological theory may aid in the assessment of these risks (Strand and Obrycki 1996; Lonsdale *et al.* 2001). However, conservation biological control most commonly involves the enhancement of populations of natural enemies that are already present in the system, so assessment of the potential for host-range expansion is not imperative. Other non-target effects of conservation biological control must nonetheless be considered before floral resource subsidies can be applied responsibly to an agricultural setting.

Possibly the greatest environmental threat posed by non-crop resource subsidies is the potential for exotic plant species to become invasive (e.g., Cheesman 1998). This is a very real risk associated with the shotgun approach (above), as even a plant that is not invasive in its home range must be considered a potential threat. For example, wild teasel, *Dipsacus fullonum* (Dipsacaceae), often occurs in field boundaries in the United Kingdom, where it provides floral resources for potential crop pollinators and natural enemies of cereal pests. Although this plant may even require conservation in the United Kingdom, it has become invasive in North America, where it threatens to displace rare native species (Cheesman 1998). This type of risk can be alleviated by a more directed approach, including knowledge of the ecology of the plant species concerned, as is advocated in weed risk assessment for plant introductions to new localities (Pheloung *et al.* 1999). For example, buckwheat has been used to enhance populations of natural enemies of leafrollers on apples in the South Island of New Zealand (Stephens *et al.* 1998; Irvin *et al.* 2000). Winter frosts prevent the persistence and spread of buckwheat as a weed by killing existing adults and preventing flowering of plants that develop from overwintered seeds (Bowie *et al.* 1995). One approach for formalizing such risk assessments used a graded-weighted checklist to select optimal (low risk) food plants in a quantitative fashion (Gurr *et al.* 1998).

There are also several potential indirect effects that may reduce the effectiveness of a conservation biological control program, or contribute to environmental harm. Although intraguild competition and predation have received considerable attention with regards to the introduction of exotic species for biological control (Ehler and Hall 1982; Tallamy 1983; Mills 1992, 1994; Janssen *et al.* 1998), these phenomena may also influence the success of natural enemy enhancement using floral resources. Many species of control agent express some degree of omnivory but floral feeding is by no means ubiquitous among insect natural enemies. Species that are solely carnivorous throughout their entire life history may suffer competitive exclusion when populations of other members of the guild are enhanced by resource subsidies. Enhancement of one species within a guild may lead to declining populations of guild members that do not receive a subsidy (Polis 1994) and impair overall control of the pest. Moreover, enhancement of populations of a predator or parasitoid using floral resource subsidies may lead to suppression of alternative hosts or prey. Such risks highlight the desirability of avoiding a shotgun approach to providing food plants but also illustrate that achieving adequate suppression of multiple pest species within a given crop system may not always be tractable.

Another potential risk of providing plant foods for agricultural benefits relates to enhancement of species other than the natural enemy. It is reasonable to assume that while natural enemies are attracted to flowers and benefit from resource subsidies, so too may predators and hyperparasitoids of the natural enemies, as well as the pests themselves. The vast majority of studies focus on enhancement of primary parasitoids by floral feeding; however, several species of hyperparasitoid have also been observed to visit flowers (Jervis *et al.* 1993). Although the effects of floral feeding on the efficiency of higher trophic levels like hyperparasitoids remain to be determined, a parasitoid (*Anacharis* sp.) of the brown lacewing has been shown to benefit from floral buckwheat resources (Stephens *et al.* 1998; see above). Such enhancement of higher-order parasitoids or predators by floral resource subsidies is potentially inimical to conservation biological control. However, if it is possible to select plants of which floral morphology or nectar quality are able to be utilized by natural enemies and not pests (Baggen and Gurr 1998; Wäckers 2001; Van Rijn *et al.* 2002; see also Sabelis *et al.*, Chapter 4), it may be possible to select plants that can be utilized by natural enemies and not higher-order predators.

The potential interactions between non-crop resource subsidies and competitors or predators of a hypothetical natural enemy are presented in Fig. 11.1. These not only illustrate the redundancy of simple food-chain models, but also highlight potential interactions that may be the subject of future research (e.g., the indirect effects of a hyperparasitoid on rates of primary host reproduction: Van Veen *et al.* 2001). Further research into the effects of floral resources on higher-order predators and hyperparasitoids may prove beneficial in reducing the risk of failure of conservation biological control.

Use of mathematical modeling to guide practice

Modeling constitutes another means by which habitat manipulation to enhance natural enemies via plant foods may be made more directed.

Population models have been widely used in classical biocontrol, as a framework for understanding pest–enemy interactions (e.g., Hassell 1980), determining the important characteristics of enemies for control of particular pests (e.g., Kean and Barlow 2001), designing optimal release strategies (e.g., Godfray and Waage 1991), and predicting pest suppression levels (e.g., Beddington *et al.* 1978). In contrast, the potential for population models to help understand and optimize conservation biological control practices has been largely ignored. This may reflect the relative youth of conservation biological control as a discipline, or perhaps practitioners have been wary of making the very specific assumptions and simplifications required to construct the models, especially since many of the simplest enemy–host models have notoriously unstable dynamics (May 1981).

Nevertheless, a handful of recent papers have shown how population models can be useful tools for exploring the effects of natural enemy augmentation. Van Rijn *et al.* (2002) used a combined approach of population modeling and experiments to show how providing pollen increases the reproductive rate of a predatory mite and thereby reduces the numbers of damaging herbivorous thrips on greenhouse cucumbers. The value of the modeling was in integrating the biological knowledge to provide a rigorous explanation for the observed effects, and then being able to explore the implications of different management actions beyond the scope of what was experimentally tested. These authors use population models to further explore the effects of food subsidies on more general predator–prey interactions in Chapter 8 of this volume. A generalized approach was also adopted by Kean *et al.* (2003), who asked: "Given that different types of food subsidies affect different aspects of natural enemy biology, what level of enhancement do we need to aim for in order to achieve the required reduction in pest density?" They explored simple models to conclude that enemy search rate or prey conversion efficiency have a bigger effect on prey density than predator consumption rate or parasitoid fecundity, but the effect of enemy longevity depends on how this interacts with other parameters. The degree of enhancement required for each natural enemy parameter depended only on the desired reduction in pests and on the maximum net reproductive rate of the dominant enemy, which is relatively easily measured.

Although these models necessarily sacrifice some biological detail for the sake of conceptual simplicity, they do demonstrate the ability of population modeling to help target conservation biological control practice and research. A real strength of models is that, unlike much of the other theory discussed in this chapter, they lead to quantitative as well as qualitative conclusions. Therefore, models allow us to ask, and answer, questions about "how much?" and "when?" in relation to pest management. With few studies to date, there is considerable potential for further use of population models in conservation biological control, both to help tease out the complex ecological interactions involved and as powerful tools for targeting management and research to where it is likely to be most effective.

Combining theory and practice: how, where, when

Floral resource subsidies have great potential to alleviate the limitations placed on insect natural enemies by modern agricultural practices. The potential for improved biological control by predators or parasitoids that receive resource subsidies is predicted by food-web theory, but care must be taken. The shotgun approach of diverse planting for its own sake may result in non-crop plants becoming weedy or competing with one another, or the enhancement of populations of herbivorous pests or higher-order predators/hyperparasitoids. This could lead to disillusionment of agricultural and horticultural practitioners with conservation biological control, and hinder further, informed plantings. In order for a conservation biological control strategy to be effective, its implementation must be guided by empirical and theoretical research.

Essentially, farmers will be concerned with practical questions such as "what?", "where?", and "when?" Researchers are increasingly able to answer these with guidance on issues such as *what* food plants should be used, *where* they should be positioned in relation to the crop for maximum benefit, and *when* to sow or slash the plants to ensure nectar and pollen are available over the desired period(s). Such research will require further rigorous empirical studies as the level of interest in conservation biological control grows in farming communities. However, to fully meet the potential for food plants in pest management, the underutilized population modeling and ecological theory approaches need to be developed. This will require researchers to more consistently address the other key questions: "how?" and "why?" Developing general theories of *how* floral subsidies affect food webs and *why* only a minority of cases of natural enemy enhancement translates into reduced crop damage (Gurr *et al.* 2000) will be critical.

Population modeling in particular has a great deal of unrealized potential to help quantify conservation biological control aims, and to target research and management effort towards those aspects of pest and enemy ecology that have the greatest influence over the success or failure of pest management. We have considered simple predator–prey models in relation to how floral resources can affect pest densities by attracting natural enemies or by enhancing their efficacy, but much work needs to be done to experimentally verify and refine such models.

Finally, the broader context of agriculture must be considered, as any techniques for natural enemy enhancement that conflict with practical farming will remain solely theoretical. In order for this theory to be put into practice, economic requirements of farmers must be met. For example, if one-third of a crop must be replanted with non-crop floral resources before a significant level of natural enemy enhancement can be achieved, this method (irrespective of its theoretical benefits) will never be utilized. Fortunately, such levels of agronomic disruption are unlikely to be necessary, as improved pest management may require as little as 1 in 20 rows to be planted with floral resources (Grossman and Quarles 1993) or for the crop itself to provide key resources (Hossain *et al.* 2002).

References

Arthur, D. R. 1944. *Aphidius granarus*, Marsh., in relation to its control of *Myzus kaltenbachi*, Schout. *Bulletin of Entomological Research* **35**: 257–270.

Ashley, T. R. and D. Gonzalez. 1974. Effect of various food substances on longevity and fecundity of *Trichogramma*. *Environmental Entomology* **3**: 169–171.

Baggen, L. R. and G. M. Gurr. 1998. The influence of food on *Copidosoma koehleri* (Hymenoptera: Encyrtidae), and the use of flowering plants as a habitat management tool to enhance biological control of potato moth, *Phthorimaea operculella* (Lepidoptera: Gelechiidae). *Biological Control* **11**: 9–17.

Baggen, L. R., G. M. Gurr, and A. Meats. 1999. Flowers in tri-trophic systems: mechanisms allowing selective exploitation by insect natural enemies for conservation biological control. *Entomologia Experimentalis et Applicata* **91**: 155–161.

2000. Field observations on selective food plants in habitat manipulation for biological control of potato moth by *Copidosoma koehleri* Blanchard (Hymenoptera: Encyrtidae). In A. D. Austin and M. Dowton (eds.) *Hymenoptera: Evolution, Biodiversity and Biological Control.* Collingwood, Australia: CSIRO, pp. 388–395.

Balmelli, L., W. Nentwig, and J. P. Airoldi. 1999. Food preferences of the common vole *Microtus arvalis* in the agricultural landscape with regard to nutritional components of plants. *Zeitschrift für Saugetierkunde* **64**: 154–168. (In German)

Barbosa, P. (ed.) 1998. *Conservation Biological Control.* San Diego, CA: Academic Press.

Beddington, J. R., C. A. Free, and J. H. Lawton. 1978. Characteristics of successful natural enemies in models of biological control of insect pests. *Nature* **273**: 513–519.

Begum, M., G. M. Gurr, and S. D. Wratten, 2004a. Flower colour affects tri-trophic biocontrol interactions. *Biological Control* **30**: 584–590.

Begum, M., G. M. Gurr, S. D. Wratten P. Hedberg, and H. I. Nicol. 2004b. The effect of floral nectar on the grapevine leafroller parasitoid, *Trichogramma carverae*. *International Journal of Ecology and Envrionmental Sciences* **30**: 3–12.

Berndt, L. A., S. D. Wratten, and P. G. Hassan. 2002. Effects of buckwheat flowers on leafroller (Lepidoptera: Tortricidae) parasitoids in a New Zealand vineyard. *Agricultural and Forest Entomology* **4**: 39–45.

Berryman, A. A. 1999. The theoretical foundations of biological control. In B. A. Hawkins and H. V. Cornell (eds.) *Theoretical Approaches to Biological Control.* Cambridge, UK: Cambridge University Press, pp. 3–21.

Bowie, M. H., G. M. Gurr, Z. Hossain, L. R. Baggen, and C. M. Frampton. 1999. Effects of distance from field edge on aphidophagous insects in a wheat crop and observations on trap design and placement. *International Journal of Pest Management* **45**: 69–73.

Bowie, M. H., S. D. Wratten, and A. J. White. 1995. Agronomy and phenology of "companion plants" of potential for enhancement of insect biological control. *New Zealand Journal of Crop and Horticultural Science* **23**: 423–427.

Bugg, R. L. and C. Waddington. 1994. Using cover crops to manage arthropod pests of orchards: a review. *Agriculture Ecosystems and Environment* **50**: 11–28.

Bugg, R. L., J. D. Dutcher, and P. J. McNeill. 1991. Cool-season cover crops in the pecan orchard understory: effects on Coccinellidae (Coleoptera) and pecan aphids (Homoptera: Aphididae). *Biological Control* **1**: 8–15.

Chaney, W. E. 1998. Biological control of aphids in lettuce using in-field insectaries. In C. H. Pickett and R. L. Bugg (eds.) *Enhancing Biological Control: Habitat Management to Promote Natural Enemies of Agricultural Pests.* Berkeley, CA: University of California Press, pp. 73–83.

Cheesman, O. D. 1998. The impact of some field boundary management practices on the development of *Dipsacus fullonum* L. flowering stems and implications for conservation. *Agriculture, Ecosystems and Environment* **68**: 41–49.

Colley, M. R. and J. M. Luna. 2000. Relative attractiveness of potential beneficial insectary plants to aphidophagous hoverflies (Diptera: Syrphidae). *Environmental Entomology* **29**: 1054–1059.

Costello, M. J. and M. A. Altieri. 1995. Abundance, growth-rate and parasitism of *Brevicoryne brassicae* and *Myzus persicae* (Homoptera, Aphididae) on broccoli growing in living mulches. *Agriculture, Ecosystems and Environment* **52**: 187–196.

Cottrell, T. E. and K. V. Yeargan. 1998. Effect of pollen on *Coleomegilla maculata* (Coleoptera: Coccinellidae) population density, predation, and cannibalism in sweet corn. *Environmental Entomology* **27**: 1402–1410.

Dale, V. H., R. V. O'Neill, F. Southworth, and M. Pedlowski. 1994. Modeling effects of land management in the Brazilian Amazonian settlement of Rondonia. *Conservation Biology* **8**: 196–206.

DeBach, P. and D. Rosen. 1991. *Biological Control by Natural Enemies*, 2nd edn. Cambridge, UK: Cambridge University Press.

Doncaster, C. P., T. Micol, and S. P. Jensen. 1996. Determining the minimum habitat requirements in theory and practice. *Oikos* **75**: 335–339.

Dyer, L. E. and D. A. Landis. 1996. Effects of habitat, temperature, and sugar availability on longevity of *Eriborus terebrans* (Hymenoptera: Ichneumonidae). *Environmental Entomology* **25**: 1192–1201.

Ehler, L. E. 1994. Parasitoid communities, parasitoid guilds, and biological control. In B. A. Hawkins and W. Sheehan (eds.) *Parasitoid Community Ecology*. Oxford, UK: Oxford University Press, pp. 418–436.

Ehler, L. E. and R. W. Hall. 1982. Evidence for competitive exclusion of introduced natural enemies in biological control. *Environmental Entomology* **11**: 1–4.

Gilbert, F. S. 1985. Ecomorphological relationships in hoverflies (Diptera: Syrphidae). *Proceedings of the Royal Society of London Series B* **224**: 91–95.

Godfray, H. C. J. and J. K. Waage. 1991. Predictive modelling in biological control: the mango mealy bug (*Rastrococcus invadens*) and its parasitoid. *Journal of Applied Ecology* **28**: 434–453.

Goller, E., L. Nunnenmacher, and H. E. Goldbach. 1997. Faba beans as a cover crop in organically grown hops: influence on aphids and aphid antagonists. *Biological Agriculture and Horticulture* **15**: 279–284.

Gorman, M. L. and P. Reynolds. 1993. The impact of land-use change on voles and raptors. *Mammal Review* **23**: 121–126.

Grossman, J. and W. Quarles. 1993. Strip intercropping for biological control. *The IPM Practitioner* **15**: 1–11.

Gurr, G. M., H. F. Van Emden, and S. D. Wratten. 1998. Habitat manipulation and natural enemy efficiency: implications for the control of pests. In P. Barbosa (ed.) *Conservation Biological Control*. San Diego, CA: Academic Press, pp. 155–183.

Gurr, G. M. and S. D. Wratten. 1999. "Integrated biological control": a proposal for enhancing success in biological control. *International Journal of Pest Management* **45**: 81–84.

2000. Preface. In G. M. Gurr and S. D. Wratten (eds.) *Biological Control: Measures of Success*. Dordrecht, the Netherlands: Kluwer Academic Publishers, pp. 1–3.

Gurr, G. M., S. D. Wratten, and M. A. Altien (eds.). 2004. *Ecological Engineering: Advances in Habitat Management for Arthropods*. Melbourne, Australia: CSIRO Publishing.

Gurr, G. M., S. D. Wratten, and P. Barbosa. 2000. Success in conservation biological control of arthropods. In G. M. Gurr and S. D. Wratten (eds.) *Biological Control: Measures of Success*. Dordrecht, the Netherlands: Kluwer Academic Publishers, pp. 105–132.

Gurr, G. M., S. D. Wratten, and J. Luna. 2002. Multi-function agricultural biodiversity: pest management and other benefits. *Basic and Applied Ecology* **4**: 107–116.

Harmon, J. P., A. R. Ives, J. E. Losey, A. C. Olson, and K. S. Rauwald. 2000. *Coleomegilla maculata* (Coleoptera: Coccinellidae) predation on pea aphids promoted by proximity to dandelions. *Oecologia* **125**: 543–548.

Hassell, M. P. 1980. Foraging strategies, population models and biological control: a case study. *Journal of Animal Ecology* **49**: 603–628.

Hawkins, B. A. 1992. Parasitoid–host food webs and donor control. *Oikos* **65**: 159–162.

Hawkins, B. A. and H. V. Cornell (eds.). 1999. *Theoretical Approaches to Biological Control*. Cambridge, UK: Cambridge University Press.

Hawkins, B. A., N. J. Mills, M. A. Jervis, and P. W. Price. 1999. Is the biological control of insects a natural phenomenon? *Oikos* **86**: 493–506.

Heimpel, G. E., J. A. Rosenheim, and D. Kattari. 1997. Adult feeding and lifetime reproductive success in the parasitoid *Aphytis melinus*. *Entomologia Experimentalis et Applicata* **83**: 305–315.

Hickman, J. M. and S. D. Wratten. 1996. Use of *Phacelia tanacetifolia* strips to enhance biological control of aphids by hoverfly larvae in cereal fields. *Journal of Economic Entomology* **89**: 832–840.

Hickman, J. M., G. L. Lövei, and S. D. Wratten. 1995. Pollen feeding by adults of the hoverfly *Melanostoma fasciatum* (Diptera: Syrphidae). *New Zealand Journal of Zoology* **22**: 387–392.

Hopper, K. R. 2001. Research needs concerning non-target impacts of biological control introductions. In E. Wajnberg, J. K. Scott, and P. C. Quimby (eds.) *Evaluating Indirect Ecological Effects of Biological Control*. Wallingford, UK: CAB International, pp. 39–56.

Horn, D. J. 1981. Effect of weedy backgrounds on colonization of collards by green peach aphid, *Myzus persicae*, and its major predators. *Environmental Entomology* **10**: 285–289.

Horton, D. R., D. A. Broers, R. R. Lavis, *et al.* 2003. Effects of moving frequency on densities of natural enemies in three Pacific Northwest pear orchards. *Entomologia Experimentalis et Applicata* **106**: 135–145.

Hossain, Z., G. M. Gurr, and S. D. Wratten. 1999. Effects of harvest on survival and dispersal of insect predators in hay lucerne. *Biological Agriculture and Horticulture* **17**: 339–348.

2002. Habitat manipulation in lucerne (*Medicago sativa* L.): arthropod population dynamics in harvested and "refuge" crop strips. *Journal of Applied Ecology* **39**: 445–454.

Howarth, F. G. 2000. Non-target effects of biological control agents. In G. M. Gurr and S. D. Wratten (eds.) *Biological Control: Measures of Success*. Dordrecht, the Netherlands: Kluwer Academic Publishers, pp. 369–404.

Hulshof, J. and O. Jurchenko. 2000. *Orius laevigatus* in a choice situation: thrips or pollen. *Mededelingen Faculteit Landbouwkundige en Toegepaste Biologische Wetenschappen Universiteit Gent* **65**: 351–358.

Idris, A. B. and E. Grafius. 1997. Nectar-collecting behaviour of *Diadegma insulare* (Hymenoptera: Ichneumonidae), a parasitoid of diamondback moth (Lepidoptera: Plutellidae). *Biological Control* **26**: 114–120.

Irvin, N. A., S. D. Wratten, and F. M. Frampton. 2000. Understorey management for the enhancement of the leafroller parasitoid *Dolichogenidea tasmanica* (Cameron) in orchards at Canterbury, New Zealand. In A. D. Austin and M. Dowton (eds.) *Hymenoptera: Evolution, Biodiversity and Biological Control*. Collingwood, Australia: CSIRO, pp. 396–403.

Jacob, H. S. and E. W. Evans. 2000. Influence of carbohydrate foods and mating on longevity of the parasitoid *Bathyplectes curculionis* (Hymenoptera: Ichneumonidae). *Environmental Entomology* **29**: 1088–1095.

Janssen, A., A. Pallini, M. Venzon, and M. W. Sabelis. 1998. Behaviour and indirect interactions in food webs of plant-inhabiting arthropods. *Experimental and Applied Acarology* **22**: 497–521.

Jervis, M. A., N. A. C. Kidd, M. G. Fitton, T. Huddleston, and H. A. Dawah. 1993. Flower-visiting by hymenopteran parasitoids. *Journal of Natural History* **27**: 67–105.

Jervis, M. A., J. C. Lee, and G. E. Heimpel. 2004. Conservation biological control using arthropod predators and parasitoids: the role of behavioural and life-history studies. In G. M. Gurr, S. D. Wratten, and M. A. Altieri (eds.) *Ecological Engineering: Advances in Habitat Manipulation for Arthropods*. Melboune, Australia: CSIRO Publishing, pp. 65–100.

Johanowicz, D. L. and E. R. Mitchell. 2000. Effects of sweet alyssum flowers on the longevity of the parasitoid wasps *Cotesia marginiventris* (Hymenoptera: Braconidae) and *Diadegma insulare* (Hymenoptera: Ichneumonidae). *Florida Entomologist* **83**: 41–47.

Kareiva, P.1996. Contributions of ecology to biological control. *Ecology* **77**: 1963–1964.

Kean, J. M. and N. D. Barlow. 2001. A spatial model for the successful biological control of *Sitona discoideus* by *Microctonus aethiopoides*. *Journal of Applied Ecology* **38**: 162–169.

Kean, J. M., S. D. Wratten, J. Tylianakis, and N. D. Barlow. 2003. The population consequences of natural enemy enhancement, and implications for conservation biological control. *Ecology Letters* **6**: 604–612.

Kruess, A. and T. Tscharntke. 1994. Habitat fragmentation, species loss, and biological control. *Science* **264**: 1581–1584.

2002. Contrasting responses of plant and insect diversity to variation in grazing intensity. *Biological Conservation* **106**: 293–302.

Kuiper, J. 1997. Organic mixed farms in the landscape of a brook valley: how can a co-operative of organic mixed farms contribute to ecological and aesthetic qualities of a landscape? *Agriculture, Ecosystems and Environment* **63**: 121–132.

Landis, D. A., S. D. Wratten, and G. M. Gurr. 2000. Habitat management to conserve natural enemies of arthropod pests in agriculture. *Annual Review of Entomology* **45**: 175–201.

Lavandero, B., S. D. Wratten, and J. Tylianakis. 2004. Marking and tracking techniques for insect predators and parasitoids in ecological engineering. In G. M. Gurr, S. D. Wratten, and M. A. Altieri (eds.) *Ecological Engineering: Advances in Habitat Manipulation for Arthropods*. Melbourne, Australia: CSIRO Publishing, pp. 117–132.

Lewis, W. J., J. O. Stapel, A. M. Cortesero, and K. Takasu. 1998. Understanding how parasitoids balance food and host needs: importance to biological control. *Biological Control* **11**: 175–183.

Liang, W. and M. Huang. 1994. Influence of citrus orchard ground cover plants on arthropod communities in China: a review. *Agriculture, Ecosystems and Environment* **50**: 29–37.

Lonsdale, W. M., D. T. Briese, and J. M. Cullen. 2001. Risk analysis and weed biological control. In E. Wajnberg, J. K. Scott, and P. C. Quimby (eds.) *Evaluating Indirect Ecological Effects of Biological Control*. Wallingford, UK: CAB International, pp. 185–210.

Lövei, G. L., D. McDougall, G. Bramley, D. J. Hodgson, and S. D. Wratten. 1992. Floral resources for natural enemies: the effect of *Phacelia tanacetifolia* (Hydrophyllaceae) on within-field distribution of hoverflies (Diptera: Syrphidae). Proc. 45th New Zealand Plant Protection Conference, pp. 60–61.

Lynch, L. D. and A. R. Ives. 1999. The use of population models in informing non-target risk assessment in biocontrol. *Aspects of Applied Biology* **53**: 181–188.

May, R. M. 1981. Models for two interacting populations. In R. M. May (ed.) *Theoretical Ecology: Principles and Applications*. Oxford, UK: Blackwell Scientific Publications, pp. 78–104.

Mensah, R. K. 2002. Development of an integrated pest management programme for cotton. I. Establishing and utilizing natural enemies. *International Journal of Pest Management* **48**: 87–94.

Mills, N. J. 1992. Parasitoid guilds, life-styles and host ranges in the parasitoid complexes of tortricoid hosts (Lepidoptera: Tortricoidea). *Environmental Entomology* **21**: 230–239.

1994. Parasitoid guilds: a comparative analysis of the parasitoid communities of tortricids and weevils. In B. A. Hawkins and W. Sheehan (eds.) *Parasitoid Community Ecology*. Oxford, UK: Oxford University Press, pp. 30–46.

Patt, J. M., G. C. Hamilton, and J. H. Lashomb. 1997a. Foraging success of parasitoid wasps on flowers: interplay of insect morphology, floral architecture and searching behavior. *Entomologia Experimentalis et Applicata* **83**: 21–30.

1997b. Impact of strip-insectary intercropping with flowers on conservation biological control of the Colorado potato beetle. *Advances in Horticultural Science* **11**: 175–181.

Peng, R. K., L. D. Incoll, S. L. Sutton, C. Wright, and A. Chadwick. 1993. Diversity of airborne arthropods in a silvoarable agroforestry system. *Journal of Applied Ecology* **30**: 551–562.

Pheloung, P. C., P. A. Williams, and S. R. Halloy. 1999. A weed risk assessment model for use as a biosecurity tool evaluating plant introductions. *Journal of Environmental Management* **57**: 239–252.

Polis, G. A. 1994. Food webs, trophic cascades and community structure. *Australian Journal of Ecology* **19**: 121–136.

Polis, G. A. and D. R. Strong. 1996. Food web complexity and community dynamics. *American Naturalist* **147**: 813–846.

Rands, M. R. W. and N. W. Sotherton. 1987. The management of field margins for the conservation of gamebirds. *British Crop Protection Council Monograph* **35**: 95–104.

Root, R. B. 1972. Organization of a plant–arthropod association in simple and diverse habitats: the fauna of collards (*Brassica oleracea*). *Ecological Monographs* **43**: 95–124.

Rosenheim, J. A. 1998. Higher-order predators and the regulation of insect herbivore populations. *Annual Review of Entomology* **43**: 421–447.

Sagarra, L. A., C. Vincent, and R. K. Stewart. 2001. Body size as an indicator of parasitoid quality in male and female *Anagyrus kamali* (Hymenoptera: Encyrtidae). *Bulletin of Entomological Research* **91**: 363–367.

Secord, D. and P. Kareiva. 1996. Perils and pitfalls in the host specificity paradigm. *BioScience* **46**: 448–454.

Sheehan, W. 1994. Parasitoid community structure: effects of host abundance, phylogeny, and ecology. In B. A. Hawkins and W. Sheehan (eds.) *Parasitoid Community Ecology*. Oxford, UK: Oxford University Press, pp. 90–107.

Simberloff, D. and P. Stiling. 1996. How risky is biological control? *Ecology* **77**: 1965–1974.

Smith, D. and D. F. Papacek. 1991. Studies of the predatory mite *Amblyseius victoriensis* (Acarina: Phytoseiidae) in citrus orchards in southeast Queensland: control of *Tegolophus australis* and *Phyllocoptruta oleivora* (Acarina: Eriophyidae), effect of pesticides, alternative host plants and augmentative release. *Experimental and Applied Acarology* **12**: 195–217.

Smith, J. G. 1969. Some effects of crop background on populations of aphids and their natural enemies on brussels sprouts. *Annals of Applied Biology* **63**: 326–333.

Sotherton, N. W. and M. R. W. Rands. 1987. The environmental interest of field margins to game and other wildlife: a game conservancy view. *British Crop Protection Council Monograph* **35**: 67–75.

Stephens, M. J., C. M. France, S. D. Wratten, and C. Frampton. 1998. Enhancing biological control of leafrollers (Lepidoptera: Tortricidae) by sowing buckwheat (*Fagopyrum esculentum*) in an orchard. *Biocontrol Science and Technology* **8**: 547–558.

Strand, M. R. and J. J. Obrycki. 1996. Host specificity of insect parasitoids and predators. *BioScience* **46**: 422–430.

Strong, D. R. and R. W. Pemberton. 2001. Food webs, risks of alien enemies and reform of biological control. In E. Wajnberg, J. K. Scott, and P. C. Quimby (eds.) *Evaluating Indirect Ecological Effects of Biological Control*. Wallingford UK: CAB International pp. 57–79.

Tallamy, D. W. 1983. Equilibrium biogeography and its application to insect host–parasite systems. *American Naturalist* **121**: 244–254.

Theunissen, J., C. J. H. Booij, and L. A. P. Lotz. 1995. Effects of intercropping white cabbage with clovers on pest infestation and yield. *Entomologia Experimentalis et Applicata* **74**: 7–16.

Thomas, M. B., N. W. Sotherton, D. S. Coombes, and S. D. Wratten. 1992. Habitat factors influencing the distribution of polyphagous predatory insects between field boundaries. *Annals of Applied Biology* **120**: 197–202.

Tylianakis, J. M., R. K. Didham, and S. D. Wratten. 2004. Improved fitness of aphid parasitoids receiving resource subsides. *Ecology* **85**: 658–666.

Van Emden, H. F. 1963. Observations on the effects of flowers on the activity of parasitic hymenoptera. *Entomologists' Monthly* **98**: 265–270.

Van Rijn, P. C. J. and L. K. Tanigoshi. 1999a. The contribution of extrafloral nectar to survival and reproduction of the predatory mite *Iphiseius degenerans* on *Ricinus communis*. *Experimental and Applied Acarology* **23**: 281–296.

1999b. Pollen as food for the predatory mites *Iphiseius degenerans* and *Neoseiulus cucumeris* (Acari: Phytoseiidae): dietary range and life history. *Experimental and Applied Acarology* **23**: 785–802.

Van Rijn, P. C. J., Y. M. van Houten, and M. W. Sabelis. 2002. How plants benefit from providing food to predators even when it is also edible to herbivores. *Ecology* **83**: 2664–2679.

Van Veen, F. J. F., A. Rajkumar, C. B. Müller, and H. C. J. Godfray. 2001. Increased reproduction by pea aphids in the presence of secondary parasitoids. *Ecological Entomology* **26**: 425–429.

Vidal, S. 1997. Factors influencing the population dynamics of *Brevicoryne brassicae* in undersown brussels sprouts. *Biological Agriculture and Horticulture* **15**: 285–295.

Waage, J. K. 1990. Ecological theory and the selection of biological control agents. In M. Mackauer, L. E. Ehler, and J. Roland (eds.) *Critical Issues in Biological Control*. Andover, UK: Intercept Press pp. 135–158.

Wäckers, F. L. 2001. A comparison of nectar- and honeydew sugars with respect to their utilization by the hymenopteran parasitoid *Cotesia glomerata*. *Journal of Insect Physiology* **47**: 1077–1084.

Wäckers, F. L., A. Bjoruson, and S. Doru. 1996. A comparison of flowering herbs with respect to their nectar accessibility for the parasitoid *Pimpla turiovellae*. *Proceedings of the section Experimental and Applied Entomology of the Netherlands Entomological Society* **7**: 177–182.

Wheeler, D. 1996. The role of nourishment in oogenesis. *Annual Review of Entomology* **41**: 407–431.

Wratten, S. D., H. F. Van Emden, and M. B. Thomas. 1998. Within-field and border refugia for the enhancement of natural enemies. In C. H. Pickett and R. L. Bugg (eds.) *Enhancing Biological Control: Habitat Management to Promote Natural Enemies of Agricultural Pests*. Berkeley, CA: University of California Press, pp. 375–404.

Index